AF785925

S. Grillo · A. Leone (Eds): Physical Stresses in Plants

Proceedings of the workshop
on Genes and their products for tolerance to physical stresses in plants,
held from 24–27 September, 1995,
in Maratea, Italy

Springer
Berlin
Heidelberg
New York
Barcelona
Budapest
Hong Kong
London
Milan
Paris
Santa Clara
Singapore
Tokyo

S. Grillo · A. Leone (Eds)

Physical Stresses in Plants

Genes and Their Products for Tolerance

With 70 Figures

Springer

Dr. Stefania Grillo
Dr. Antonella Leone
Research Centre for Vegetable Breeding
National Research Council
Via Universitá, 133
I-80055 Portici (Naples)
Italy

ISBN-13: 978-3-642-64732-1 Springer-Verlag Berlin Heidelberg New York

Library of Congress Cataloging-in-Publication Data
Physical stresses in plants: genes and their products for tolerance/[edited by] S. Grillo, A. Leone.
p. cm.
"Proceedings of the Workshop on Genes and Their Products for Tolerance to Physical Stresses in Plants, held from 24–27 September 1995, in Maratea, Italy" – P.
Includes bibliographical references and indexes.

ISBN-13: 978-3-642-64732-1 e-ISBN-13: 978-3-642-61175-9
DOI: 10.1007/978-3-642-61175-9

1. Crops – Effect of stress on – Congresses. 2. Plants, Effect of stress on – Congresses. 3. Plant gene expression – Congresses.
I. Grillo, S. (Stefania), 1957– . II. Leone, A. (Antonella), 1957– . III. Workshop on Genes and Their Products for Tolerance to Physical Stresses in Plants (1995: Maratea, Italy)
SB112.5.P476 1996 96–22127
582'.05222–dc20 CIP

This work is subject to copyright. All rights are reserved, whether the whole or part of the material is concerned, specifically the rights of translation, reprinting, reuse of illustrations, recitation, broadcasting, reproduction on microfilm or in any other way, and storage in data banks. Duplication of this publication or parts thereof is permitted only under the provisions of the German Copyright Law of September 9, 1965, in its current version, and permission for use must always be obtained from Springer-Verlag. Violations are liable for prosecution under the German Copyright Law.

© Springer-Verlag Berlin Heidelberg 1996

Softcover reprint of the hardcover 1st edition 1996

The use of general descriptive names, registered names, trademarks, etc. in this publication does not imply, even in the absence of a specific statement, that such names are exempt from the relevant protective laws and regulations and therefore free for general use.

Cover design: Design & Production, Heidelberg
Cover illustration from Lee, Hübel and Schöffl (1995); Plant Journal 8:603–612, Blackwell Science, Oxford
Typesetting: camera ready by author

SPIN 10516516 31/3137-5 4 3 2 1 0 – Printed on acid-free paper

Preface

The workshop "*Genes and their products for tolerance to physical stresses in plants*" was held in Maratea, Italy, from 24-27 September 1995. As one of a series of activity launched by the European Science Foundation (ESF) Network "*Cell Stress Genes and their Protein Products*", the workshop was the only one entirely devoted to research in the plant field. Around one hundred participants from fourteen different countries enjoyed a lively meeting in an atmosphere of sunny Mediterranean weather. A total of twenty-eight speakers from universities and research institutes were invited to present the most recent results of their research.

The workshop was divided into eight sessions, namely heat, cold, salt, drought, oxidative stress, light stress, a conclusive session for recommendations and a poster session.

The results reported in this workshop broadly showed the rapid progress made in the understanding of the molecular mechanisms underlying basic aspects of the plant response to environmental stresses.

The list of new genes whose expression might be genetically manipulated to produce tolerant crop varieties is becoming longer and longer. This includes genes encoding products of the signal transduction pathway, which are filling the missing gaps in our understanding of the complex cascade of events leading from the perception of the stress signal to plant response. Tolerance to stress conditions is a quantitative trait and several reports discussed the use of RFLPs or other kind of linked molecular markers, which constitute a powerful tool for speeding up the selection of tolerant varieties by classical breeding. There was an incredibly large number of reports regarding the production of transgenic plants carrying specific genes conferring tolerance to environmental constraints, proving that recombinant DNA technology will lead to commercial varieties with improved tolerance to environmental stress in the near future.

Besides the very high quality and novelty of the papers presented, the workshop will be remembered primarily for the continuous discussion and exchanges of ideas, which were always friendly, informal and fruitful. Opportunity for discussion was limited not only to the scientific programme, but was also extended into leisure time, constituting the basis for collaboration and further exchanges among the participants.

It was decided to publish this book in order to further disseminate the results reported during the workshop, and that it might serve as a state-of-the-art overview on the molecular basis of plant response to physical stresses.

As editors, we greatly appreciated the support of the ESF, but we also wish to acknowledge the generous support of the Italian Ministry of Agriculture and Forestry, the special National Project "Resistenze Genetiche delle Piante Agrarie agli stress biotici e abiotici" and the University of Naples, without which the workshop would never have taken place. Special thanks go to Daniela Nenna for revision of the manuscript and to all the staff of the Research Centre for Vegetable Breeding, National Research Council, Portici, Italy, for help and assistance during the workshop.

Portici (Italy), March 1996

Stefania Grillo
Antonella Leone

Contents

Part 1 Heat Stress

Part 2 Cold Stress

Part 1

Heat Stress

Analysis of Multiple Classes of Soybean Heat Shock Genes and Proteins

Ron T. Nagao[1], Yuh-Ru Julie Lee[2], Peter R. LaFayette[3], Virginia H. Goekjian[1], Kevin O'Grady[1] and Joe L. Key[1]

[1] Department of Botany, University of Georgia, Athens, GA 30602, USA
[2] Department of Biology, Pennsylvania State University, University Park, PA 16802, USA
[3] Department of Biochemistry, University of Georgia, Athens, GA 30602, USA

Abstract. The influence of high temperature stress (heat shock or HS) and other environmental stress agents on gene expression of soybean seedlings has been extensively studied. The sequence analysis of HS genes has revealed a high degree of conservation among individual members of several heat shock protein (HSP) families and different classes within a family, but some interesting differences have been noted. These studies have also revealed complex patterns of regulation of expression of the HS genes and accumulation of the HSPs. Based primarily upon the deduced amino acid sequence of the HSPs, immunological cross-reactivity, and intracellular localization, the complex group of low molecular weight (LMW) HSP genes have been organized into multiple classes. In soybean several cDNA and genomic clones encoding 20 to 24 kD LMW HSPs have been isolated which represent new classes of the LMW HSP gene super family based on nucleotide/amino acid sequence and cell fractionation analyses. The mRNAs transcribed from these genes are of lower abundance than those for the 15 to 18 kD Class I and II proteins, and these genes occur as small multigene (i.e. three to four) classes or subfamilies. The mRNAs of three of these classes of LMW HSP genes are translated on ER-bound ribosomes and possess hydrophobic leader sequences. The presence of a consensus ER retention sequence on two of these proteins indicates that they probably reside within the ER. The third protein lacks the consensus ER retention signal and presumably is translocated to an as yet unidentified location. The mRNA representing a fourth LMW gene class is translated on unbound cytoplasmic ribosomes, and the predicted protein has a N-terminal sequence with properties similar to that of some proteins which are translocated into mitochondria. Early studies with soybean seedlings indicated that some 22 to 24 kD HSPs are localized in mitochondria. Differential induction by amino analog treatment indicates that genes assigned to the same class based on amino acid similarity and localization can be regulated differently. The possible role of the multiple classes on LMW 15 to 24 kD HSPs in protein protection from denaturation at high temperature (i.e. a chaperone function), based on studies from other laboratories is noted and some of these results will be summarized.

One aspect of the physiological/biochemical role(s) of HSPs in cellular function was studied by my laboratory, emphasizing the phenomenon of acquired thermotolerance. A soybean *HSP101* gene was isolated and sequenced. This soybean gene is homologous to the yeast *HSP104* gene and was used to complement a yeast *HSP104* deletion mutant in the acquisition of thermotolerance. Results of these experiments demonstrate that the soybean gene can partially restore heat tolerance in the yeast deletion mutant indicating that soybean *HSP101* is functionally similar to yeast *HSP104*. The *HSP101* gene family is again one of several groups or gene families for high molecular weight HSPs (e.g. HSP70s, HSP80s, HSP60s and HSP90/92s). Studies on these other HSPs/*HSP* genes will be reviewed along with the presentation of some of the newer results from our laboratory. Results from a number of studies in many laboratories support the view that these HSPs function as chaperones in multiple types of protein-protein.

Keywords. *Glycine max*, heat shock, heat shock proteins, soybean, thermotolerance

1 Introduction

Plants, like other organisms, respond to an increase in temperature above a threshold by synthesizing a set of proteins called the heat shock proteins (HSPs). The heat shock (HS) response has been highly conserved throughout evolution not only as a physiological phenomenon but also at the level of individual proteins. The HSPs comprise some of the most highly conserved proteins known, and this conservation has been very useful in categorizing the various HSP families within an organism and among groups of organisms.

The HS response has been the subject of numerous reviews in recent years (e.g. Craig, 1985; Key et al., 1985a; 1985b; Kimpel and Key, 1985a; Lindquist, 1986; Nagao et al., 1986; Key et al., 1987a; 1987b; Lindquist and Craig, 1988; Nagao and Key, 1989; Neumann et al., 1989; Nagao et al., 1990; Gurley and Key, 1991; Vierling, 1991). In general terms, the HS response in plants is characterized by a significant slowing of most normal (control temperature) protein synthesis as a result of a dramatic decline both in transcription of genes, and translation of the gene products, which were active prior to heat stress. Most normal mRNAs persist to varying degrees during a typical HS treatment. Concomitant with the cessation or slowing of normal mRNA and protein synthesis at the onset of HS, a rapid activation of the transcription of HS genes and accumulation of HS mRNAs occur. These HS mRNAs are preferentially translated resulting in a rapid and substantial accumulation of HSPs. Selective intracellular localization of many HSPs occurs during HS. The acquisition of thermotolerance, i.e. the ability of an organism to withstand an otherwise lethal temperature after being given a pretreatment with a non lethal HS, is correlated with HSP synthesis and accumulation. During prolonged heat treatment, HS mRNA synthesis declines, with maximum levels occurring after 1 to 2 h of HS; HSP synthesis also declines gradually with a normal pattern of protein synthesis occurring after 6 to 8 h in soybean seedlings (Kimpel et al., 1990). Some HS genes are also expressed under tight developmental controls, whereas some HS genes have counterparts that are constitutively expressed at non-HS temperatures (HS cognate genes) and that also show patterns of expression regulated during development (Vierling, 1991).

The customary nomenclature for HSPs is designation by molecular weight, e.g. HSP70 for a 70 kD HSP or HSP83 for an 83 kD HSP. This has lead to some confusion in the literature with numerous molecular weight variants and discrepancies between apparent mol wt estimated after fractionation on polyacrylamide gels and molecular weight calculated from deduced amino acid sequence. Sequence analysis of cloned genes/cDNAs demonstrates that HSPs are very highly conserved among very diverse organisms ranging from bacteria, lower eukaryotes, mammals, and plants. This conservation allows the designation of major HSP families as HSP100, HSP90, HSP70, HSP60, and low mol wt (LMW) HSPs (15-30 kD).

A number of review articles have recently addressed the subjects of the role(s) of HSPs as molecular chaperones and their involvement in stress tolerance (e.g. Hartl and Martin, 1992; Ciocca et al., 1993; Craig et al., 1993; Georgopoulos and Welch, 1993; Parsell and Lindquist, 1993; Parsell et al., 1993; Schatz, 1993; Yura et al., 1993; Alexandrov, 1994). Recent significant progress has been made on the biology of the stress response and the function of HSPs as documented by the results and perspectives of two monographs (Morimoto et al., 1990; Morimoto et al., 1994).

This paper focuses on the characterization of two HSP gene families from soybean, the HSP100 family and the LMW HSP gene family.

2 The HSP100 Family of Soybean

Genes encoding heat-inducible yeast HSP104 and plant HSP101 from soybean (Lee et al., 1994) and *Arabidopsis* (Schirmer et al., 1994) belong to a large family of highly conserved proteins known as the HSP100 or Clp protein family (Squires and Squires, 1992). Members of the HSP100/Clp family are found in both prokaryotes and eukaryotes. Some members are heat inducible while others are not. Members of the HSP100/Clp protein family share two large blocks of sequence homology (about 200 amino acids), each containing an ATP-binding domain. The ATP-binding domains are separated by a variable spacer region and are flanked by less conserved leader and trailer regions. The size of the spacer region has been used to define three subfamilies, namely *ClpA*, *ClpB*, and *ClpC* which have short, long, and intermediate spacers, respectively. The yeast HSP104 and the plant HSP101s are heat inducible and belong to the ClpB family. The ClpB homolog of *E. coli* is also heat inducible and is involved in heat tolerance in that organism (Squires et al., 1991). The involvement of yeast HSP104 and plant HSP101s in thermotolerance is discussed below. Additionally, several *Clp* genes have been characterized from higher plants, including tomato (Gottesman et al., 1990), pea (Moore and Keegstra, 1993), and *Arabidopsis* (Kiyosue et al., 1994) all of which are classified as ClpC family members owing to their intermediate spacer lengths of 60 to 70 amino acids. The ClpC proteins encoded by these plant genes contain N-terminal targeting sequences. ClpC in pea has been shown to be localized to chloroplasts *in vitro* (Moore and Keegstra, 1993). During the isolation and characterization of the soybean HSP101 (*ClpB*) clone, several additional *Clp*-like clones were isolated. Spacer length chacterization indicates that these clones should be classified as *ClpC* genes. Northern analysis showed that these *ClpC* genes of soybean are not heat inducible (Lee and Key, unpublished data).

3 The LMW HSPs of Soybean

One striking feature of the soybean HS response, and plants in general, is the relatively complex constellation of LMW HSPs that is induced by elevated temperature. The complexity of the LMW group has been reviewed for a number of plant species (Nagao et al., 1986; Mansfield and Key, 1987). The relative abundance of these HSPs and their complexity sharply contrast other eukaryotes which contain far fewer LMW HSPs. For example, *Drosophila* has genes for only four major LMW HSPs while yeast and mammals have only one (Lindquist and Craig, 1988).

3.1 Classification of LMW HSPs

Characterization of cDNA and genomic clones for soybean HSPs demonstrated that genes for soybean LMW HSPs may be grouped into several multigene families/subfamilies with domains of highly conserved sequence identity from evolutionarily distant organisms, including *Drosophila*, *Xenopus* and *Caenorhabditis elegans* (Nagao et al., 1986). In discussing the different cDNA clones, the original term group is now used synonymously with the more formal terms class and/or subfamily. Representative cDNA clones have been characterized and grouped into classes or subfamilies as summarized in Table 1. It should be noted that the group represented by cDNA clone, pCE54 (Czarnecka et al., 1984), and genomic clone, *GmHsp26A* (Czarnecka et al., 1988), is not included as HSPs even though some members of this group are induced several fold by HS above a level of constitutive expression. This group is now considered a class of general stress proteins (Nagao and Key, 1989; Vierling, 1991) responding to essentially every stress insult to which soybean seedlings have been exposed (Czarnecka et al., 1984; Edelman et al., 1988; Lee et al., 1996). Subsequently it was shown that *GmHsp26A* has significant homology to glutathione S-transferases (Droog et al., 1993). The LMW HSPs are less conserved on a percentage basis than the HMW HSPs, but comparison of amino acid sequences shows distinct domains of conservation and similar hydropathy profiles. Additional criteria such as nucleic acid cross-hybridization, immunological cross-reactivity, hybrid select/arrest translation, localization of mRNA translation, and protein localization of the LMW HSPs have been used to subdivide these HSP genes/proteins into a number of classes/groups (Table 2) which both correspond to and extend the classes defined by Vierling (1991). A major point of this paper is to use the characterization of the soybean LMW HSPs to illustrate the complexity of the family and to propose a functional relationship for this complexity.

Table 1. Classes of cDNA clones to LMW HSPs of soybean

Representative cDNA clone (Gene name)	Molecular Mass of encoded HSP(s) kD	Estimated members no.	Cellular localization
pCE75[*]	15 - 16	8	Cytoplasm
pCE53 (*GmHsp17.5E*)	15 - 18	13	Cytoplasm
pEV3 (*GmHsp20.7*)	22 and 27	2	Chloroplast
pFS2033 (*GmHsp22*)	22 - 24	4	Endomembrane (ER)
pEV2 (*GmHsp22.3*)	23 - 24	3-4	Endomembrane or ?
pEV1 (*GmHsp23.9*)	21 - 24	4	Mitochondria
pCE54[**] (*GmHsp26A*)	25 26	4-5	Cytoplasm

[*]See text for possible translation products of pCE75 other than 15 to 16 kD HSPs
[**]Considered not to be a class of HSPs, but rather a class of general stress proteins

3.2 Cytoplasmic-Localized LMW HSPs

The cytoplasmically-localized LMW HSPs represent the most abundant and the most complex class of the LMW HSPs. In soybean these HSPs are represented by cDNA clones pCE53 and pCE75 and are designated as class I and class II, respectively, in Table 2. These classes were recently reviewed (Key et al., 1994); therefore, only a summary will be presented here. The class I LMW HSPs are represented by a multigene family consisting of at least 13 proteins from hybrid-select translation data with pCE53 (Key et al., 1985c) and on immunological cross-reactivity (Hsieh et al., 1992). Recent data indicate that one or two additional proteins, not induced by HS, which immunocross-react with class I antibody were induced by treatment with the amino acid analog azetidine-2-carboxylic acid (Lee et al., 1996). This raises the interesting possibility that additional regulatory control must be operative to specifically induce additional HSPs of this class by azetidine treatment and perhaps other stress treatments.

Soybean genomic clones thus far identified as encoding LMW class I HSPs are presented in Table 2. The amino acid sequence of class I HSPs of soybean range from 87.1 to 92.2% identical and 96.1 to 97.4% similar. Sequence comparison of soybean *GmHSP17.5E* with pea *PsHSP18.1*, *Arabidopsis AtHSP17.6*, and wheat HSPC5-8 showed 85.1/92.9, 71.4/85.1, 72.3/84.5% identity/similarity, respectively. Amino acid sequence comparisons among HSPs of other classes give much lower percentages within a species or between different species.

Table 2. Amino Acid Sequence Relationship of Low Molecular Weight Heat Shock Proteins from Soybean[a]

	Percent amino acid identity (similarity) to: [b]											
	Class I					Class II		Class III	Class IV [c]		Class V [c]	Class VI
	Gmhsp17.5E	Gmhsp17.3B	Gmhsp17.5M	Gmhsp17.6L	Gmhsp18.5	Gmhsp17.9D	pCE75	Gmhsp20.7	Gmhsp22	Gmhsp22.5	Gmhsp22.3	Gmhsp23.9
Class I - Cytosolic												
Gmhsp17.5E	—											
Gmhsp17.3B	89.6 (97.4)	—										
Gmhsp17.5M	92.2 (96.1)	89.0 (96.1)	—									
Gmhsp17.6L	91.0 (96.1)	89.4 (95.4)	90.9 (95.4)	—								
Gmhsp18.5	91.6 (93.5)	89.6 (96.1)	92.2 (93.5)	87.1 (91.6)	—							
Class II - Cytosolic												
Gmhsp17.9D	38.4 (56.3)	35.1 (55.0)	33.1 (57.9)	34.9 (57.9)	35.9 (57.7)	—						
pCE75cDNA	43.8 (59.1)	38.0 (57.3)	44.8 (60.3)	40.9 (58.4)	44.5 (61.3)	92.1 (97.1)	—					
Class III - Chloroplast												
Gmhsp20.7	30.9 (52.3)	32.9 (56.6)	30.9 (52.3)	31.5 (53.7)	30.9 (53.9)	29.1 (57.4)	28.4 (58.8)	—				
Class IV - Endomembrane												
Gmhsp22	50.6 (70.4)	48.6 (67.6)	46.9 (67.1)	49.0 (69.5)	49.0 (68.9)	35.1 (61.7)	37.0 (60.4)	29.7 (62.1)	—			
Gmhsp22.5	45.7 (68.2)	45.3 (66.9)	43.6 (66.4)	42.1 (64.5)	43.7 (65.6)	34.7 (60.4)	35.4 (61.1)	29.1 (57.6)	77.9 (90.1)	—		
Class V - Endomembrane												
Gmhsp22.3	47.6 (68.0)	44.0 (66.0)	44.4 (62.9)	46.0 (64.7)	44.4 (66.7)	37.4 (56.1)	40.0 (56.8)	37.2 (60.0)	48.2 (69.9)	46.7 (65.3)	—	
Class VI - Mitochondria												
Gmhsp23.9	31.2 (58.9)	31.7 (55.6)	31.7 (59.0)	28.5 (56.3)	30.9 (55.0)	26.6 (53.8)	25.5 (53.9)	39.0 (61.6)	26.9 (57.9)	23.4 (61.4)	30.3 (59.3)	—

[a] Sequences are as follows: Gmhsp17.5E (Czarnecka et al., 1985); Gmhsp17.3B is synonymous with HS6871 (Schöffl et al., 1984); Gmhsp17.5M and Gmhsp17.6L (Nagao et al., 1985); Gmhsp18.5 and Gmhsp17.9D (Raschke et al., 1988); Gmhsp20.7, predicted mature size (Vierling et al., 1988) and pCE75 (unpublished data).
[b] Percentage identity and similarity were calculated using the GAP computer program (set with default parameters) of the Wisconsin GCG sequence analysis software.
[c] Endomembrane calculations were performed with predicted mature proteins.

The soybean class II cytoplasmic HSPs are represented by cDNA clone pCE75 and genomic clone *GmHsp17.9D* (Raschke et al., 1988) (Table 2). A comparison of the deduced amino acid sequence of pCE75 and GmHSP17.9D showed 92.1% identity/97.1% similarity. Hybrid-select translation analyses with pCE75 identified eight polypeptides estimated to be 15 to 16 kD (Key et al., 1985c). DNA sequence analysis of pCE75 indicates an open reading frame with at least five possible translation start sites predicting proteins of 15.4 to 18.0 kD. Utilizing the consensus rules of a purine at -3 and a guanine at +1 next to the initiator methionine codon, polypeptides of 18.0, 17.3 and 15.7 would be favored. Although favor of the first consensus initiator is predicted, as mentioned above, hybrid select translation analysis with pCE75 showed synthesis of polypeptides in the 15 to 16 kD range based on polyacrylamide gel electrophoresis separation. Further research is required to clarify this issue and to assess the relevance of multiple translation start sites.

In comparing the amino acid sequence of various classes of LMW HSPs, there is sequence divergence as illustrated in Table 2. However, the different classes contain highly conserved regions. The predicted amino acid sequence of a representative from each of the HSP classes I to IV of Table 2 are presented in Fig. 1.

```
              1                                                   50
GmHsp17.5E    .......... .......... ........MS LIPGFFGGRR SNVFDPFSLD
pCE75         .......... .......... .......... .......... ....MDMSED
GmHsp20.7     GDNKDNSVEV QHVSKGDQGT AVEKKPRRTA MDISPFGILD PWSPMRSMRQ
GmHsp22       .........M RLQQLNLFFL LLCVAKANGS LLP.FMDPPI TLLADLWSDR

              51                                                 100
GmHsp17.5E    MWDPFKDF.H VPTSSVSAEN S.AFVSTRVD WKETPEAHVF KADIPGLKKE
pCE75         AAGENKTYSA PTRSYVRDAK AMAATPADV. .KEYPNSYVF EIDMPGLKSG
GmHsp20.7     ILDTMDRVFE DTMTFPGRNI GGGEIRAPWD IKDEEHEIRM RFDMPGLAKE
GmHsp22       FPDPFRVLEH IPFGVDKDEA SMAMSPARVD WKETPEGHVI MLDVPGLKRE

              101                                                150
GmHsp17.5E    EVKVQIEDDR VLQISGERNV EKEDKNDTWH RVERSSGKFT RRFRLPENAK
pCE75         DIKVQVEDDN VLLISGERKR DEEKEGVKYL RMERRVGKFM RKFVLPENAN
GmHsp20.7     DVKVSVEDD. MLVIKGGHKS EQEHGGDDS. WSSRTYSSYD TRLKLPDNCE
GmHsp22       EIKVEVEENR VLRVSGERKK EEEKKGDHWH RVERSYGKFW RQFRLPQNVD
               Consensus II

              151                                                200
GmHsp17.5E    VNEVKASMEN GVLTVTVPK. ..EEVKKPDV KAIEISG... ..........
pCE75         TDAISAVCQD GVLSVTVQKL PPPEPKKPRT IEVKVF.... ..........
GmHsp20.7     KDKVKAELKN GVLYITIPKT KVE....RKV IDVQVQ.... ..........
GmHsp22       LDSVKAKLEN GVLTLTLDKL SPGKIKGPRV VSIAGEDHQQ GNLNNDGAKQ
               Consensus I

              201
GmHsp17.5E    ..
pCE75         ..
GmHsp20.7     ..
GmHsp22       EL
```

Fig. 1. Comparative alignment of the deduced amino acid sequence of representatives of four classes of soybean low mol wt heat shock protein genes. Class I (GmHsp17.5E), class II (pCE75), class III (GmHsp20.7), and class IV (GmHsp22) amino acid sequence identity of three out of four is bolded and shaded. Dots within a sequence indicate gaps inserted to optimize alignment. Consensus I and Consensus II sequences are double underlined. The derived amino acid sequences are as follows: GmHsp17.5E (Czarnecka et al., 1985), GmHsp20.7 (Vierling et al., 1988), and GmHsp22 (Helm et al., 1993)

Consensus region I (27 amino acids) and consensus region II (29 amino acids) were identified by Vierling (1991), and hydropathy analysis indicates that the less conserved sequence between consensus I and II is a highly hydrophilic domain present in all LMW HSPs. The consensus I region contains a highly conserved hydrophobic domain found in *Drosophila* LMW HSPs and in α-crystallin protein of the eye lens (Nagao et al., 1985; Vierling, 1991). Both the LMW HSPs and the α-crystallins form high mol wt complexes *in vivo*, and it has been suggested that the domain shared by these proteins is essential for complex formation (Lindquist and Craig, 1988).

3.3 Chloroplast-Localized LMW HSPs

Vierling et al., (1986) demonstrated that certain nuclear-encoded HS mRNAs of a number of species, including soybean, were translated in the cytoplasm and that the translation products were imported *in vitro* into isolated pea chloroplast. Similar results were obtained by Kloppstech et al. (1985) using pea and *Chlamydomonas*, and this phenomenon has been firmly established in diverse plant species (Vierling, 1991). Amino acid sequence comparisons show an amino-terminal 45-50 residues with no sequence conservation between these proteins from a range of species; this sequence is the transit peptide that is removed during import of the proteins into the chloroplast. The mature chloroplast-localized HSPs compared from various plant species have three highly conserved regions designated as consensus regions I, II, and III. Regions I and II delimit the regions of homology with other LMW HSPs, and consensus region III contains a highly conserved domain unique to chloroplast HSPs (Chen and Vierling, 1991). The soybean chloroplast HSPs were originally estimated as 22 and 27 kD proteins, but the deduced amino acid sequence of the 22 kD protein has a predicted mature size of 20.7 kD (*GmHSP20.7*, Table 2).

3.4 Endomembrane/Endoplasmic Reticulum-Localized LMW HSPs

The localization of LMW HSPs to the chloroplasts is an obvious unique feature to the plant HS response. Additionally, cell fractionation studies indicate that some of the LMW HSPs are localized in other plant organelles, e.g. mitochondria (Lin et al., 1984; Lenne and Douce, 1994; Key et al., unpublished data), endoplasmic reticulum (ER) (Cooper and Ho, 1987). While localization of specific HMW HSPs (e.g. HSP70 and HSP-related proteins) has been demonstrated in a number of eukaryotic systems (Brodsky and Schekman, 1994; Gething et al., 1994), only recently have LMW HSPs been shown to be specifically associated with the ER in any organism (Helm et al., 1993). DNA sequence analysis of four LMW HSP clones has identified open reading frames encoding polypeptides of 22.0, 22.3, 22.5, and 23.9 kD for *GmHsp22*, *GmHsp22.3*, *GmHsp22.5* and *GmHsp23.9*, respectively (Helm et al., 1993; LaFayette et al., 1996). Comparison of predicted amino acid sequences encoded by these four LMW HSP clones revealed a high degree of conservation at the carboxyl-terminus with greater divergence at the amino-terminus of these proteins. Sequences similar to consensus region I and II of other LMW HSPs were identified (Fig. 2).

```
Gmhsp22      .......MRLqqLnLfflLLcVAKAnGSLLPFmDpPITLLADLWSDRFPD
Gmhsp22.5    ...MrHfLvLVpLiLLVfagfpSKAkGSLLPFTNhPnTLLADLWSNHFPD
Gmhsp22.3    ...MaRTItIaAMcLLLsLVsLcvTTnALMPYT..rsTLwdmMl..qseD
Gmhsp23.9    massliAkRfLSssLLsrsLlrpaASAShrsFdtnaMrqydNradDHstD

Gmhsp22      ......................PFRVLEhIPF..GVDKDEAsmAMSp.AR
Gmhsp22.5    ......................PFRVLEQIPF..GVDKDETfTALSShAR
Gmhsp22.3    ......................PFgILEQnPFn.nIpnirGGAetlAlAR
Gmhsp23.9    idrhsersfpstarrddiflrcvgsIfsdseFepGsEHDgpGhGqSvplR

Gmhsp22      VDW.............KETPEGHVIMLDVPGLKREEIKVEVEENRVLRVS
Gmhsp22.5    VDW.............KETPEGHVIMLDVPGLKRDEIKIEVEgNRVLRVS
Gmhsp22.3    aDW.............KETPsAHVIVLDLPGMKKkDVKIEVEEsRVLRIS
Gmhsp23.9    VardrswrwsgrgwdaRETeDAlhLrVDMPGLaKEDVKIsVEQNtLI.Ik
                                                  II

Gmhsp22      GERKKEEEK.....kGDHWHRVERSYGKFWRQFRLPqNV.DLDSVKAKLE
Gmhsp22.5    GERKREEEK.....EGDHWHRVERSYGKsWRQFKVPDNV.DLDSVKAKME
Gmhsp22.3    GERKgEEEeeeeevEGEKWHRaERTnGKFmRQFRLPvNa.DLEkVtARLE
Gmhsp23.9    GEgaKEgDe.............EeSarRYtsridLPDkLykIDqIRAeMk
                                                    I

Gmhsp22      NGVLTLTLDKLSPgKIKGPRVVSIAGEDhQqgNLnnDgaKQEL
Gmhsp22.5    NGVLTLTMNKLSPDKVKGPRLVSIAGDDeQApkLKgNedKQEL
Gmhsp22.3    NGVLrITVgKfgeDKkRqPKVIdIAqrDsaAeNVKptkpqm..
Gmhsp23.9    NGVLkVvVpKMkeEerKdvisVkVe..................
```

Fig. 2. Comparative alignment of the deduced amino acid sequence corresponding to *GmHSP22*, *GmHSP22.5*, *GmHSP22.3*, and *GmHSP23.9*. All sequences represent the complete coding region (LaFayette et al., 1996). Identical residues are shown as uppercase letters on a gray background. Conserved residues are shown as uppercase letters on a white background. Dots within a sequence indicate gaps inserted to optimize alignment. The underlined regions show the Consensus I and Consensus II domains of LMW HSPs; the double underlined region shows the highly conserved GVL sequence characteristic of the HS domain I of LMW HSPs (Vierling, 1991)

Hydropathy profiles and analysis of the deduced amino-terminal translated sequence showed a hydrophobic central core characteristic of signal peptides (von Heijne, 1983). This was true across species for both pea *PsHSP22.7* and soybean *GmHSP22* and subsequent fractionation studies showed ER localization of these HSPs (Helm et al., 1993). ER localization was also predicted for *GmHSP22.3* and *GmHSP22.5* since the polysomal mRNAs of each of these genes were preferentially associated with ER (LaFayette et al., 1996) (Fig. 3).

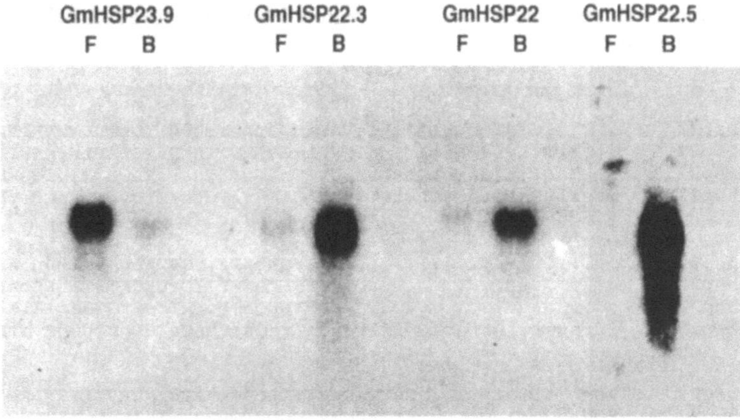

Fig. 3. Northern analysis of the distribution of LMW HSP mRNAs between free and membrane-bound polysomes. Total RNA from free (F) and membrane-bound (B) fractions from heat-stressed soybean seedlings were hybridized with cDNA probe for the indicated HSPs (from LaFayette et al., 1996)

Based upon localization of *in vitro* mRNA translation products, peptide structure, and amino acid sequence comparisons, GmHSP22 and GmHSP22.5 are categorized in the endomembrane class designated class IV (endomembrane) in Table 2. Both of these proteins by virtue of their carboxyl-terminal tetrapeptide (KQEL), which is similar to the consensus ER retention signal (KDEL) are likely retained in the lumen on the ER (LaFayette et al., 1996). *GmHSP22.3*, although translated on ER-bound polysomes, lacks an ER retention signal. These results indicate that the GmHSP22.3 polypeptide may be targeted to a different compartment of the endomembrane system or even transported to the cell wall. Additionally, amino acid sequence comparisons show that *GmHSP22.3* has no more sequence identity (similarity) to the Class IV endomembrane LMW HSPs than to any of the other designated classes of LMW HSPs; thus it is proposed to represent an additional class of endomembrane-associated LMW HSP (Class V in Table 2; LaFayette et al., 1996).

3.5 Mitochondrial-Localized LMW HSPs

The first evidence suggesting that some HSPs are localized to mitochondria came from *in vivo* labeling experiments where radiolabeled proteins were extracted from isolated mitochondria after heat treatment and shown to co-migrate with 22 to 24 kD HSPs (Lin et al., 1984). These proteins were not just mitochondrial-associated (i.e. external to the mitochondrial membrane) because protease treatment did not remove them. In contrast, the 15 to 18 kD HSPs (i.e. class I and class II HSPs) which were associated with the mitochondrial fraction were digested by the same protease treatment. Additionally, the 15 to 18 kD radiolabeled HSPs chased from the mitochondrial fraction during a 30°C (non-

HS) non-radioactive amino acid chase while the 22 to 24 kD HSPs remained associated with the mitochondria (Lin et al., 1984). Subsequently a number of soybean HS cDNAs have been cloned which by hybrid-select translation experiments encode 22 to 24 kD proteins. The full-length extension of one of these clones (GmHsp23.9) identified an amino-terminal sequence containing amino acids typical of some proteins which are imported into mitochondria (Gavel and von Heijne, 1990). GmHsp23.9 mRNA is translated on free polysomes, and the deduced amino acid sequence does not contain an ER-retention sequence (LaFayette et al., 1996). These criteria along with the amino acid sequence analysis and dissimilarity to class I through class V HSPs, grouped GmHsp23.9 into a different class of HSP, designated class VI (mitochondrial) in Table 2. This additional class is proposed in part because sequence comparisons show that the percentage amino acid identity and similarity among LMW HSPs of the same class from different species is much greater than different gene classes from the same species; this criterion has been used as part of the basis of LMW HSP classification. The amino acid sequence of GmHSP23.9 has as low or lower amino acid identity/similarity to chloroplast and endomembrane HSPs of similar mol wt than to any of the other designated classes of HSPs (Table 2). A comparison of GmHSP23.9 with the recently characterized pea mitochondrial PsHSP22 revealed 66.5% identity and 76.5% similarity (C. Lenne, personal communication; Lenne et al., 1995). Although the intra-mitochondrial localization of the soybean LMW HSPs needs to be confirmed by the development of antibodies and their application in immunolocalization experiments, this evidence suggests a functional relationship between the demonstrated thermotolerance of isolated mitochondria (Chou et al., 1989) and the temperature-dependent localization of 22 and 24 kD HSPs in the mitochondria.

4 The Function of HSPs in Stress Thermotolerance

Over the past few years a number of excellent reviews have summarized the significant strides made toward understanding the roles of the major HSPs in normal cellular physiology (Ellis and van der Vies, 1991; Welch, 1991; Gething and Sambrook, 1992; Hartl et al., 1992) and during stress conditions (Craig et al., 1993; Georgopoulos and Welch, 1993; Parsell and Lindquist, 1993). In plants, similar functions of HSPs as molecular chaperones and agents involved in stress tolerance or thermotolerance are implied (Key et al., 1985a; 1985b; Kimpel and Key, 1985b). Some specific data on molecular chaperone-like function in plants are available (Jinn et al., 1989; 1995; Lee et al., 1995). More specific data on involvement in thermotolerance come from studies with the HSP100 class. When yeast cells grown at 25°C are pretreated at 37°C to induce thermotolerance, they survive exposure to 50°C 1,000- to 10,000-fold better than non-pretreated cells. This thermotolerance is not exhibited by an Hsp104 gene deletion mutant of yeast (*Δhsp104*). Complementation of *Δhsp104* cells with *HSP104* restores the ability to acquire thermotolerance demonstrating that HSP104 plays a critical role in cell survival at extreme temperatures

(Sanchez and Lindquist, 1990). Complementation of yeast *Δhsp104* cells with a heat-inducible soybean HSP101 gene (*GmHsp101*) partially restored the defect of the mutant in the acquisition of thermotolerance indicating that soybean HSP101 performs a function similar to if not exactly the same as yeast HSP104 (Fig. 4; Lee et al., 1994). Schirmer et al. (1994) isolated one heat-inducible HSP101 cDNA clone from *Arabidopsis* and found that it was also able to partially complement the yeast *Δhsp104* mutant in acquiring thermotolerance.

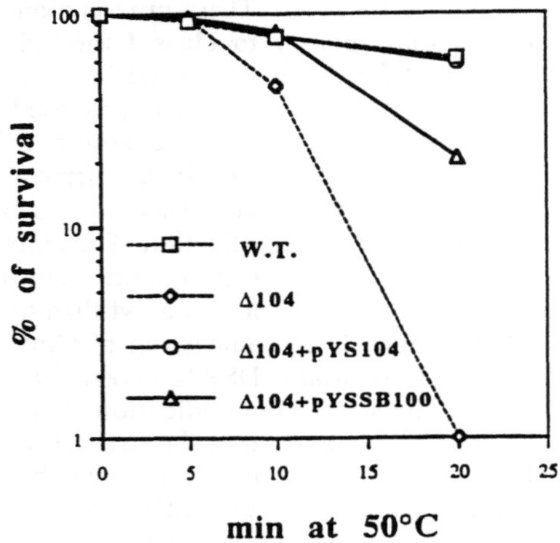

Fig. 4. Thermotolerance assay of the *HSP104* deletion mutant transformed with pYSSB101. Prior to the 50°C heat treatment, yeast cells were incubated at 37°C for 30 min in induce HSP synthesis. After heat treatment at 50°C for the period of time indicated, cells were diluted in ice-cold media and plated. Colonies that survived the 50°C treatment were counted 3 d after the treatment. Values of relative survival percentage were averaged from three independent transformation experiments. W.T., the wild type yeast cells; *Δ104,* the *Δhsp104* cells; *Δ*104 + pYS104, the *Δhsp104* cells transformed with the yeast HSP104 genomic clone pYS104; *Δ*104 +pYSSB101, the *Δhsp104* cells transformed with pYSSB101 that contains the *GmHsp101* coding sequence under the control of the yeast *HSP104* promoter (from Lee et al., 1994)

While the exact mechanism of HSP100 function(s) in thermotolerance is not known, recent work suggests that HSP104 in yeast functions in thermotolerance by promoting the reactivation of aggregated, heat-damaged proteins after high temperature stress (Parsell et al., 1994). Additionally, increased expression of Hsp104 can confer thermotolerance to yeast cells in the absence of synthesis of other HSPs; further, over-expression of Hsp104 is not toxic to yeast cells (S. Lindquist, personal communication). Thus it may be possible to manipulate the expression of HSP100 proteins in plants to engineer greater stress tolerance. It should be noted that HSP100 apparently is not present in *Drosophila*; therefore, the functional role(s) of HSP100 must be provided by other HSPs in this organism. It is known that, while HSPs (e.g. HSP70) have a considerable degree of specialization, they also have overlapping and/or compensating functions, as evidenced in the case of deletion mutations) (Parsell and Lindquist, 1993). While evolution may emphasize the use of particular proteins for specific purposes, functional equivalency must exist in other proteins when a particular type is not present in all organisms. This apparent redundancy of function may provide a selective safeguard through evolution and may apply in particular to the LMW HSPs of plants.

The function of LMW HSPs is less well understood than for HMW HSPs. Transfection of human *HSP27* into rodent cells conferred heat resistance (Landry et al., 1989), but on the other hand over-expression of the one major LMW *Hsp* of yeast (*Hsp26*) had very little effect on thermotolerance (Susek and Lindquist, 1989). Moreover, deletion of the yeast *Hsp26* gene had no effect on growth or thermotolerance at any temperature (Petko and Lindquist, 1986). A number of cell-type specific and system-specific small proteins have been investigated from different laboratories, but only recently has immunological and molecular biology techniques identified these proteins as HSP27. The biological and clinical implications of HSP27 have recently been reviewed (Ciocca et al., 1993), but the functional difference(s) between organisms remains unclear.

Minton et al. (1982) originally proposed a model of non-specific stabilization of stress-susceptible proteins as the biological role of HSPs. Jinn et al. (1989) reported that a soybean HSP-enriched ammonium sulfate fraction (containing a high percentage of all the HSPs) significantly protected control proteins from heat denaturation. The degree of protection was proportional to the amount of HSP-enriched fraction added. While no mechanism was presented, stabilization of proteins by maintaining solubility is somewhat analogous to the role of chaperones proposed by Hemmingsen et al. (1988). Further characterization of an ammonium sulfate-enriched fraction of HSPs by non-denaturing polyacrylamide gel electrophoresis revealed the presence of a high molecular mass complex (280 kD) consisting primarily of class I LMW HSPs. A purified class I HSP complex from soybean thermostabilized proteins against denaturation *in vitro* (Jinn et al., 1995). Mammalian small HSPs and α-crystallin have been shown to possess molecular chaperone activity *in vitro* by suppressing protein aggregation (denaturation) and promoting renaturation of denatured proteins (Horwitz, 1992; Jakob et al., 1993; Merck et al., 1993). Recombinant class I HSP18.1 and class II HSP17.7 from pea form discrete sized homo-oligomeric complexes as estimated by PAGE to be approximately about 240 and 320 kD, respectively. Sedimentation equilibrium analytical ultracentrifugation revealed that each complex consists of approximately 12 subunits (Lee et al., 1995); hetero-oligomeric complexes were not detected. Both proteins (or complexes) enhance the refolding of chemically denatured citrate synthase and lactate dehydrogenase at stechiometric levels in an ATP-independent manner, and both HSPs similarly prevented denaturation or aggregation (Lee et al., 1995).

The demonstration of a chaperone like-function for LMW HSPs *in vitro* is interesting because it provides a functional basis for the LMW HSPs that is consistent with the correlation of functional conservation and may be used to explain the abundance and complexity of LMW HSP synthesis in plants. Although the LMW HSPs have lower overall sequence identity than the HMW

HSPs among organisms, the identification of conserved regions in the carboxyl region of the proteins suggests that these regions might be conserved functional domains. The conservation of these domains across kingdoms and in different compartments within plant cells strongly suggests this to be the case. The demonstration that recombinant class I and class II complexes of pea LMW HSPs can prevent denaturation/aggregation *in vitro* (Lee et al., 1995) supports this idea. Some of the complexity and abundance of LMW HSPs in plants as compared to animals, can be accounted for the evolution of HSPs targeted to specific plant organelles (e.g. chloroplasts), or other organelles like mitochondria and ER. Some of the divergence arises in that different classes of LMW HSPs have evolved specialized domains, so that a class could be more related among species than different classes within the same species (e.g. class III chloroplastic LMW HSPs have been shown to have a conserved chloroplast domain not found in other classes). The less conserved amino terminal portion of these HSPs is in part expected based on this scenario, since the N-terminal sequences are involved in targeting the functional protein to specific locations within the cell (e.g. chloroplast, mitochondria, ER) where the HSPs function. The divergent targeting sequences contribute in part to the complexity and abundance of the LMW HSPs in plants. Thus, plants have evolved the complex and abundant LMW HSPs to protect cellular organelles and their contents as well as cytoplasmic components of the cell. The equivalent function in animal systems is provided by the few LMW HSPs present and by compensation by other HSPs (e.g. HSP70s). The latter must be the case in yeast particularly where as mentioned above, the deletion of the one major LMW HSP did not affect thermotolerance. The fact that the relative amount of HSP70 is so much higher in non-plant organisms, being the major group of HSPs synthesized in animal systems, is also consistent with the idea of functional compensation. The evolution of specialized LMW HSPs in plants may be a reflection of the additional demands placed on plants as sessile organisms in an ever changing environment. The specifically-localized LMW HSPs in plants could provide additional protection/functions or perhaps could have evolved to replace the major thermoprotectant used by many other organisms (e.g. HSP70), perhaps because LMW HSP can be synthesized more rapidly. Additional research is required to determine whether the LMW HSPs of plants function independently or whether they function also in conjunction with other HSPs.

The biochemical and physiological complexity of LMW HSP involvement in the HS response is only beginning to be unravelled. Many observations and questions remain unresolved. For example the cytosolic LMW HSPs show a complex and baffling pattern of localization and relocalization in response to HS and during recovery from HS. The relative importance of misfolding, refolding, and degradation in the cellular response to and recovery from stress is a major unresolved issue. These and possibly other issues are likewise unresolved for LMW HSPs present during certain stages of development under non-stress conditions. While some of the diversity of HSPs employed in stress tolerance may reflect the different sensitivities of different targets to damage in different organisms, a clear correlation between HSP abundance and complexity with heat tolerance is not apparent.

References

Alexandrov VY (1994) Functional aspects of cell response to heat shock. Int Rev Cytol **148**: 171-227

Brodsky JL, Schekman R (1994) Heat shock cognate proteins and polypeptide translocation across the endoplasmic reticulum membrane. *In* RI Morimoto, A Tissières, C Georgopoulos, eds, The Biology of Heat Shock Proteins and Molecular Chaperones. Cold Spring Harbor Monograph Series 26, Cold Spring Harbor Laboratory Press, Plainview, NY, pp 85-109

Chen Q, Vierling E (1991) Analysis of conserved domains identifies a unique structural feature of a chloroplast heat shock protein. Mol Gen Genet **226**: 425-431

Chou M, Chen Y-M, Lin C-Y (1989) Thermotolerance of isolated mitochondria associated with heat shock proteins. Plant Physiol **89**: 617-621

Ciocca DR, Oesterreich S, Chamness GC, McGuire WL, Fuqua SAW (1993) Biological and clinical implications of heat shock protein 27000 (Hsp27): a review. J Natl Cancer Inst **85**: 1558-1570

Cooper P, Ho THD (1987) Intracellular localization of heat shock proteins in maize. Plant Physiol **84**: 1197-1203

Craig EA (1985) The heat shock response. CRC Critical Rev Biochem **18**: 239-280

Craig EA, Gambill BD, Nelson RJ (1993) Heat shock proteins: molecular chaperones of protein biogenesis. Microbiol Rev **57**: 402-414

Czarnecka E, Edelman L, Schöffl F, Key JL (1984) Comparative analysis of physical stress responses in soybean seedlings using cloned heat shock cDNAs. Plant Mol Biol **3**: 45-58

Czarnecka E, Gurley WB, Nagao RT, Mosquera LA, Key JL (1985) DNA sequence and transcript mapping of a soybean gene encoding a small heat shock protein. Proc Natl Acad Sci USA **82**: 3726-3730

Czarnecka E, Nagao RT, Key JL, Gurley WB (1988) Characterization of *Gmhsp26-A*, a stress gene encoding a divergent heat shock protein of soybean: heavy-metal-induced inhibition of intron processing. Mol Cell Biol **8**: 1113-1122

Droog FNJ, Hooykaas PJJ, Libbenga KR, van der Zaal EJ (1993) Proteins encoded by an auxin-regulated gene family of tobacco share limited but significant homology with glutathione *S*-transferases and one member indeed shows *in vitro* GST activity. Plant Mol Biol **21**: 965-972

Edelman L, Czarnecka E, Key JL (1988) Induction and accumulation of heat shock-specific poly(A^+) RNAs and proteins in soybean seedlings during arsenite and cadmium treatments. Plant Physiol **86**: 1048-1056

Ellis RJ, van der Vies SM (1991) Molecular chaperones. Annu Rev Biochem **60**: 321-347

Gavel Y, von Heijne G (1990) Cleavage-site motifs in mitochondrial targeting peptides. Protein Engineering **4**: 33-37

Georgopoulos C, Welch WJ (1993) Role of the major heat shock proteins as molecular chaperones. Annu Rev Cell Biol **9**: 601-634

Gething M-J, Sambrook J (1992) Protein folding in the cell. Nature **355**: 33-45

Gething M-J, Blond-Elguindi S, Mori K, Sambrook, JF (1994) Structure, function, and regulation of the endoplasmic reticulum chaperone, BiP. *In* RI Morimoto, A Tissières, C Georgopoulos, eds, The Biology of Heat Shock Proteins and Molecular Chaperones. Cold Spring Harbor Monograph Series 26, Cold Spring Harbor Laboratory Press, Plainview, NY, pp 111-135

Gottesman S, Squires C, Pichersky E, Carrington M, Hobbs M, Mattick JS, Clark WP, Ross B, Squires CL, Maurizi MR (1990) Conservation of the regulatory subunit for the Clp ATP-dependent protease in prokaryotes and eukaryotes. Proc Natl Acad Sci USA **87**: 3513-3517

Gurley WB, Key JL (1991) Transcriptional regulation of the heat-shock response: a plant perspective. Biochemistry **30**: 1-12

Hartl FU, Martin J (1992) Protein folding in the cell: the role of molecular chaperones Hsp70 and Hsp60. Annu Rev Biophys Biomol Struct **21**: 293-322

Helm KW, LaFayette PR, Nagao RT, Key JL, Vierling E (1993) Localization of small heat-shock proteins to the higher-plant endomembrane system. Mol Cell Biol **13**: 238-247

Hemmingsen SM, Woolford C, van der Vies SM, Tilly K, Dennis DT, Georgopoulos CP, Hendrix RW, Ellis RJ (1988) Homologous plant and bacterial proteins chaperone oligomeric protein assembly. Nature **333**:330-334

Horwitz J (1992) α-Crystallin can function as a molecular chaperone. Proc Natl Acad Sci USA **89**: 10449-10453

Hsieh M-H, Chen J-T, Jinn T-L, Chen Y-M, Lin C-Y (1992) A class of soybean low molecular weight heat shock proteins. Plant Physiol **99**: 1279-1284

Jakob U, Gaestel M, Engel K, Buchner J (1993) Small heat shock proteins are molecular chaperones. J Biol Chem **268**: 1517-1520

Jinn T-L, Yeh Y-C, Chen Y-M, Lin C-Y (1989) Stabilization of soluble proteins *in vitro* by heat shock proteins-enriched ammonium sulfate fraction from soybean seedlings. Plant Cell Physiol **30**: 463-469

Jinn T-L, Chen Y-M, Lin C-Y (1995) Characterization and physiological function of class I low-molecular-mass, heat-shock protein complex in soybean. Plant Physiol **108**: 693-701

Key JL, Kimpel JA, Lin CY, Nagao RT, Vierling E, Czarnecka E, Gurley WB, Roberts JK, Mansfield MA, Edelman L (1985a) The heat shock response in soybean. *In* JL Key, T Kosuge, eds, Cellular and Molecular Biology of Plant Stress. Vol. 22, Alan R. Liss, New York, pp 161-179

Key JL, Kimpel J, Vierling E, Lin C-Y, Nagao RT, Czarnecka E, Schöffl F (1985b) Physiological and molecular analyses of the heat shock response in plants. *In* BG Atkinson, DB Walden, eds, Changes in Eukaryotic Gene Expression in Response to Environmental Stress. Academic Press, New York, pp 327-348

Key JL, Gurley WB, Nagao RT, Czarnecka E, Mansfield MA (1985c) Multigene families of soybean heat shock proteins. *In* L van Vloten-Doting, GSP Groot, TC Hall, eds, Molecular Form and Function of the Plant Genome. Plenum Press, New York, pp 81-100

Key JL, Kimpel J, Nagao RT (1987a) Heat shock gene families of soybean and the regulation of their expression. *In* JL Key, L McIntosh, eds, Plant Gene Systems and Their Biology. 62th edn. Alan R. Liss, New York, pp 87-97

Key JL, Nagao RT, Czarnecka E, Gurley WB (1987b) Heat stress: expression and structure of heat shock protein genes. *In* D von Wettstein, NH Chua, eds, Plant Molecular Biology. Plenum Publishing Corp, New York, pp 385-397

Key JL, Lee Y-R, Goekjian V, Nagao RT (1994) The low molecular weight heat shock proteins of soybean seedlings. *In* JH Cherry, ed, Biochemical and Cellular Mechanisms of Stress Tolerance in Plants. NATO ASI Series, Vol H 86, Springer-Verlag, Berlin, pp 115-140

Kimpel JA, Key JL (1985a) Heat shock in plants. Trends Biochem Sci **10**: 353-357

Kimpel JA, Key JL (1985b) Presence of heat shock mRNAs in field grown soybeans. Plant Physiol **79**: 672-678

Kimpel JA, Nagao RT, Goekjian V, Key JL (1990) Regulation of the heat shock response in soybean seedlings. Plant Physiol **94**: 988-995

Kiyosue T, Yamaguchi-Shinozaki K, Shinozaki K (1994) Cloning of cDNAs for genes that are early-responsive to dehydration stress (ERDs) in *Arabidopsis thaliana* L.: identification of three ERDs as HSP cognate genes. Plant Mol Biol **25**: 791-798

Kloppstech K, Meyer G, Schuster G, Ohad I (1985) Synthesis, transport and localization of a nuclear coded 22-kD heat-shock protein in the chloroplast membranes of peas and *Chlamydomonas reinhardi*. EMBO J **4**: 1901-1909

LaFayette PR, Nagao RT, O'Grady K, Vierling E, Key JL (1996) Molecular characterization of cDNAs encoding low-molecular-weight heat shock proteins of soybean. Plant Mol Biol **30**: 159-169

Landry J, Chrétien P, Lambert H, Hickey E, Weber LA (1989) Heat shock resistance conferred by expression of the human HSP27 gene in rodent cells. J Cell Biol **109**: 7-15

Lee GJ, Pokala N, Vierling E (1995) Structure and *in vitro* molecular chaperone activity of cytosolic small heat shock proteins from pea. J Biol Chem **270**: 10432-10438

Lee Y-RJ, Nagao RT, Key JL (1994) A soybean 101-kD heat shock protein complements a yeast *HSP104* deletion mutant in acquiring thermotolerance. Plant Cell **6**: 1889-1897

Lee Y-RJ, Nagao RT, Lin C-Y, Key JL (1996) Induction and regulation of heat-shock gene expression by an amino acid analog in soybean seedlings. Plant Physiol (in press)

Lenne C, Douce R (1994) A low molecular mass heat-shock protein is localized to higher plant mitochondria. Plant Physiol **105**: 1255-1261

Lenne C, Block MA, Garin J, Douce R (1995) Sequence and expression of the mRNA encoding HSP22, the mitochondrial small heat-shock protein in pea leaves. Biochem J **311**: 805-813

Lin CY, Roberts JK, Key JL (1984) Acquisition of thermotolerance in soybean seedlings: synthesis and accumulation of heat shock proteins and their cellular localization. Plant Physiol **74**: 152-160

Lindquist S (1986) The heat-shock response. Annu Rev Biochem **55**:1151-1191

Lindquist S, Craig EA (1988) The heat shock proteins. Annu Rev Genet **22**: 631-677

Mansfield MA, Key JL (1987) Synthesis of the low molecular weight heat shock proteins in plants. Plant Physiol **84**: 1007-1017

Merck KB, Groenen JTA, Voorter CEM, de Haard-Hoekman WA, Horwitz J, Bloemendal H, de Jong WW (1993) Structural and functional similarities of bovine α-crystallin and mouse small heat-shock protein. J Biol Chem **268**: 1046-1052

Minton KW, Karmin P, Hahn GH, Minton AP (1982) Nonspecific stabilization of stress-susceptible proteins by stress-resistant proteins: a model for the biological roles of heat shock proteins. Proc Natl Acad Sci USA **79**: 7101-7111

Moore T, Keegstra K (1993) Characterization of a cDNA clone encoding a chloroplast-targeted Clp homologue. Plant Mol Biol **21**: 525-537

Morimoto RI, Tissières A, Georgopoulos C (1990) Stress Proteins in Biology and Medicine. Cold Spring Harbor Monograph Series 19, Cold Spring Harbor Laboratory Press, Cold Spring Harbor, NY

Morimoto RI, Tissières A, Georgopoulos C (1994) The Biology of Heat Shock Proteins and Molecular Chaperones. Cold Spring Harbor Monograph Series 26, Cold Spring Harbor Laboratory Press, Plainview, NY

Nagao RT, Czarnecka E, Gurley WB, Schöffl F, Key JL (1985) Genes for low-molecular-weight heat shock proteins of soybeans: sequence analysis of a multigene family. Mol Cell Biol **5**: 3417-3428

Nagao RT, Kimpel JA, Vierling E, Key JL (1986) The heat shock response: a comparative analysis. *In* BJ Miflin, ed, Oxford Surveys of Plant Molecular & Cell Biology, Vol 3, Oxford Univ Press, London, pp 384-438

Nagao RT, Key JL (1989) Heat shock protein genes of plants. *In* IK Vasil, J Schell, eds, Cell Culture and Somatic Cell Genetics of Plants, Academic Press, New York, pp 297-328

Nagao RT, Kimpel JA, Key JL (1990) Molecular and cellular biology of the heat-shock response. Adv Genet **28**: 235-274

Neumann D, Nover L, Parthier B, Rieger R, Scharf K-D, Wollgiehn R, Nieden UZ (1989) Heat shock and other stress response systems of plants. Biol Zentralbl **108**: 1-156

Parsell DA, Lindquist S (1993) The function of heat-shock proteins in stress tolerance: degradation and reactivation of damaged proteins. Annu Rev Genet **27**: 437-496

Parsell DA, Taulien J, Lindquist S (1993) The role of heat-shock proteins in thermotolerance. Phil Trans R Soc Lond **339**: 279-286

Parsell DA, Kowal AS, Singer MA, Lindquist S (1994) Protein disaggregation mediated by heat-shock protein Hsp104. Nature **372**: 475-478

Petko L, Lindquist S (1986) Hsp26 is not required for growth at high temperatures, nor for thermotolerance, spore development, or germination. Cell **45**: 885-894

Raschke E, Baumann G, Schöffl F (1988) Nucleotide sequence analysis of soybean small heat shock protein genes belonging to two different multigene families. J Mol Biol **199**: 549-557

Sanchez Y, Lindquist S (1990) HSP104 required for induced thermotolerance. Science **248**: 1112-1115

Schatz G (1993) The protein import machinery of mitochondria. Protein Sci **2**: 141-146

Schirmer EC, Lindquist S, Vierling E (1994) An *Arabidopsis* heat shock protein complements a thermotolerance defect in yeast. Plant Cell **6**: 1899-1909

Schöffl F, Raschke E, Nagao RT (1984) The DNA sequence analysis of soybean heat-shock genes and identification of possible regulatory promoter elements. EMBO J **3**: 2491-2497

Squires CL, Pedersen S, Ross BM, Squires C (1991) ClpB is the *Escherichia coli* heat shock protein F84.1. J Bacteriol **173**: 4254-4262

Squires C, Squires CL (1992) The Clp proteins: proteolysis regulators or molecular chaperones? J Bacteriol **174**: 1081-1085

Susek RE, Lindquist SL (1989) hsp26 of *Saccharomyces cerevisiae* is related to the superfamily of small heat shock proteins but is without a demonstrable function. Mol Cell Biol **9**: 5265-5271

Vierling E, Mishkind ML, Schmidt GW, Key JL (1986) Specific heat shock proteins are transported into chloroplasts. Proc Natl Acad Sci USA **83**: 361-365

Vierling E, Nagao RT, DeRocher AE, Harris LM (1988) A heat shock protein localized to chloroplasts is a member of a eukaryotic superfamily of heat shock proteins. EMBO J **7**: 575-581

Vierling E (1991) The roles of heat shock proteins in plants. Annu Rev Plant Physiol Plant Mol Biol **42**: 579-620

von Heijne G (1983) Patterns of amino acids near signal-sequence cleavage sites. Eur J Biochem **133**: 17-21

Welch WJ (1991) The role of heat-shock proteins as molecular chaperones. Curr Opin Cell Biol **3**: 1033-1038

Yura T, Nagai H, Mori H (1993) Regulation of the heat-shock response in bacteria. Annu Rev Microbiol **47**: 321-350

Stress-Induced Activation of Chaperone Genes Implies the Operation of a Novel Transcriptional Regulatory Mechanism in the Cyanobacterium, *Synechocystis* PCC6803

Attila Glatz, Ibolya Horváth, Viktória Varvasovszki, Eszter Kovács, Zsolt Török and László Vigh

Institute of Biochemistry, Biological Research Centre, Hungarian Academy of Sciences, H-6701 Szeged, Hungary

Abstract. Transcriptional analysis of the chaperone genes of *Synechocystis* PCC6803 have been carried out. Data obtained from northern experiments suggest that the mechanism of the transcriptional control of the *dnaK* gene is different from that of shown for the *groEL*-type chaperonins. Furthermore, it was also possible to demonstrate a distinct regulation of the two chaperonin genes. Primer extension analysis revealed, that the transcriptional startpoint of both chaperonin mRNA is located within CIRCE, a common potential regulatory element, tested either under normal or heat shock (HS) conditions. The above findings implying a novel transcriptional regulatory pathway for molecular chaperones in *Synechocystis* will be discussed.

Keywords. Chaperone genes, *Cyanobacterium*, heat stress, *Synechocystis* PCC6803

1 Introduction

There is unequivocal evidence that in higher plant cells exposed to HS the photosynthetic apparatus is irreversibly damaged prior to the impairment of other cellular functions (Berry and Bjorkman, 1980). The general assembly of the photosynthetic membranes in cyanobacteria is similar to that of higher plants, therefore *Synechocystis* PCC6803 might serve as a powerful model for studying the molecular mechanism of the stress response and long-term adaptation (Lehel et al., 1993a; Vigh et al., 1994, 1995).

Chaperonins, a subclass of molecular chaperones, belong to a highly conserved family of proteins and are probably present in all living organism.

They assist in the correct folding of newly synthesized proteins, and assembly of oligomer structures (see Ellis and van der Vies, 1991; Hendrick and Hartl, 1993, for reviews). As an implication of such effects, chaperonins could provide protection against different environmental stresses (Gatenby et al., 1993). In a previous study it has been shown that sublethal heat stress enhances the synthesis of at least four type of heat shock proteins in *Synechocystis* PCC6803 (Lehel et al., 1992). It has also been demonstrated that a GroEL-type chaperonin is associated to the thylakoid membrane upon elevated temperatures, indicating that these proteins might be involved directly in the protection of the photosynthetic apparatus of cyanobacteria (Kovács et al., 1994a, b).

Transcriptional regulation of the HS genes are best characterized in *E. coli.* A temperature up shift transiently induces the HS genes by activating transcription from promoters specifically recognized by RNA polymerase containing σ^{32} (see Bukau, 1993; Yura et al., 1993, for reviews). Recently, a novel mechanism of chaperone transcriptional regulation has emerged. It was demonstrated, that several bacteria contain a highly conserved inverted repeat ("CIRCE") preceding their *groE* and *dnaK* genes. Mutational analysis of this element in *Bacillus subtilis* revealed that CIRCE could act as a negative cis regulator (Zuber and Schumann, 1994). It has been reported that the first member of the operon containing the *orf39-grpE-dnaK-dnaJ* genes codes for the repressor acting on CIRCE (Shulz et al., 1995; Yuan and Wong, 1995a). It has also been documented, that a rapid turnover of the *groE* mRNA requires the presence of the IR. Without CIRCE the *groE* transcript displayed a longer half-life in *Bacillus subtilis* (Yuan and Wong, 1995a). Similarly, CIRCE can be found in the upstream region of both *groESL* and *dnaK* operon of *Clostridium acetobutilycum* (Narberhaus and Bahl, 1992; Narberhaus et al., 1992) and *orf A*, encoded by the *dnaK* gene shows striking homology with *orf* 39 (Narberhaus et al.,1992; Yuan and Wong, 1995b).

Synechocystis PCC6803, in contrast to most of bacteria, contains two *groEL* type genes: a *groESL* operon (Lehel et al., 1993b) and an additional *cpn60* gene, possessing no *groES* in the neighbouring region (Chitnis and Nelson; 1991). Whereas the CIRCE element is present upstream of both chaperonin genes, no similar element was documented in the putative promoter region of the *dnaK* (Chitnis and Nelson, 1991). The major *dnaK* transcript seems to be monocistronic, although the presence of a larger mRNA species cannot be ruled out (Chitnis and Nelson, 1991). Due to the fact that the transcriptional regulatory mechanism of chaperone families in *Synechocystis* hitherto unrevealed, the mode of stress-induced mRNA formation of individual chaperone genes of this cyanobacterial model was investigated.

2 Materials and Methods

2.1 Growth Conditions

Synechocystis PCC6803 cells were grown photoautotrophically at 30 °C in BG-11 medium buffered with HEPES-NaOH (pH 7.5) under illumination with cool white fluorescent light as described (Lehel et al., 1992) . Cultures at exponential growth phase were used for all experiments.

2.2 DNA Manipulation

Standard molecular genetic procedures were followed unless otherwise stated (Ausubel at al.,1987). Plasmid pUC18 was chosen as a general cloning vehicle (Yanisch-Perron et al., 1985). *Synechocystis groESL* and *cpn60* specific probes for northern analysis were generated by PCR (Lehel et al., 1993b). *dnaK* probe was prepared with the same method using primers 5' TGGCGATCGCCTAGTGGGGC 3' (positions from 1307 to 1326 of the sequence deposited in Gen Bank, under accession number M57518) and 5'TGGCCCGGGACAGGGTGGTGT3' (positions from 2068 to 2047). The amplified fragments were purified on agarose gel and ligated into the HincII site of pUC18.

Transformation of *E. coli* strain DH5αF' were carried out according to Inoue et al. (1990). Recombinant plasmids were verified by partial sequencing of the inserts.

Plasmid pAN4 containing the ribosomal RNA genes of *Anacystis nidulans* were kindly provided by M. Sugiura (Tomioka et al., 1981). Probes were labeled with α^{32}P-dCTP using Multiprime DNA labeling kit (Amersham).

Synthetic oligonucleotides C_1 (5'AGTAGAAACATTAATGGAAAT3') and C_2 (5'TGGATTCATCCTTAAAGGAA3') were used to identify the transcriptional start point of the *groESL* (C_1) and *cpn60* (C_2) genes.

Plasmids pESL (containing the XmnI fragment of the *groESL* operon) and pCPN (carrying the BclI-KpnI fragment of the *cpn60* gene) served for generating sequencing ladder for primer extension.

2.3 Northern Hybridization

RNA were extracted as described (Lehel et al., 1993b). Five microgram of total RNA were run on 1% agarose gel containing 6% formaldehyde and blotted onto nylon membrane (ZetaProbe GT, BioRad). Filters were washed, dried and heated in a 80 °C oven for 1 h. High stringency hybridization, washing and stripping were carried out according to the manufacturer's directions.

2.4 Primer Extension

C_1 and C_2 primers were labeled with T_4 polynucleotide kinase (USB) as recommended by the manufacturer. Five picomoles of labeled oligonucleotides were co-precipitated with 25µg of total RNA. After denaturation at 95°C for 2 min, annealing were carried out at room temperature for 20 min. Extension reactions were performed at 42°C in total volume of 15 µl of reaction buffer containing 50 mM Tris-HCl pH 8.3, 8 mM $MgCl_2$, 50 mM NaCl, 5 mM DTT, 1 mM each of dNTPs, 20 U RNasin (Promega) and 12 U AMV reverse transcriptase (USB). Reactions were stopped after 30 min, and loaded onto a standard sequencing gel.

3 Results and Discussion

Synechocystis cells were cultured at 30°C, exposed to different temperatures and the extracted total RNA was subjected to northern analysis. Hybridization with *groES*-specific probe revealed that the *groESL* mRNA level increased gradually up to 44°C (Fig. 1A).

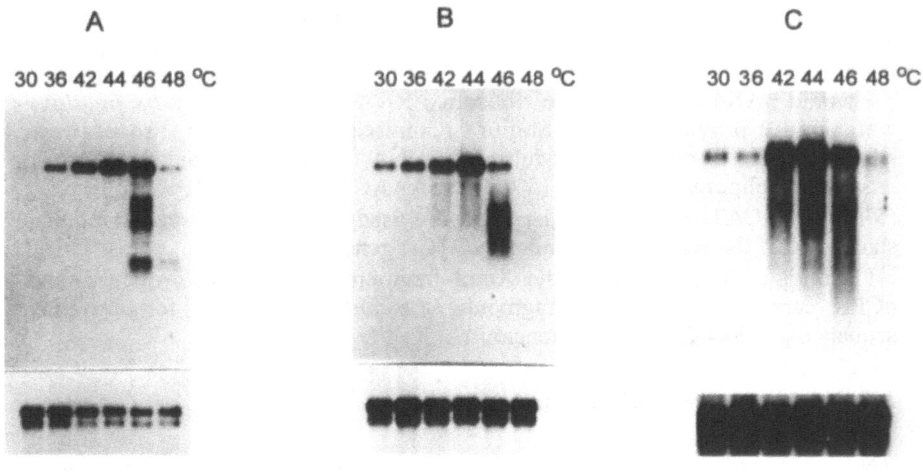

Fig. 1. Effect of heat stress on the level of chaperone mRNAs. Cells were incubated at temperatures indicated and total RNA were subjected to northern analysis using *groESL* (A), *cpn60* (B) and *dnaK* (C) specific probes (upper panels). Hybridizations with rRNA were used as loading controls (lower panels)

At higher temperatures, e.g. 46 and 48°C, transcript of lower mol wt were detected and the amount of the *groESL* transcript decreased dramatically. The expression pattern of the *cpn60* gene proved to be very similar to that of *groESL* except that the *cpn60* message started to decline at 46°C (Fig. 1B). Since both PSII activity and *de novo* protein synthesis were shown to be impaired at this temperatures even in cells adapted to 36°C (Lehel et al., 1993), presence of shorter mRNA species detectable at 46 and 48°C could be the result of degradation.

The *dnaK* gene has also proved to be strongly heat-inducible. While chaperonin genes displayed a significantly elevated transcriptional activity upon exposure of cells to mild heat stress (36°C), *dnaK* remained to be uninduced under identical conditions (Fig. 1C). This findings indicate that the mechanism of transcriptional activation of *dnaK* differs from that shown for the operation of *groEL*-analog genes. Differential regulation of two chaperone families is further supported by data obtained with heavy metal stress. Cells were incubated with different concentrations of $CdCl_2$ for 30 min, and the chaperone mRNA levels were determined. At a higher concentration of cadmium (Fig. 2) the level of activation obtained for *dnaK* was comparable with the heat induction. In contrast, the two *groEL*-type genes did not reveal such a strong induction. These results strongly suggest that the chaperone gene families possess distinct transcriptional controlling mechanisms in *Synechocystis* PCC6803.

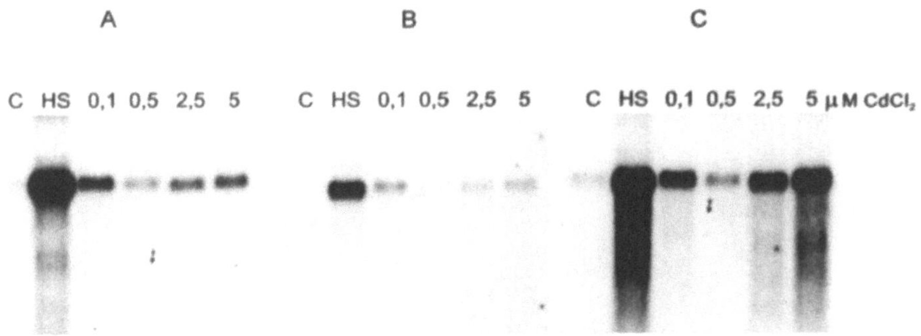

Fig. 2. Expression of chaperone genes under heavy metal stress. Total RNA were purified from cells exposed to different concentrations of $CdCl_2$ for 30 min and analyzed by northern hybridization. RNA extracted from cells incubated at 30 (lanes C) and 42°C (lanes HS) for the same time were used as controls. Filters were hybridized with probes complementary to *groESL* (panel A), *cpn60* (panel B) and *dnaK* (panel C) mRNA. The same filter was used in all three cases

In order to obtain further information on the transcriptional regulation of the two *groEL*-analog chaperonin genes, effects of various drugs has been tested. Prior to heat stress, cells were incubated with metabolic inhibitors for 30 min in order to ensure the proper level of the penetration of these compounds (Fig 3). As expected, addition of rifampicin completely inhibited the transcription of both chaperonin genes. As an indication that protein factor(s) are required for the transcriptional activation and/or mRNA stabilization, chloramphenicol treatment abolished also the formation of intact chaperonin mRNAs. A transcript of lower mol wt was observed and this might be a consequence of degradation and/or early termination. Reprobing the filter with a fragment representing the 3' end of the *groESL* operon revealed a single band corresponding to the full length mRNA, suggesting an early termination process (data not shown). The treatment with the photosyntheting inhibitor diuron had no significant effect on *groESL* activation. In contrary, the formation of *cpn60* mRNA was reduced dramatically upon diuron treatment: the transcript level was even somewhat lower than that seen under basal conditions. To our knowledge this is the first evidence showing that manipulation of the rate of photosynthetic electron transport modulates chaperonin gene transcription. This difference found in the regulation of the two chaperonin genes might support our hypothesis, that GroEL and Cpn60 proteins fulfill physiologically distinct roles.

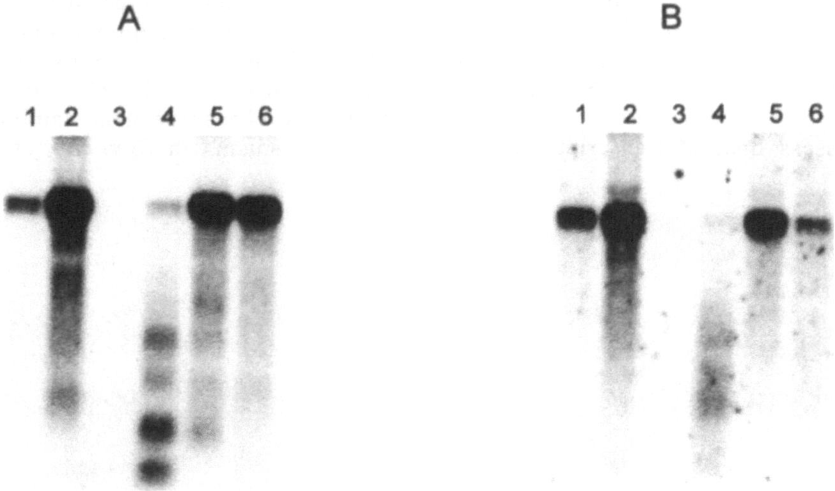

Fig. 3. The combined effect of various drugs and heat stress on transcriptional activity of the two *groEL*-analog genes. Cells were incubated at 30 °C in the absence (lanes 1-2) or in the presence of antibiotics (lane 3: 20μg/ml rifampicin; lane 4: 100μg/ml chloramphenicol; lane 5: 100μg/ml novobiocin and lane 6: 3.5 μg/ml diuron) for 30 min in and then shifted to 42°C for 30 min (lanes 2-6). Total RNA was hybridized to *groESL* specific fragment (A). The filter was stripped and then rehybridized with radiolabeled *cpn60* probe (B)

Finally, primer extention experiments were carried out in order to identify the promoter region of the chaperonin genes. Since the transcriptional startpoints. proved to be identical for both chaperonins irrespective of temperature treatment, our results clearly suggest that no additional heat shock promoter is present in front of chaperonin genes in *Synechocystis* PCC6803 (Fig. 4). The first nucleotide of the mRNAs is located within the CIRCE element, similarly to *Synechococcus* PCC7942 reported by Webb et al. (1990).

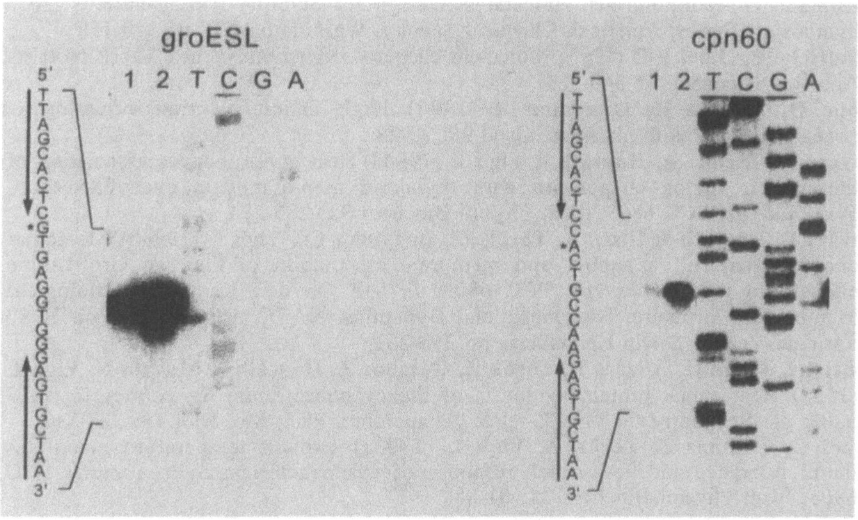

Fig. 4. Mapping of the 5' end of mRNAs of both chaperonin genes by primer extension analysis. ^{32}P radiolabeled C_1 (*groESL*) and C_2 (cpn60) primers were hybridized to 25µg of total RNA from *Synechocystis* before (lane 1) and 30 min after a heat shock from 30 to 42°C (lane 2). The primer extension products were analyzed on a sequencing gel. T, C, G, and A are products of the sequencing reactions obtained by using the same oligonucleotides as primers. The depicted sequences represent the noncoding strands. The converging arrows mark CIRCE elements. Potential transcriptional start points are indicated by asterisks

In conclusion, operation of distinct transcriptional regulatory mechanisms was evidenced for both the two chaperone families and the individual chaperonins, also, in *Synechocystis*.

References

Ausubel FM, Brent R, Kingston RE, Moore DD, Smith JA, Struhl K (1987) Current Protocols in Molecular Biology. John Wiley & Sons USA

Berry JA, Bjorkman O (1980) Photosynthetic response and adaptation to temperature in higher plants. Annu Rev Plant Physiol **31**: 491-553

Bukau B (1993) Regulation of the *Escherichia coli* heat shock response. Mol Microbiol **9**: 671-680

Chitnis PR, Nelson M (1991) Molecular cloning of the genes encoding two chaperone proteins of the cyanobacterium *Synechocystis* sp. PCC 6803. J Biol Chem **266**: 58-65

Ellis RJ, van der Vies S (1991) Molecular chaperones. Annu Rev Biochem **60**: 321-347

Gatenby AA, Donaldson GK, Baneyx F, Lorimer GH, Viitanen PV, van der Vies S (1993) Participation of GroE heat shock proteins in polypeptide folding. *In* ME

Himmel, G Georgiou, eds, Biocatalyst Design for Stability and Specificity. ACS Symposium Series, American Chemical Society, Washington DC, pp 140-150

Hendrick JP, Hartl F-U (1993) Molecular chaperone functions of heat shock proteins. Ann Rev Biochem **62**: 349-384

Inoue H, Nojima H, Okayama H (1990) High efficiency transformation of *Escherichia coli* with plasmids. Gene **96**: 23-28

Kovács E, Török Zs, Horváth I, Vigh L (1994a) Heat stress induces association of the GroEL analog chaperonin with thylakoid membranes in cyanobacterium, *Synechocystis* PCC 6803. Plant Physiol Biochem **32**: 285-293

Kovács E, Horváth I, Glatz A, Török Zs, Bagyinka Cs, Vigh L (1994b) Molecular characterisation , assembly and membrane association of the two GroEL-type chaperonin in *Synechsystis* PCC 6803. *In* JAF Op den Kamp, ed, Biological Membranes: Structure, Biogenesis and Dynamics. NATO ASI Series, Vol. H 82 Springer-Verlag, Berlin Heidelberg, pp 253-261

Lehel Cs, Wada H, Kovács E, Török Z, Gombos Z, Horváth I, Murata N, Vigh L (1992) Heat shock protein synthesis of the cyanobacterium *Synechocystis* PCC 6803: purification of the GroEL-related chaperonin. Plant Mol Biol **18**: 327-336

Lehel C, Gombos Z, Török Z, Vigh L (1993a) Growth temperature modulates thermotolerance and heat shock response of cyanobacterium *Synechocystis* PCC 6803. Plant Physiol Biochem **31**: 81-88

Lehel C, Los D, Wada H, Györgyei J, Horváth I, Kovács E, Murata N, Vigh L (1993b) A second groEL-like gene, organized in a groESL operon is present in the genome of *Synechocystis* sp. PCC 6803. J Biol Chem **268**: 1799-1804

Narberhaus F, Bahl H (1992) Cloning, sequencing and molecular analysis of the groESL operon of *Clostridium acetobutilycum*. J Bacteriol **174**: 3282-3289

Narberhaus F, Giebeler K, Bahl H (1992) Molecular of the dnak gene region of *Clostridium acetobutilycum*, including grpE, dnaJ, and a new heat shock gene. J Bacteriol **174**: 3290-3299

Schulz A, Tzschaschel B, Schumann W (1995) Isolation and analysis of mutants of the dnak operon of *Bacillus subtilis*. Mol Miocrobiol **15**: 421-429

Tomioka N, Shinozaki K, Sugiura M (1981) Molecular cloning and characterization of ribosomal RNA genes from a blue-green alga, *Anacystis nidulans*. Mol Gen Genet **184**: 359-363

Vigh L, Török Z, Kovács E, Glatz A, Balogh N, Horváth I (1994) Thermal acclimation and heat stress response of *Synechsystis* PCC 6803: The possible role of thylakoid physical state, lipid saturation and molecular chaperones. *In* JH Cherry, ed, Biochemical and Cellular Mechanisms of Stress Tolerance in Plants. NATO ASI Series, Vol. H 86 Springer-Verlag, Berlin Heidelberg, pp 77-95

Vigh L, Los D, Murata N, Glatz A, Kovács E, Horváth I (1995) Is the membrane the primary target in the biological perception of temperature? Effect of membrane physical state on the expression of stress-defence genes. *In* J-C Kader, P Mazliak, eds, Plant Lipid Metabolism, Kluwer Academic Publishers, The Netherlands, pp 369-371

Webb R, Reddy KJ, Sherman L (1990) Regulation and sequence of the *Synechococcus* sp. strain PCC 7942 groESL operon encoding a cyanobacerial chaperonin. J Bacteriol **172**: 5079-5088

Yanisch-Perron C, Viera J, Messing J (1985) Improved M13 phage cloning vectors and host strains: nucleotide sequence of the M13mp18 and pUC19 vectors. Gene **33**: 103-119

Yuan G, Wong SL (1995a) Isolation and characterization of *Bacillus subtilis* groE regulatory mutants : Evidence for orf39 in the dnaK operon as a repressor gene in regulating the expression of both groE and dnaK. J Bacteriol **177**: 6462-6468

Yuan G, Wong SL (1995b) Regulation of groE expression in *Bacillus subtilis*: The involvement of the σ^A-like promoter and the roles of the inverted repeat sequence (CIRCE). J Bacteriol **177**: 5427-5433
Yura T, Nagai H, Mori H, (1993) Regulation of the heat shock response in bacteria. Annu Rev Microbiol **47:** 321-350
Zuber U, Schumann W (1994) CIRCE, a novel heat shock element involved in regulation of heat shock operon dnaK of *Bacillus subtilis*. J Bacteriol **176**: 1359-1363

Genetic Dissection of Thermotolerance in Maize

Carla Frova

Dipartimento di Genetica e di Biologia dei Microrganismi, I-20133 Milano, Italy

Abstract. High temperature tolerance is one of the main components of yield stability in crop plants. The genetic mechanisms controlling variability within and between species remain largely unknown, even though their genetical basis is expected to be complex. Therefore the genetic dissection of the character requires adequate methods of analysis. A two step strategy was followed: i) character dissection into physiological components, such as cellular membrane stability (CMS), heat shock protein (HSP) synthesis, pollen germination and tube growth, root elongation; ii) genetic dissection of these components by means of molecular markers, such as RFLPs, used to detect QTLs (Quantitative Trait Loci) controlling each trait. All analyses were performed on a population of 45 recombinant inbred lines (RI) derived by two contrasting inbreds. In each case, except for HSPs, thermotolerance was evaluated as degree of injury of stressed *versus* non stressed material. HSP expression in stressed seedling roots was measured quantitatively by reading each HSP band after lD electrophoresis with a densitometric scanner. CMS was evaluated as electrolyte leakage from leaf disks. For all traits a significant difference among the RI genotypes was found. By RFLP analysis from 3 to 7 QTLs were identified for the different components. A comparison between all the QTLs detected revealed that some of them appear to be in common between the physiological components of thermotolerance, thus opening the way to marker assisted selection procedures.

Keywords. Heat shock protein, maize, quantitative molecular markers, thermotolerance, trait loci, *Zea mays*

1 Introduction

High temperature stress is one of the main factors accounting for unpredictable and often severe yield losses worldwide. Accordingly thermotolerance, i.e. the ability to maintain and protect basic plant metabolic and developmental processes on exposure to high temperatures, is an important component of yield stability. Improvement of this trait is a major target within the general objective of breeding genotypes characterized by a stable yield also in limiting environmental conditions. The task, however, is not easy for several reasons: i) thermotolerance is a very complex trait resulting from the optimization of several complex physiological processes at both the cellular and organismic level and at different stages of plant development; ii) due to its complexity and to a strong genotype-environment interaction the trait shows a low heritability and is difficult to measure, so that the classification of genotypes as tolerant or susceptible is often uncertain; iii) the genetical basis of the trait is almost unknown even though a large number of genes is expected to be involved.

A possible strategy to by-pass these problems is to dissect the trait into physiological components that generally show a higher heritability, for which accurate and reliable laboratory tests are available, and to analyse them separately. This allows to perform the analyses in controlled conditions and on traits characterized by simpler genetics. The genetic dissection of the components can be performed by means of molecular markers (RFLP, RAPD, AFLP, microsatellites). These markers have several advantages: i) they are genetically neutral, i.e. they are independent of gene expression, of environmental effects or of interaction with other genes and show no temporal or tissue specificity; ii) they are usually highly polymorphic; iii) their number is almost unlimited, so as to saturate the whole genome. The availability for many species of densely saturated maps allows, at least theoretically, the identification and chromosomal localization of near all the quantitative trait loci (QTL) controlling a quantitative trait, such as thermotolerance components, on the basis of the association between the trait expression and the allelic composition of a segregating population at each of the marker loci included in the map (Burr et al., 1988; Tanksley, 1993). The data so obtained can be subsequently pooled together for the evaluation of each genotype global tolerance and for the reconstruction of the genomic architecture of the trait. Moreover, the information can be exploited for marker assisted selection (MAS): the choice of the individuals to be included in the next generation is made in a non destructive way on the basis of the molecular alleles without the need to evaluate the character, and at a very early stage of the plant development.

2 Materials and Methods

2.1 Genetic Material

All analyses were performed on a set of 46 recombinant inbred lines (RI), derived from repeated selfing of the F_1 between two highly diverse lines, T232 and CM37. The population was already characterized for approximately 200 RFLP loci. This type of material is a very powerful tool for genetic analysis since RIs are a highly homozygous and permanent population; this allows a very accurate evaluation of each trait, and also to compare data obtained for different components, in different years and environments.

2.2 Physiological Components

The physiological components of thermotolerance analyzed in this work, their main characteristics and the evaluation tests are summarized in Table 1.

Table 1. Thermotolerance component traits included in the analysis

Traits	Level	Variability	Evaluation
CMS (leaves)	cell	**	Injury (electrolyte leakage)
HSP expression (root)	cell	* /**	Relative band intensity
Root growth	organ	*	Injury
Pollen germination	organ	**	Injury
Tube growth	organ	*	Injury

* and ** indicate genetic differences among 46 RI genotypes, significant at $P<0.05$ and $P< 0.01$, respectively

a) Cellular Membrane Stability (CMS). It is considered one of the main components of cellular tolerance to temperature and water stress. Its reliability as an index of heat stress tolerance is supported, in several plant species, by a good correlation between CMS and plant performance in the field under high temperature stress (Martineau et al., 1979; Sullivan and Ross, 1979). The trait was evaluated as membrane injury by measuring the electrolyte leakage in controlled and stressed leaf disks cut from the fully expanded second leaf of 10-day-old seedlings. $I = 1- \{1 - (T_0/T_1) / 1 - (C_0/C_1)\} \times 100$ (Ottaviano et al., 1991).

b) Quantitative expression of heat shock proteins (HSP). Among the molecular factors involved in the development of thermotolerance, HSPs are assumed to play a prominent role. Their involvement in the short term adaptation of plants to high temperature stress is supported by a large body of correlative evidence (Lin et al., 1989; Nagao et al., 1990), and the chaperone-like function reported for some of them (Vierling, 1991) suggests a possible role in maintaining cellular membrane stability. The reason for a quantitative assay of HSP expression is that in plants intraspecific qualitative variability, evaluated as presence/absence of specific HSP bands on a protein gel, is very rare, while quantitative variability in band intensities is high. Temperature treatment (3 h at 41°C) was applied *in vivo* to intact 4-day-old seedlings, whose primary root was immersed in dH_2O in the presence of labeled methionine. After the treatment, root tips were excised, the proteins extracted, subjected to 1D PAGE and autoradiographed. Each lane on the autoradiographs was analyzed with a densitometric scanner and the intensity of each HSP band, expressed as percentage of the total protein intensity of the lane, was used as the measure of quantitative HSP expression (Frova and Sari Gorla, 1993).

c) Radicle growth in stress conditions was chosen as a general index of seedling thermotolerance. Germination and radicle elongation in maize is highly influenced by temperature, being strongly inhibited over 35°C (Blacklow, 1972).

For the test, seeds were germinated in the dark at 27°C. When the primary root was 1 cm long, seedlings were heat-stressed at 45°C for 3 h and subsequently returned to 27°C for recovery. Four d later root length was recorded. Thermotolerance was evaluated as rate of injury (IRG) by comparing root growth of treated (T) and control (C) samples (I = 1 - T/C).

d) Pollen germination ability and tube growth under stress. Pollen thermotolerance is a major factor contributing to yield stability, since it relates to the ability of the plant to protect reproduction processes under high temperature stress. This is particularly relevant for those crops, as maize, for which the main product are seeds and for which flowering and fertilization occur on the hottest days of the year. In fact, in maize a reduction in seed set is consistently observed when temperatures at flowering time exceed 38°C. The phenomenon is mainly due to a drop in pollen germination and tube growth, while the female components of the system are much less affected (Herrero and Johnson, 1980; Binelli et al., 1985; Shoper et al., 1986; Dupuis and Dumas, 1990). Freshly shed pollen from field grown plants was grown on a solid germination medium at 27°C for 180 min (control), at 41°C for 120 min followed by 60 min at 27°C (T1), at 27°C for 30 min followed by 120 min at 41°C and by 30 min at 27°C (T2). T1 and T2 are specific treatments for the evaluation of injury of germination ability (IPGG) and tube growth (IPTG), respectively. Germination percentage and tube growth were recorded at the end of the treatment. Thermotolerance is expressed as percent injury of treated *versus* control pollen (I = 1 - T/C) (Frova and Sari Gorla, 1994).

2.3 Linkage Analysis between RFLPs and Thermotolerance Components

Detection of the chromosome regions where loci controlling the studied characters are located was carried out by regression analysis. The mean value of the character for each line is the dependent variable, and the allelic composition of each molecular marker is the regressor, assuming the value of 1 or 2 according to the presence of the allele from P1 (T232) or P2 (CM37). A significant regression is interpreted as an indication that a QTL controlling the trait is closely linked to that RFLP locus. The sign of the coefficient of regression indicates which parental line carries the allele that increases the trait value. In order to avoid false assignements, the correlation matrices between all significant loci were computed and for each character only the significant non correlated loci were included in a multiple regression model, on the basis of which the cumulative contribution of these loci to the expression of the trait (multiple R^2) was evaluated.

3 Results and Discussion

Analysis of variance indicated significant differences between the 46 recombinant inbred lines for CMS, IRG, IPGG, IPTG and 5 HSPs (84, 82, 27, 22 and 18 kD). Heritability estimates were generally high, h^2 ranging from 0.48 (IRG) to 0.73 (CMS).

Linkage analysis detected a minimum of 3 (HSP 84 and HSP18) and a maximum of 8 (HSP 27) QTLs for each trait. The results are summarized in Table 2: the number of QTLs identified and the proportion of genetic variability explained (expressed as the ratio between the multiple R^2 and h^2) are indicated for each component.

Table 2. QTLs detected by linkage analysis with molecular markers for different trait components of thermotolerance

Trait	Loci no.	Genetic Variability Explained (R^2/h^2) %
CMS	6	54
HSP84	3	35
HSP82	4	52
HSP27	8	60
HSP22	6	55
HSP18	3	50
IRG	4	86
IPGG	5	80
IPTG	6	88

On average the number of QTLs detected was relatively small and the high R^2/h^2 values indicate that most of them were identified by the analysis.

The use of the same genetic material for the dissection of all components allows additional information to be obtained by comparison of the data. For HSPs, for instance, the comparison identifies a few clusters of loci that are significant for three of the five HSPs included in the analysis, thus indicating a possible regulatory role for the associated QTLs. Three significant loci (on chromosomes 1, 8 and 9) were common to HSPs and CMS, and the sign of the coefficient of regression shows that the alleles associated with an increased HSP synthesis are also associated with a lower level of injury of cell membrane. Thus, even though the correlation between CMS and the different HSPs was not

Table 3. Coefficient of regression of RFLP loci associated to QTLs involved in the control of different thermotolerance components

Chromosome	Characters									
	Locus	CMS	HSP84	HSP82	HSP27	HSP22	HSP18	IPGG	IPTG	IRG
					b values					
1	YNH20			0.57	-0.73			-1.00	-5.07	
	7.21	**-0.10**			**0.86**					
2	NPI271						0.62	1.26		
	Dia1				0.68			1.16		
	5.21B	**0.10**							**4.99**	
3	**6.06**					**-0.27**		**1.25**		
7	15.21			0.52				1.27		
	6.27				0.73			1.00		
	4.24						0.51	1.31		
8	**NPI220**	**0.136**				**-0.24**		**1.56**		
	9.11			0.83		-0.38		1.07		
	Mdh1		-0.67	0.51					-7.27	
	10.39		-0.69						-8.02	
	9.44L		-0.61						-0.67	6.42
	7.08			**0.54**					**-5.83**	
9	**C1**		**0.64**					**-1.02**		
	Wx1	-0.133	0.94						4.60	

significant, it is possible to postulate that HSPs play a minor role in maintaining CMS. In the case of the two pollen traits, the map position of the QTLs identified for IPGG and IPTG is very different from that of the QTLs that control germination ability and tube growth in non stress conditions. This means that different sets of genes are involved and therefore pollen performance at normal temperatures is not necessarily predictive of its performance under stress.

These results can be exploited for breeding purposes. Analysis of all data together reveals that the favourable alleles for the different components are not carried all by T232, on the whole the most thermotolerant parental line. In several cases the sign of the coefficient of regression indicates that they are those of the susceptible line CM37. It should thus be possible to select for genotypes that maximize the number of favourable alleles. Monitoring all loci involved in the control of each component is not feasible. However, it is possible to choose key chromosomal regions where factors with appreciable effects on more components are localized. This multiple association can be due either to the presence of more QTLs, each relative to one trait component, or to pleiotropic effects or to the presence of a factor with regulatory function. Whatever the biological basis, the choice of the adequate marker allele could help selecting for the favourable QTL(s). For instance, in this case (Table 3, bold figures), by selecting the allele of line T232 (P1) for the loci 5.21B (Chr 2), 6.06 (Chr 3) and NPI220 (Chr 8), one would select QTL alleles that increase cellular membrane stability, tolerance of pollen and HSP synthesis at the same time. For the loci 7.21 (Chr 1), 7.08 (Chr 8) and C1 (Chr 9) instead, the "good" allele to be chosen would be that of line CM37 (P2).

Acknowledgements. This research was supported in part by Ministero Agricoltura e Foreste, Italy, Programma Nazionale Resistenze Genetiche delle Piante Agrarie agli Stress Biotici e Abiotici.

References

Binelli G, Vieira de Manincor E, Ottaviano E (1985) Temperature effects on pollen germination and pollen tube growth in maize. Genet Agr **39**: 269-281

Blacklow WM (1972) Influence of temperature on germination and elongation of the radicle and shoot of corn (*Zea mays* L). Crop Sci **12**: 647-650

Burr B, Burr FA, Thompson KH, Albertson CM, Stuber CW (1988) Gene mapping with recombinant inbreds in maize. Genetics **118**: 519-526

Dupuis I, Dumas C (1990) Influence of temperature stress on *in vitro* fertilization and heat shock protein synthesis in maize (*Zea mays* L) reproductive tissues. Plant Physiol **94**: 665-670

Frova C, Sari Gorla M (1993) Quantitative expression of maize HSPs: genetic dissection and association with thermotolerance. Theor Appl Genet **86**: 213-220

Frova C, Sari Gorla M (1994) Quantitative trait loci (QTLs) for pollen thermotolerance detected in maize. Mol Gen Genet **245**: 424-430

Herrero HP, Johnson RR (1980) High temperature stress and pollen viability in maize. Crop Sci **20**: 796-800

Lin CY, Roberts JK, Key JL (1984) Acquisition of thermotolerance in soybean seedlings. Plant Physiol **74**: 152-160

Martineau JR, Specht JE, Williams JH, Sullivan CY (1979) Temperature tolerance in soybeans I. Evaluation of a technique for assessing cell membrane thermostability. Crop Sci **19**:75-78

Nagao RI, Kimpel JA, Key JL (1990) Molecular and cellular biology of the heat shock response. Adv Genet **28**: 235-274

Ottaviano E, Sari Gorla M, Pè ME, Frova C (1991) Molecular markers (RFLPs and HSPs) for the genetic dissection of thermotolerance in maize. Theor Appl Genet **81**: 713-719

Shoper JB, Lambert RJ, Vasilas BL (1986) Pollen viability, pollen shedding and combining ability for tassel heat tolerance in maize. Crop Sci **27**: 27-31

Sullivan CY, Ross WM (1979) Selecting for drought and heat resistance in grain sorghum. *In* H Mussell, R Staples, eds, Stress Physiology in Crop Plants. John Wiley and Sons, NY, pp 263-281

Tanksley SD (1993) Mapping polygenes. Annu Rev Genet **27**: 205-233

Vierling E (1991) The roles of heat shock proteins in plants. Annu Rev Plant Physiol Plant Mol Biol **42**: 579-620

Derepression of the Heat Shock Response and Genetic Engineering of Thermotolerance in *Arabidopsis* using Recombinant Heat Shock Factor

Fritz Schöffl, Anja Hübel and Jeong Hee Lee

Universität Tübingen, Allgemeine Genetik, D-72076, Tübingen, Germany

Abstract. The co-ordinate expression of HSPs is regulated by plant heat shock (HS) transcription factor (HSF) which recognizes the conserved HSE promoter sequences in a temperature-dependent fashion and subsequently stimulates transcription of HS genes. The activation of HSF is the key to an alteration of the HS response by genetic engineering, e.g. for constitutive synthesis of HSPs and increase in thermotolerance. Using *Arabidopsis* as a model for higher plants we have isolated *Athsf1*, the prime HSF gene responsible for the control of transcription of HS genes. This gene is constitutively expressed but its activation for trimerization, DNA-binding, and transcriptional stimulation of HS genes depends on heat stress. Ectopically over-expressed ATHSF1 is regulated in the same way in transgenic *Arabidopsis*. Derepression of HSF activity was achieved by transgenic expression of gene fusions between *Athsf1* and a glucuronidase reporter gene. The recombinant HSF/GUS proteins are constitutively active at normal temperature as demonstrated by trimerization, DNA-binding and a constitutive synthesis of HSPs. Transgenic plants exhibit a significantly higher level of basic thermotolerance. These data have implications for the mechanism of HSF regulation and provide evidence for a fundamental role of HSPs in plant stress tolerance.

Keywords. *Arabidopsis thaliana*, glucuronidase enzyme, heat shock factor, heat shock proteins, thermotolerance

1 Introduction

Each organism exhibits basic thermotolerance that allows survival at high temperatures. The basic level can be significantly increased by a sublethal heat treatment. This is the acquisition or induction of a higher level of

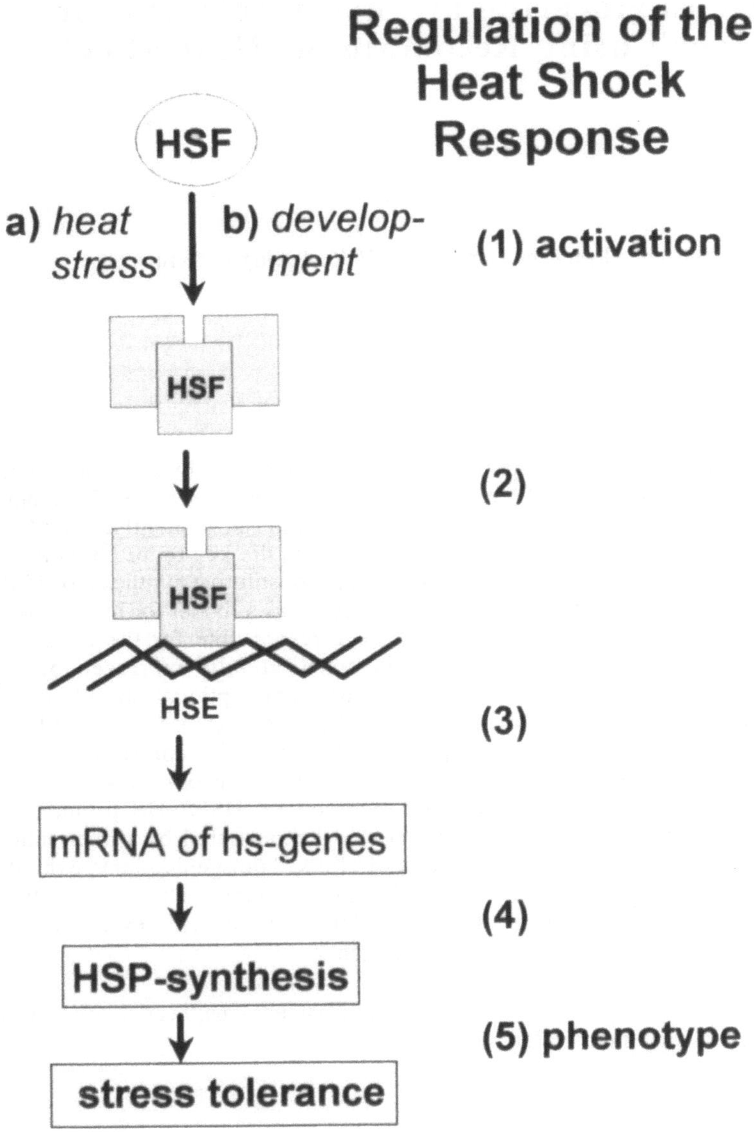

Fig. 1. Molecular model of HSF activation and regulation of the HS response

thermotolerance that protects the cells and organisms from a subsequent, otherwise lethal heat treatment (Yarwood, 1961; Lin et al., 1984). The synthesis of HSPs correlates with the acquisition of thermotolerance and the HS response is also induced by some other environmental stressors (Lin et al., 1984; see Vierling, 1991, for review). The expression of HSPs is primarily regulated at the transcriptional level, the activation of HSF plays a central role in the control in the expression of HSPs and the regulation of the HS response. HSF genes have been isolated from tomato (Scharf et al., 1990) and *Arabidopsis* (Hübel and Schöffl, 1994) and it was demonstrated that recombinant HSFs were able to recognize the conserved DNA binding site, the heat shock element (HSE) located in the promoter regions of HS genes. ATHSF1, the HSF of *Arabidopsis*, is constitutively expressed (Hübel and Schöffl, 1994) but its activation for DNA binding is strictly dependent on heat stress and requires a transition from monomers to trimers (Hübel and Schöffl, 1994; Hübel et al., 1995). The DNA binding domain is highly conserved among organisms (Vuister et al., 1994), efficient DNA binding requires a second domain which is involved in oligomerization. These domains, including leucine zipper motifs, are less conserved than the DNA binding regions.

The keys to the regulation of the HS response are the molecular signals and events involved in the control of HSF trimerization and consequently DNA binding (see Fig. 1). Expression of recombinant *Arabidopsis* ATHSF1 in *E. coli, Drosophila* and human cells resulted in a constitutive DNA binding activity (Hübel and Schöffl, 1994; Hübel et al., 1995), the derepression of transcriptional activity of ATHSF1 was demonstrated in *Drosophila* by DNA footprinting to a *Dhsp70* promoter and by ATHSF1-dependent expression of a *Dhsp70-CAT* reporter gene (Hübel et al., 1995). However, over-expression of ATHSF1 in stably transformed transgenic *Arabidopsis* was not sufficient to derepress HSF activity at normal temperature (Hübel et al., 1995).

The goal of our recent investigation was to break the negative regulation of HSF that should consequently alter the HS response in *Arabidopsis*. Here we summarize our results about a very special and efficient way of genetic engineering of ATHSF1 by N- and C-terminal fusions with the GUS reporter enzyme. Stably transformed HSF/GUS plants exhibit both, constitutive HSPs and significantly enhanced levels of basic thermotolerance.

2 Results and Discussion

2.1 Derepression of the Activity of HSF/GUS Fusion Proteins In Transgenic *Arabidopsis*

In *Arabidopsis* wild type (wt) and transgenic ATHSF1 over-expressing lines DNA binding and transcriptional activation of HS genes by HSF requires still activation by HS (Hübel et al., 1995). High levels of CaMV 35S-driven ATHSF1 in transgenic lines (HSF↑) suggest that the negative regulation of HSF activity cannot be broken by an approximately 10-fold over-expression of

HSF, this higher concentration was not sufficient to titrate a potential trans-active negative regulator of HSF activity in *Arabidopsis*. There were no obvious changes in the HS response and in thermotolerance phenotype in these lines (Lee et al., 1995). In order to generate a constitutively active HSF another attempt was devoted to the modification of HSF by genetic engineering. A number of different *Athsf1* constructs containing deletions, modifications, or representing gene fusions were generated. Surprisingly, the most pronounced effects in transgenic *Arabidopsis* were obtained using protein fusion constructs between HSF and GUS proteins (Lee et al., 1995).

2.1.1 Expression of Chimeric HSF/GUS Gene Constructs

Two types of gene fusions, derived from CaMV promoter-driven HSF, were constructed (Fig. 2).

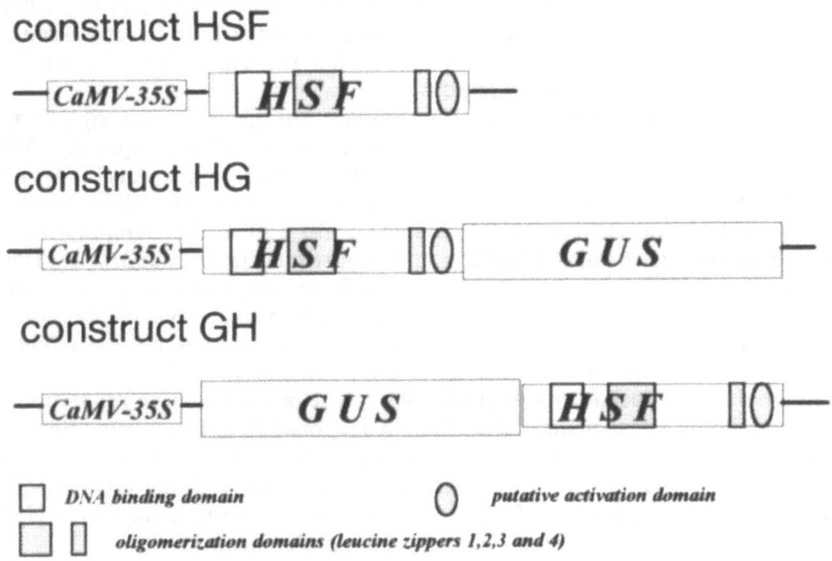

Fig. 2. Chimeric HSF/GUS constructs for transformation of *Arabidopsis*

Both types, HSF-GUS (HG) and GUS-HSF (GH), contain almost the complete amino acid sequences of both proteins with only minor changes at the fusion points. In transgenic *Arabidopsis* plants generated via *Agrobacterium tumefaciens*-mediated transformation of BIN19-derived vectors, both genes were expressed to about the same constitutive level as shown by Northern hybridizations and determination of GUS activities of fusion proteins.

The mRNA levels of the chimeric HSF genes were significantly higher than the mRNA level of the authentic wt HSF.

The oligomeric state of the chimeric HSF-GUS and GUS-HSF proteins was determined by cross-linking experiments following expression in *Arabidopsis* and in *E.coli*. For expression in *E. coli*, the chimeric genes were cloned into pQE-32 vector that allowed the purification of the recombinant proteins by affinity chromatography. In both organisms HSF-GUS and GUS-HSF proteins formed trimer complexes of approximately 540 kD (monomer: 180 kD). Owing to the intrinsic GUS activity of the chimeric proteins it was possible to visualize the proteins in gels using a novel GUS activity staining protocol (Lee and Schöffl, 1995). This method is highly sensitive and will have a wide range of applications for studying protein:protein and protein:DNA interactions.

The formation of HSF-GUS trimers at normal temperature (25°C) indicated that at least the initial steps in HSF activation were derepressed in transgenic plants.

2.1.2 Constitutive DNA Binding of HSF-GUS Trimers in *Arabidopsis*

The molecular and functional properties for HSF-GUS and GUS-HSF genes and proteins in *Arabidopsis* were identical, only the data for HSF-GUS are further exemplified in this summary. Using gel retardation assay it was shown that the constitutive HSF-GUS trimers (present in total protein extracts from transgenic plants grown at 25°C) had the capacity to bind DNA (containing the consensus HSE sequences of HS promoters). The authentic wt HSF remained as an inactive monomer, unable to bind DNA under these conditions. However, following HS (37°C) the HSF-GUS complexes, HSF trimers and two forms of HSF/HSF-GUS quasi-heterotrimers bound to DNA. Using recombinant proteins purified after expression in *E. coli*, heterotrimers could be reconstituted *in vitro*. The term heterotrimers is used operationally and does not mean an interaction between different types (species) of HSF, it refers only to the interaction of the ATHSF1 with its GUS-tagged HSF-GUS variants.

The formation of heterotrimeric HSF:DNA binding complexes of 340 and 440 kD (the HSF runs as an 80 kD monomer, HSF-GUS as an 180 kD monomer) is also a highly sensitive and powerful marker of the activation of the wt HSF in transgenic plants. We were able to demonstrate that the endogenous wt HSF is activated by temperatures between 30 and 45°C, with a maximum between 35 and 40°C, but not at 25 or 50°C.

2.1.3 Constitutive Transcription of HS Genes and Synthesis of HSPs in HSF-GUS Plants

The capacity of HSFs to bind DNA (HS promoters) is a prerequisite to transcription of HS genes and to the synthesis of HS proteins (HSP). An exception is the HSF of yeast, which binds constitutively to DNA, but in a second step that requires heat shock, becomes competent for transcriptional activation (Sorger and Pelham, 1988; Wiederrecht et al., 1988). By the criterion of constitutive DNA binding, a similar situation was generated in *Arabidopsis* containing HSF-GUS and it was interesting to check whether HS was required for transcriptional activation in these plants. Using northern and western blot techniques it was shown that transcription of HS genes and synthesis of HSPs (exemplified for the low mol wt HSP18 family) was constitutive at normal temperature (25°C) in transgenic *Arabidopsis*. Similar results were also obtained for HSP70 and other HSPs indicating that the entire HS response was derepressed in these plants. Quantification of the constitutive levels of HSP18 proteins revealed approximately 20% constitutive HSPs present at normal temperature compared to the maximum heat-inducible level in plants. These experiments provide evidence that heat shock is not essential for either activation of HSF, transcription of HS genes or translation of HSPs in plants. It is not known whether the different processes are as efficient under normal conditions as after heat shock.

2.1.4 Enhanced Levels of Basic Thermotolerance

The relatively high level of constitutive HSPs offered a good chance for identifying changes in thermotolerance in transgenic HSF-GUS plants compared to *Arabidopsis* wt and transgenic HSF↑ over-expressing lines. One-week old seedlings (grown at 25°C under controlled dark/light conditions) subjected to different HS regimes exhibited a dramatic change in the basic but not in the acquired thermotolerance. Survival temperature was increased by about 5°C up to 52°C after 1 h treatment compared to wt and HSF↑ plants. The acquired (or acclimated) level of thermotolerance, tested after 2 h pretreatment at 35°C, was not affected, all plants survived the subsequent heat treatments up to 54°C.

The increase in basic thermotolerance is the first evidence for determining HSPs as the important constituents of the cell and the organism in thermotolerance. It is noteworthy that HSP synthesis in these line was constitutive and uncoupled from the heat stress treatment.

3 Conclusions

Arabidopsis is the first plant species which has been altered in thermotolerance properties by genetic engineering. This is also the first example showing alteration of a complex pathway via manipulation of the activity of a central regulator in plants. HSF/GUS fusion proteins were uncoupled from heat-dependent activation and fulfilled all criteria of a constitutively derepressed HSF trimer formation, DNA binding, transcriptional activation of HS genes. So far there was no evidence that a trans-active negative regulator is involved in a repression HSF activity at normal temperature. Our results obtained with HSF-GUS suggest that conformational changes may be involved in formation of an active HSF trimer.

The physiological and phenotypic consequences of constitutive HSP synthesis and an enhanced level of basic thermotolerance are consistent with the model that HSPs are the important determinants of stress tolerance. The maximum level of the HS response or acquired thermotolerance seems to be dependent on stress dosage. The equal levels of acquired thermotolerance in wt and transgenic lines are consistent with equal levels of HSPs induced by HS in these plants. These findings can be extrapolated to other plant species in order to improve the stress tolerance phenotype.

The formation of HSF/HSF-GUS quasi-heterotrimers (Fig. 3) is a very sensitive indicator of HSF activity that can be tested in wt and transgenic plants. The technology will be applicable to other regulatory proteins, particularly to the investigation of multimeric complexes.

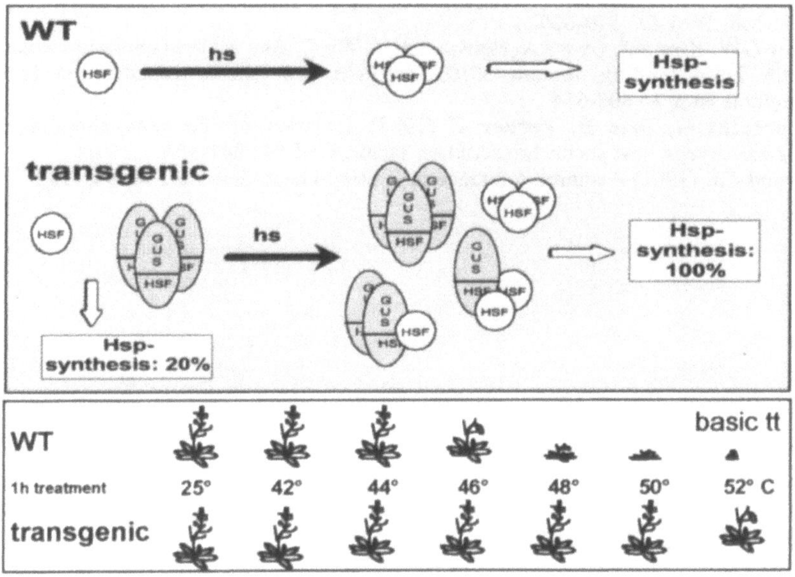

Fig. 3. Model of HSF activation and generation of enhanced basic thermotolerance in transgenic HSF-GUS *Arabidopsis* plants

Acknowledgements. This work was supported by grants of the Deutsche Forschungsgemeinschaft.

References

Hübel A, Schöffl F (1994) *Arabidopsis* heat shock factor: isolation and characterization of the gene and the recombinant protein. Plant Mol Biol **26**: 353-362

Hübel A, Lee JH, Wu C, Schöffl F (1995) *Arabidopsis* heat shock factor is constitutively active in *Drosophila* a human cells. Mol Gen Genet **348**: 136-141

Lee JH, Hübel A, Schöffl F (1995) Derepression of the activity of genetically engineered heat shock factor causes constitutive synthesis of heat shock proteins and increased thermotolerance in transgenic *Arabidopsis*. Plant J **8**: 603-612

Lee JH, Schöffl F (1995) GUS activity staining in gels - a powerful tool for studying protein interactions in plants. Plant Mol Biol Rep **13**: 365-373

Lin C-Y, Roberts JR, Key JL (1984) Acquisition of thermotolerance in soybean seedlings. Plant Physiol **74**: 152-160

Scharf K-D, Rose S, Zott W, Schöffl F, Nover L (1990) Three tomato genes code for heat stress transcription factors with a region of remarkable homology to the DNA binding domain of yeast HSF. EMBO J **9**: 4495-4501

Sorger PK, Pelham HRB (1988) Yeast heat shock factor is an essential DNA binding protein that exhibits temperature dependent phosphorylation. Cell **54**: 855-864

Vierling E (1991) The role of heat shock proteins in plants. Annu Rev Plant Physiol Plant Mol Biol **42**: 579-620

Vuister GW, Kim S-J, Orosz A, Marquardt J, Wu C, Bax A (1994) Solution structure of the DNA binding domain of *Drosophila* heat shock transcription factor. Structural Biol **1**: 605-614

Wiederrecht G, Seto D, Parker C (1988) Isolation of the gene encoding the *Saccharomyces* heat shock transcription factor. Cell **54**: 841-853

Yarwood CE (1961) Acquired tolerance in leaves to heat. Science **134**: 941-942

Part 2

Cold Stress

Physiological and Molecular Response of Barley to Cold and Drought Stress

Luigi Cattivelli, Cristina Crosatti, Maria Grossi, Carla Murelli, Fulvia Rizza and A. Michele Stanca

Istituto Sperimentale per la Cerealicoltura, Sezione di Fiorenzuola d'Arda, I-29017 Fiorenzuola d'Arda (PC), Italy

Abstract. Many physiological and molecular changes have been shown to occur at low temperatures, but the role and the contribution of each to frost resistance are still poorly understood. These changes include variations in enzyme activity, accumulation of soluble components, alterations in membrane composition and induction of specific genes.

The protein encoded by the cDNA clone *pt59*, induced in barley by cold treatment, was over-expressed in *E. coli* to produce the matching antibody, which *in vivo* recognizes a cold-induced protein of 14 kD (COR14). In the cold resistant cv. "Onice" the COR14 is stored in amounts only slightly greater than those found in the susceptible "Gitane", although the former has a higher induction-temperature threshold of COR14 than the latter. This feature may confer an evolutionary advantage enabling the resistant varieties in the field to be prepared for the cold well ahead the susceptible ones.

Two other cDNA clones, *paf93* and *cdr29*, are regulated by low temperature and drought stress, but not by exogenous ABA treatment. Indeed during the early stage of dehydration, when only a small increase occurs in ABA content, the mRNAs corresponding to these two genes are expressed before the induction of other known ABA-regulated genes, such as dehydrins. The sequence analysis revealed that *paf93* encodes a protein homologous to the cold-regulated protein COR47 of *Arabidopsis*, whereas *cdr29* represents a plant gene homologous to yeast and mammalian sequences coding for acyl-Coenzyme A oxidase.

Keywords. Barley, cold-regulated genes, cold stress, drought stress, gene expression, *Hordeum vulgare*

1 Introduction

Barley is well known for its wide adaptability to a range of different ecological situations Nevertheless, the ability of barley to growth in stressful environments can be limited by extreme stress situations which determine crop injury and yield loss. Studies of genetic resistance have intensely focused on the selection of those plant genotypes less subjected than others to environmental stress factors so as to increase yields and yield stability. Thus, genetic variability plays a primary role in determining positive adaptation to environmental stresses and, hence, in supporting the spread of various barley genotypes to extreme climatic conditions (Stanca et al., 1992). Frost resistance in barley is an important agronomic character because barley sowing has been shifted from spring to autumn, as winter varieties have been found to give higher yield than spring cultivars. The tolerance of winter cereals to low temperatures depends on the physiological process known as hardening or acclimation, that occurs when plants are exposed to temperatures ranging from 0 to 5°C prior to winter freezing. An increase in membrane stability has been associated with the adaptation of plant tissues to cold. The level of membrane damage after freezing in standard conditions can therefore be used as a test to assess the genetic variability for frost tolerance. Studies for assessing the influence of hardening on cold tolerance of barley have been carried out by evaluating 30 barley genotypes in the field and laboratory conditions. The data from the field experiments indicate that most of the tested cultivars show a degree of cold tolerance in a narrow range. High cold injury adversely affected yield (Rizza et al., 1994).

2 Metabolic Changes Associated with Cold Acclimation

Many physiological and molecular changes have been shown to occur at low temperatures, but the role and the contribution of each to frost resistance is still poorly understood. These changes include variations in enzyme activity, accumulation of soluble components, alterations in membrane composition and induction of specific genes (Cattivelli and Bartels, 1992). In a preliminary experiment two cultivars of barley (*Hordeum vulgare* L.), the frost-resistance Onice and the frost-sensitive Gitane, were acclimated under controlled conditions (8 h photoperiod at +4°C -light- and +2°C -dark- for 21 d) and used to monitor the metabolic changes associated with the developing of frost-resistance. Exposure of barley plants to low temperature resulted in an increase in proline, equal in both cultivars. During the first days of cold acclimation, ABA levels showed a peak in the frost-resistant cultivar, lasting about 24 h, followed by a decrease. It has been recently reported that also other plant species respond to low temperatures with a transient ABA accumulation at the beginning of acclimation which correlate with frost resistance (Lång et al., 1994; Ryu and Li, 1994). The water soluble carbohydrates reached their highest content after 3 d of hardening, although after 14 to 21 d of acclimation the carbohydrate content was similar to that of unhardened plants. In the frost-resistant cultivar Onice, 86% of free fatty acids was represented by unsaturated molecular species, linolenic acid alone being 71%. In contrast, in the frost-sensitive cultivar only 31% of free fatty acids was unsaturated and a large amount of 9-oxo-nonanoic acid, a product present in the linolenic acid cascade, was also detected. The ABA content after 2 d of hardening and the free fatty acids composition were clearly different between the two cultivars and may explain, at least in part, the different frost-resistant capacities of Onice and Gitane (Murelli et al., 1995).

3 Molecular Cloning of Cold-Regulated Genes

It is well known that low temperature conditions lead to changes in several physiological parameters as well as to modifications in gene expression (Cattivelli and Bartels, 1989). A set of cDNA clones homologous to cold-regulated mRNAs have been isolated in our laboratory from barley plants exposed to low temperatures. The analyses of their expression pattern indicate that the transcripts accumulate to different levels during cold-treatment regardless the plant developmental stage. The mRNAs homologous to cDNA clone *pt59* were expressed only in leaves whereas other cold-regulated transcripts were induced in all plant tissues during exposure to temperatures between 0° and +4°C. All the mRNAs corresponding to cold-regulated clones were found to disappear a few hours after increasing the temperature up to 22°C (Cattivelli and Bartels, 1990). An additional set of barley cold-regulated genes have been isolated and characterized by Dunn et al. (1994 and other references herein). The finding that, with only one exception, none of these sequences share common domains with those isolated in our lab suggests that probably a large number of cold-regulated genes are involved in barley cold acclimation.

The cold-regulated cDNA clone *pt59* encodes for a cold-induced protein of 14 kD (COR14). The accumulation of COR14 occurs only at low temperatures after even brief exposure of the plants to light. Plants grown and fully hardened in the dark accumulate a reduced amount of *pt59*-corresponding mRNA and only traces of COR14. Light exposure for as short as 5 min is enough to normalize the expression of *pt59*-corresponding mRNA and increases the accumulation of COR14. These findings indicate that one or more light-dependent factors are involved in gene transcription and in protein accumulation. The accumulation of COR14 was monitored by using a specific antibody throughout hardening and subsequent dehardening in the cold resistant Onice and the cold-susceptible Gitane barley cultivars. Analysis of the proteins in the subcellular fractions (nucleus, endoplasmic reticulum, intact chloroplasts, thylakoids) indicates that COR14 is present in the proteins extracted from intact chloroplasts but not in the thylakoid fraction. This suggests that COR14 is localized in the stroma of the chloroplasts (Crosatti et al., 1995). A protein with traits similar to those of COR14 in barley is COR15 in *Arabidopsis* (Lin and Thomashow, 1992). While COR15 is also localized in the chloroplast and has also a high alanine content, the nucleic and/or amino acid sequences of COR14 and COR15 do not correspond substantially.

The protein COR14 was detected in both cultivars after about 7 d of exposure to low temperatures and gradually accumulated over a period of up to four weeks. When the plants were subsequently transferred to 20°C, COR14 persisted for at least 6 d. The resistant cultivar Onice and the susceptible cultivar Gitane were monitored under hardening at 4, 6, 8 and 10°C. The amount of COR14 accumulated by Onice plants over one week was limited at 10°C and higher at 8, 6 and 4°C. By contrast, COR14 was synthesized in Gitane only at the 4°C hardening temperature. These data indicate that the difference between the resistant and susceptible cultivar is more related to the regulation of hardening than to the hardening process itself. An early induction of hardening is an evolutionary advantage because under field conditions of progressively decreasing temperatures it enables the resistant cultivar to be prepared for the cold well before the susceptible cultivar. These results also show that the COR14 antibody might be a potential tool for selecting cultivars with a greater capacity of resistance to low temperatures.

4 Genes Regulated by Low Temperatures and Expressed under Desiccation Link the Molecular Adaptation to Frost and Drought Conditions

It has been widely demonstrated that desiccation causes changes in gene expression, and many genes induced under these conditions have been isolated and characterized (Skriver and Mundy, 1990; Bray, 1991). Most of these genes are ABA-regulated and can be induced by exogenous application of the hormone. It is well known that ABA is involved in plant adaptation to environmental stress. In plant species such as *Arabidopsis* cold acclimation can be triggered by exogenous ABA treatment. In addition, many ABA-regulated genes are expressed upon low temperatures and desiccation conditions (Heino et al., 1990). Nevertheless, our experience indicates that in barley only few ABA-regulated mRNAs and proteins are accumulated under both stress conditions (Grossi et al., 1992). An example is the protein named COR75. This protein is the predominant COR protein in barley and immunologically related to the high mol wt COR proteins of wheat. Yet the extent of frost resistance of a given genotype was not found to be related to the amount of COR75 accumulated in the tissues. The expression pattern of COR75 in response to various environmental stresses was monitored in plants at the first-leaf stage. COR75, which is expressed both in leaf and root tissues, can be induced by exogenous treatment with ABA or drought stress, as well as by low temperatures (Crosatti et al., 1994). It has also been postulated that proteins inducible by such environmental factors (low temperatures and drought) may be the plant response to variations in tissue moisture status, conditions brought about by both these stresses (Guy et al., 1992).

Although COR75 is the only example of ABA-regulated protein accumulated upon cold and desiccation in barley, other ABA-independent genes are known to be involved in plant adaptation to both these environmental factors. We have recently reported the characterization of cold and drought-regulated cDNA clones, *paf93* and *cdr29*, which respond rapidly to dehydration before a great increase in ABA content occurs. In a time-course experiment made with barley leaves, the induction of known ABA-regulated mRNAs, as detected by hybridization with the *Dhn1* probe, occurred after 5h of desiccation (4.3% loss of fresh weight), when the ABA level had already increased to 466 ng/g fresh weight, whereas the expression of the mRNAs corresponding to *paf93* was detected after 30 min (0.8% fresh weight loss), when the ABA level was only 20.5 ng/g fresh weight (1.5 times the basal level). A transcript of 1.8 kb, homologous to clone *cdr29*, was induced in the same conditions. These finding suggest that two mechanisms are involved in the barley molecular response to drought. Genes such as dehydrins are induced only after a great increase of ABA, whereas the genes rapidly induced by dehydration are either ABA-independent or triggered by very small variation in ABA content. Drought-induced genes whose expression is not controlled by ABA have also been isolated in *Arabidopsis*. An analysis of the promoter region of the *Arabidopsis* gene *rd29A* has indeed made possible to identify a *cis*-acting element responsive to drought and low temperature but not to ABA (Yamaguchi-Shinozaki and Shinozaki, 1994).

Sequence analysis of the barley cDNA clones revealed that *paf93* encodes for a protein with some similarity to Group II LEA proteins *paf93* represents a rare example of a gene coding for a protein homologous to a class of ABA-regulated genes whose induction is ABA-independent. The comparison of CDR29 with other known protein sequences revealed homologies with several peroxisome localized acyl-Coenzyme A oxidases (Grossi et al., 1995). These results lead to the hypothesis that *cdr29* encodes for a potential acyl-Coenzyme A oxidase, an enzyme involved in the fatty acids metabolism. By using a physiological approach we were able to demonstrate not only that changes in the fatty acid metabolism occur during cold acclimation, but also that such modifications could in part explain the difference between frost-resistant and frost-sensitive cultivars. The interrelation between the physiological and the molecular components involved in cold-acclimation should provide in the near future a better understanding of the mechanism leading to cold tolerance.

Acknowledgements. This research was supported by National Research Council of Italy, Special Project RAISA, Sub-project N. 2, Paper N. 2460.

References

Bray EA (1991) Regulation of gene expression by endogenous ABA during drought stress. *In* WJ Davies, HG Jones, eds, Abscisic Acid: Physiology and Biochemistry. ßios Scientific Publishers, Oxford, pp 81-98

Cattivelli L, Bartels D (1989) Cold-induced mRNAs accumulate with different kinetics in barley coleoptiles. Planta **178**: 184-188

Cattivelli L, Bartels D (1990) Molecular cloning and characterization of cold-regulated genes in barley. Plant Physiol **93**: 1504-1510

Cattivelli L, Bartels D (1992) Biochemistry and molecular biology of cold-inducible enzymes and proteins in higher plants. *In* JL Wray, ed, Society for Experimental Biology Seminar Series 49: Inducible Plant Proteins. Cambridge University Press, pp 267-288

Crosatti C, Rizza F, Cattivelli L (1994) Accumulation and characterization of the 75 kDa protein induced by low temperature in barley. Plant Sci **97**: 39-46

Crosatti C, Soncini C, Stanca AM, Cattivelli L (1995) The accumulation of a cold-regulated chloroplastic protein is light-dependent. Planta **196**: 458-463

Dunn MA, Goddard NJ, Zhang L, Pearce RS, Hughes MA (1994) Low-temperature-responsive barley genes have different control mechanisms. Plant Mol Biol **24**: 879-888

Grossi M, Cattivelli L, Terzi V, Stanca AM (1992) Modification of gene expression induced by ABA, in relation to drought and cold stress in barley shoots. Plant Physiol Biochem **30** (1): 97-103

Grossi M, Gulli M, Stanca AM, Cattivelli L (1995) Characterization of two barley genes that respond rapidly to dehydration stress. Plant Sci **105**: 71-80

Guy C, Haskell D, Neven L, Klein P, Smelser C (1992) Hydration-state-responsive proteins link cold and drought stress in spinach. Planta **188**: 265-270

Heino P, Sandman G, Lång V, Nordin K, Palva ET (1990) Abscisic acid deficiency prevents development of freezing tolerance in *Arabidopsis thaliana* (L.) Heynh. Theor Appl Genet **79**: 801-806

Lång V, Mantyla E, Welin B, Sundberg B, Palva ET (1994) Alterations in water status, endogenous abscisic acid content, and expression of *Rab18* gene during the development of freezing tolerance in *Arabidopsis thaliana*. Plant Physiol **104**: 1341-1349

Lin C, Thomashow M (1992) DNA sequence analysis of a complementary DNA for cold-regulated *Arabidopsis* COR15 and characterization of the CPR15 polypeptide. Plant Physiol **99**: 519-525

Murelli C, Rizza F, Marinone Albini F, Dulio A, Terzi V, Cattivelli L (1995) Metabolic changes associated with cold-acclimation in contrasting cultivars of barley. Physiol Plant **94**: 87-93

Rizza F, Crosatti C, Stanca AM, Cattivelli L (1994) Studies for assessing the influence of hardening on cold tolerance of barley genotypes. Euphytica **75**: 131-138

Ryu SB, Li PH (1994) Potato cold hardiness development and abscisic acid. 1. Conjugated abscisic acid is not the source of the increase in free abscisic acid during potato (*Solanum commersonii*) cold acclimation. Physiol Plant **90**: 15-20

Skriver K, Mundy J (1990) Gene expression in response to abscisic acid and osmotic stress. Plant Cell **2**: 503-512

Stanca AM, Terzi V, Cattivelli L (1992) Biochemical and molecular studies of stress tolerance in barley. *In* PR Shewry, ed, Barley: Genetics, Biochemistry, Molecular Biology and Biotechnology. C.A.B. International, Wallingford, UK, pp 277-288

Yamaguchi-Shinozaki K, Shinozaki K (1994) A novel *cis*-acting element in an *Arabidopsis* gene is involved in responsiveness to drought, low temperature, or high-salt stress. Plant Cell **6**: 251-264

Genes for Fatty Acid Desaturases and Choline Oxidase are Responsible for Tolerance to Low-Temperature and Salinity Stresses in Cyanobacteria and Plants

Norio Murata[1], Hajime Wada[2], Toshio Sakamoto[3], Yasushi Tasaka[1], Zoltan Gombos[4], Byoung Yong Moon[5], Patcharaporn Deshnium[1], Dmitry A. Los[1] and Hidenori Hayashi[6]

1 National Institute for Basic Biology, Okazaki 444, Japan
2 Faculty of Science, Kyushu University, Ropponmatsu, Fukuoka 810, Japan
3 Department of Biology, College of Arts and Sciences, the University of Tokyo, Komaba, Meguro-ku Tokyo 153, Japan
4 Department of Plant Biology, Biological Research Center, H-6701 Szeged, Hungary
5 Department of Biology, Inje University, Kimhae, Korea
6 Department of Chemistry, Ehime University, Matsuyama 790, Japan

Abstract. Most cyanobacteria and plants contain high levels of polyunsaturated fatty acids of membrane lipids. We isolated the *desA*, *desB* and *desC* genes of *Synechocystis* sp. PCC 6803, which encode the Δ12, Δ15 (ω3) and Δ9 desaturases, respectively, of the acyl-lipid type. We disrupted the *desA* gene and the *desD* gene (for Δ6 desaturase) in *Synechocystis* sp. PCC 6803 by insertion of antibiotic resistance gene cartridges. These mutations greatly modified the extent of unsaturation of the fatty acids of membrane lipids. These changes in the unsaturation of membrane lipids altered the tolerance of cyanobacterial cells to low temperature. In higher plants, we isolated a cDNA for acyl-ACP: glycerol-3-phosphate acyltransferase from squash and introduced it into tobacco plants. The transformation modified the level of unsaturation of chloroplastic phosphatidylglycerol, and altered the tolerance of tobacco plants to low temperature. Glycinebetaine, a compatible solute, is synthesized in some halophilic plants. We cloned, from the soil bacterium *Arthrobacter globiformis*, the *codA* gene for choline oxidase which converts choline into glycinebetaine. When we introduced this gene into *Synechococcus* sp. PCC 7942, the resultant transformants accumulated glycinebetaine to some extent and became tolerant to salinity stresses.

Keywords. Choline oxidase, cyanobacterium, fatty acid desaturase, low temperature tolerance, *Nicotiana tabacum*, salt stress, *Synechococcus*, *Synechocystis*, transformation, transgenic plants

1 Introduction

Physiological ability of plants is limited by various stresses such as low temperature, high temperature, drought, salinity, strong light and UV light. However, some plants have genetically adapted to non-optimal conditions so that they can sustain high levels of physiological ability in various environments. Recently genes for tolerance to various stresses have been identified, and it is now possible to genetically manipulate these genes for improving the photosynthetic capability of plants and algae. Some years ago, we could genetically manipulate the level of unsaturation of glycerolipids of thylakoid membranes and modified the chilling sensitivity of cyanobacteria (Wada et al., 1990) and higher plants (Murata et al., 1992a). In the present study we present the molecular mechanism for the low temperature tolerance in relation to alteration in unsaturation of membrane lipids. We also present the effect of a compatible solute, glycinebetaine, on photosynthesis and the gene-engineered improvement of salinity tolerance of cyanobacteria.

2 Genetic Alteration of Membrane Lipids and Low temperature Tolerance in Cyanobacteria

2.1 Acyl-Lipid Desaturases and their Genetic Manipulation

The cyanobacterium, *Synechocystis* sp. PCC 6803, contains high levels of polyunsaturated fatty acids of membrane lipids, which are important in their tolerance to low temperature. From this cyanobacterium we have isolated the *desA*, *desB*, and *desC* genes, which encode, respectively, the Δ12, Δ15 (ω3) and Δ9 desaturase of acyl-lipid type (Wada et al., 1990; Sakamoto et al., 1994a; Sakamoto et al., 1994b), and Reddy et al. (1993) have isolated the *desD* gene which encodes the Δ6 desaturase (Table 1). The deduced amino acid sequences of the four desaturases together with those in other cyanobacterial strains indicated that the Δ6, Δ9 and Δ12 desaturases have about 350-360 amino acid residues, but that the Δ9 desaturases are smaller than the other desaturases (Murata and Wada, 1995). The hydropathy profiles are likely to suggest that all these desaturases span the membrane four times and have hydrophilic domains exposed to the cytoplasm. Three histidine clusters, located in the hydrophilic domains, are regarded as ligands to Fe atoms which constitute the catalytic center of these enzymes.

Table 1. Cloned genes for cyanobacterial acyl-lipid desaturases

Strain	Residues	Reference
desA (for Δ12 desaturase)		
Synechocystis sp. PCC 6803	351	Wada et al. 1990
Synechocystis sp. PCC 6714	349	Sakamoto et al. 1994a
Synechococcus sp. PCC 7002	347	Sakamoto et al 1994a
Anabaena variabilis	350	Sakamoto et al 1994a
Spirulina platensis	351	Murata et al. 1995
desB (for Δ15 or ω3 desaturase)		
Synechocystis sp. PCC 6803	359	Sakamoto et al. 1994b
Synechococcus sp. PCC 7002	350	Bryant, personal commun.
desC (for Δ9 desaturase)		
Synechocystis sp. PCC 6803	319	Sakamoto et al. 1994c
Anabaena variabilis	272	Sakamoto et al. 1994c
desD (for Δ6 desaturase)		
Synechocystis sp. PCC 6803	359	Reddy et al. 1993
Spirulina platensis	368	Murata et al. 1995

We disrupted the *desD* and *desA* genes, in a step-wise manner, by inserting, respectively, chloramphenicol and kanamycin resistance gene cartridges. These mutations greatly modified the extent of unsaturation of the fatty acids of membrane lipids. In the wild type strain, each of the saturated, monounsaturated, diunsaturated and triunsaturated lipid molecules contributes to about 15% of the total membrane lipids; in the Δ*desD* mutant, the triunsaturated lipid molecular was replaced by the diunsaturated lipid molecules; in the Δ*desA*/Δ*desD* mutant, monounsaturated lipid molecules accounted for 80% of the total lipid species with a 15% contribution of saturated molecules. In some experiments, however, we used the Fad6 mutant (an ethylmethanesulfonate-induced mutant of the *desD* gene, Wada and Murata, 1989) and the Fad6/Δ*desA* strain in which the *desA* gene of the Fad6 mutant strain was insertionally disrupted with the kanamycin-resistance gene cartridge (Wada et al., 1992).

2.2 Effects of Unsaturation of Membrane Lipids on Photosynthesis

Our previous study (Gombos et al., 1992) demonstrated that the elimination of unsaturation of membrane lipids in *Synechocystis* sp. PCC 6803 decreased the tolerance to low temperature photoinhibition of the photosystem II complex. However, the extent of the photoinhibition in intact cells reflects a balance between the light-induced inactivation of the photosystem II complex and the recovery of the complex from the inactivated state. Thus, we separately determined the two processes by employing several methods such as inhibition of protein synthesis with antibiotics to block the recovery process. The results demonstrated that the unsaturation of the membrane lipids did not affect the light-induced inactivation, but accelerated the recovery of the photosystem II complex from the inactivated state (Gombos et al., 1994).

The inactivation and the recovery of the photosystem II complex under the photoinhibitory conditions are explained in terms of the turnover of the D1 protein in the photosystem II complex. The immunochemical analysis with antibodies raised against the D1 protein indicated that the unsaturation of membrane lipids did not affect the light-induced damage to the D1 protein and the proteolysis of the damaged D1 protein (Kanervo et al., 1995). It is likely that the unsaturation of membrane lipids accelerates the processing of the pre-D1 protein into the mature D1 protein in the photosystem II complex.

3 Genetic Alteration of Membrane Lipids and Low temperature Tolerance in Tobacco Plants

3.1 Acyltransferases and their Genetic Manipulation

The sensitivity of higher plants to chilling is closely correlated with the level of unsaturation of the fatty acids of the thylakoid membrane lipids of their chloroplasts (Murata and Nishida, 1989). We demonstrated that both unsaturation of thylakoid membrane lipids and chilling sensitivity were significantly affected upon transformation of tobacco plants with cDNAs for glycerol-3-phosphate acyltransferases from squash and *Arabidopsis* (Murata et al., 1992a). In particular, the extent of unsaturation of phosphatidylglycerol was most effectively modified, and this change appears to be responsible for modification of the ability to tolerate low temperatures.

To study the relationship between the unsaturation of thylakoid membrane lipids and the low temperature photoinhibition in higher plants, we employed transgenic tobacco plants in which the extent of unsaturation of thylakoid membrane lipids was decreased as a result of over-expression of the glycerol-3-phosphate acyltransferase from squash under the control of the cauliflower mosaic virus 35S promoter (Moon et al., 1995). Of the four major lipid classes isolated from the thylakoid membranes, phosphatidylglycerol showed the most conspicuous decrease in the level of unsaturation in the transformed plants (Table 2).

3.2 Effect of Unsaturation of Phosphatidylglycerol on Photosynthesis

Photosynthesis at low temperature is impeded when plants are exposed to light, and this phenomenon is known as low temperature photoinhibition. The main target for photoinhibition is the photosystem II protein complex (Powles, 1984). Impairment of electron transport is caused by the irreversible damage to the D1 protein, which is one of the heterodimeric polypeptides of the photosystem II reaction center complex (Aro et al., 1990).

The isolated thylakoid membranes from wild type and transgenic plants did not significantly differ from each other in terms of the sensitivity of photosystem II to low temperature photoinhibition. However, leaves of the transformed plants were more sensitive to photoinhibition than those of wild type plants. Moreover, the recovery of photosynthesis from photoinhibition in leaves of wild type plants was faster than that in leaves of the transgenic tobacco plants. These results suggest that unsaturation of fatty acids of phosphatidylglycerol in thylakoid membranes protects the photosynthetic machinery against low temperature photoinhibition by accelerating the recovery of the photosystem II complex in a way similar to the case in cyanobacteria (Moon et al., 1995).

Table 2. Major molecular species of phosphatidylglycerol in leaves of wild type tobacco and tobacco transformed with the cDNA for glycerol-3-phosphate acyltransferase from squash (p-SQ)

Molecular species			Strain	
			Wild type	p-SQ
			%	
16:0 16:0 PG	+	16:0 16:1t PG	30	64
18:0 16:0 PG	+	18:0 16:1t PG	4	11
18:1 16:0 PG	+	18:1 16:1t PG	8	4
18:2 16:0 PG	+	18:2 16:1t PG	26	6
18:3 16:0 PG	+	18:3 16:1t PG	32	15

Abbreviations: 16:0, Hexadecanoic acid; 16:1t, Δ3-*trans*-hexadecenoic acid; 18:0, octadecanoic acid; 18:1, Δ9-*cis*-octadecenoic acid; 18:2, Δ9,12-*cis*-octadecadienoic acid; 18:3, Δ9,12,15-*cis*-octadecatrienoic acid; PG, phosphorylglycerol.

4 Genetic Manipulation of Glycinebetaine Accumulation and Salinity Tolerance

4.1 Protective Effects of Glycinebetaine

Glycinebetaine (betaine hereinafter) is a compatible solute, which is accumulated in chloroplasts of halotolerant plants and microorganisms. Our previous studies demonstrated that betaine has a strong protective effect on the oxygen-evolving activity of the photosystem II complex under high-salt conditions (reviewed in Papageorgiou and Murata, 1995). When the oxygen-evolving PS II complex is exposed to 1 M NaCl, the extrinsic 23-kD and 18-kD proteins are dissociated from the PS II complex (Miyao and Murata, 1983). When it is exposed to 1 M $CaCl_2$ or $MgCl_2$, the extrinsic 33-kD protein, in addition to the extrinsic 23-kD and 18-kD proteins, is dissociated (Ono and Inoue, 1984). In the presence of 1 M betaine, these dissociations are remarkably prevented and the oxygen-evolving activity is preserved (Papageorgiou et al., 1991; Murata et al., 1992b). Betaine has also a remarkable protective effect on the oxygen-evolving activity of the thylakoid membranes isolated from *Synechocystis* sp. PCC 6803 (Mamedov et al., 1991). Betaine is also effective in protecting the oxygen-evolving activity at high temperature, whereas betaine has no protective effect on the photochemical reactions I and II and the other electron transport reactions such as reactions mediated by the photosystem I complex and the cytochrome b_6/f complex (Mamedov et al., 1993). Probably, betaine stabilizes the structure of the photosystem II complex and, thereby, establishes the stable oxygen-evolving activity. It is strongly recommended to supplement betaine to all preparation, storage and reaction media to obtain the maximum activity and to keep the activity at high levels (Papageorgiou and Murata, 1995).

4.2 Improvement of Salinity Tolerance by *in vivo* Accumulation of Glycinebetaine.

Choline oxidase from the soil bacterium, *Arthrobacter globiformis*, is an enzyme which converts choline into betaine (Fig. 1).

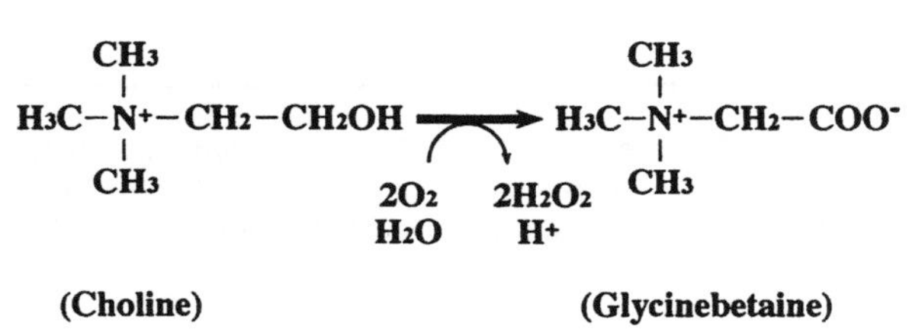

Fig. 1. The reaction catalyzed by choline oxidase from *Arthrobacter globiformis*

This reaction of betaine synthesis contrasts to that in higher plants and *E. coli* in which choline is converted into betaine by two enzyme reactions (Landfald and Strøm, 1986; Weretilnyk et al., 1989). However, the choline oxidase produces hydrogen peroxide as a by-product. We isolated the gene for choline oxidase, designated *codA*, from the soil bacterium, and introduced it into *Synechococcus* sp. PCC 7942 with use of a platform vector plasmid, pAM1044, which was kindly provided by Dr. S. Golden (Deshnium et al., 1995). The PCR analysis and the western blot analysis indicated that the introduced gene was incorporated into all copies of the chromosome and was well expressed to synthesize choline oxidase in cells of the cyanobacterium. The transformed cells accumulated betaine in the cytoplasm at the level of about 80 mM.

The wild type cells and the transformed cells were compared in terms of growth, accumulation of chlorophyll, and photosynthetic activity at various concentrations of NaCl (Deshnium et al., 1995). In a medium containing 400 mM NaCl, the transformed cells grew well after a lag period, whereas the wild type cells hardly grew. The chlorophyll accumulation in the wild type and the transformed cells showed profiles similar to those of growth. The photosynthetic activity declined for two d in both wild type and transformed cells after the cells were transferred from 0 mM to 400 mM NaCl, and then the activity of the transformed cells started to increase whereas that of the wild type cells remained at the declined level. These observations indicate that betaine synthesized in the cells transformed with the *codA* gene enabled the cells to become tolerant to salinity stress.

As indicated in the reaction catalyzed by choline oxidase, the biosynthesis of betaine in the cytoplasm simultaneously produces hydrogen peroxide, which is toxic to the cells. However, when the transformed cells were grown under non-salinity conditions, they grew at the same rate as the wild type cells (Deshnium et al., 1995). These observations suggest that the produced hydrogen peroxide must have been efficiently scavenged and did not apparently produce an apparent toxic effect.

Acknowledgements. This work was supported, in part, by Grants-in-Aid for Scientific Research on Priority Areas (04273102 and 04273103) from the Ministry of Education, Science, Sports and Culture, Japan.

References

Aro E-M, Hundal T, Carbery I, Andersson B (1990) *In vitro* studies on light-induced inhibition of Photosystem II and D_1-protein degradation at low temperatures. Biochim Biophys. Acta 1019: 269-275

Deshnium P, Los DA, Hayashi H, Mustardy L, Murata N (1995) Transformation of *Synechococcus* with a gene for choline oxidase enhances tolerance to salt stress. Plant Mol Biol **29**: 897-907

Gombos Z, Wada H, Murata N (1992) Unsaturation of fatty acids in membrane lipids enhances tolerance of the cyanobacterium *Synechocystis* PCC6803 to low-temperature photoinhibition. Proc Natl Acad Sci USA **89**: 9959-9963

Gombos Z, Wada H, Murata N (1994) The recovery of photosynthesis from low-temperature photoinhibition is accelerated by the unsaturation of membrane lipids: a mechanism of chilling tolerance. Proc Natl Acad Sci USA **91**: 8787-8791

Kanervo E, Aro E-M, Murata N (1995) Low unsaturation level of thylakoid membrane lipids limits turnover of the D1 protein of photosystem II at high irradiance. FEBS Lett **364**: 239-242

Landfald B, Strøm AR (1986) Choline-glycine betaine pathway confers a high level of osmotic tolerance in *Escherichia coli*. J Bacteriol **165**: 849-855

Mamedov MD, Hayashi H, Wada H, Mohanty PS, Papageorgiou GC, Murata N (1991) Glycinebetaine enhances and stabilizes the evolution of oxygen and the synthesis of ATP by cyanobacterial thylakoid membranes. FEBS Lett **294**: 271-274

Mamedov MD, Hayashi H, Murata N (1993) Effects of glycinebetaine and unsaturation of membrane lipids on heat stability of photosynthetic electron-transport and phosphorylation reactions in *Synechocystis* PCC6803. Biochim Biophys Acta **1142**: 1-5

Miyao M, Murata N (1983) Partial disintegration and reconstitution of the photosynthetic oxygen evolution system: binding of 24 kilodalton and 18 kilodalton polypeptides. Biochim Biophys Acta **725**: 87-93

Moon BY, Higashi S, Gombos Z, Murata N (1995) Unsaturation of chloroplastic membrane lipids protects photosynthesis against low-temperature photoinhibition in transgenic tobacco plants. Proc Natl Acad Sci USA **92**: 6219-6223

Murata N, Nishida I (1989) Lipids in relation to chilling sensitivity of plants. *In* CY Wang, ed, Chilling Injury of Horticultural Crops. CRC Press, Boca Raton, FL, pp 181-199

Murata N, Ishizaki-Nishizawa O, Higashi S, Hayashi H, Tasaka Y, Nishida I (1992a) Genetically engineered alteration in the chilling sensitivity of plants. Nature **356**: 710-713

Murata N, Mohanty PS, Hayashi H, Papageorgiou GC (1992b) Glycinebetaine stabilizes the association of extrinsic proteins with the photosynthetic oxygen-evolving complex. FEBS Lett **296**: 187-189

Murata N, Wada H (1995) Acyl-lipid desaturases and their importance in the tolerance and acclimatization to cold of cyanobacteria. Biochem J **308**: 1-8

Ono T, Inoue Y (1984) Ca^{2+}-dependent restoration of O_2-evolving activity in $CaCl_2$-washed PS II particles depleted of 33, 24 and 16 kDa protein. FEBS Lett **168**: 281-286

Papageorgiou GC, Fujimura Y, Murata N (1991) Protection of the oxygen-evolving photosystem II complex by glycinebetaine. Biochim Biophys Acta **1057**: 361-366

Papageorgiou GC, Murata N (1995) The unusually strong stabilizing effect of glycinebetaine on the structure and function of the oxygen-evolving photosystem II complex. Photosyn Res **44**: 243-252

Powles SB (1984) Photoinhibition of photosynthesis induced by visible light. Annu Rev Plant Physiol **35**: 15-44

Reddy AS, Nuccio ML, Gross LM, Thomas TL (1993) Isolation of a Δ6-desaturase gene from the cyanobacterium *Synechocystis* sp. strain PCC 6803 by gain-of-function expression in *Anabaena* sp. strain PCC 7120. Plant Mol Biol **27**: 283-300

Sakamoto T, Wada H, Nishida I, Ohmori M, Murata N (1994a) Identification of conserved domains in the Δ12 desaturase of cyanobacteria. Plant Mol Biol **24**: 643-650

Sakamoto T, Los DA, Higashi S, Wada H, Nishida I, Ohmori M, Murata N (1994b) Cloning of ω3 desaturase from cyanobacteria and its use in altering the degree of membrane-lipid unsaturation. Plant Mol Biol **26**: 249-263

Sakamoto T, Wada H, Nishida I, Ohmori M, Murata N (1994c) Δ9 acyl-lipid desaturase of cyanobacteria: molecular cloning and substrate specificities in terms of fatty acids, *sn*-positions, and polar head groups. J Biol Chem **269**: 25576-25580

Wada H, Murata N (1989) *Synechocystis* PCC6803 mutants defective in desaturation of fatty acids. Plant Cell Physiol **30**: 971-978

Wada H, Gombos Z, Murata N (1990) Enhancement of chilling tolerance of a cyanobacterium by genetic manipulation of fatty acid desaturation. Nature **347**: 200-203

Wada H, Gombos Z, Sakamoto T, Murata N (1992) Genetic manipulation of the extent of desaturation of fatty acids in membrane lipids in the cyanobacterium *Synechocystis* PCC6803. Plant Cell Physiol **33**: 535-540

Weretilnyk EA, Bednarek S, McCue KF, Rhodes D, Hanson AD (1989) Comparative biochemical and immunological studies of the glycine betaine synthesis pathway in diverse families of dicotyledons. Planta **178**: 342-352

Marker-Assisted Genetic Analysis of Winter Hardiness in Barley

Tony H.H. Chen[1], Patrick M. Hayes[2], Aihong Pan[2], Fu Qiang Chen[2], Karen van Zee[1], Thomas K. Blake[3], Timothy J. Close[4] and Ildiko Karsai[5]

[1] Department of Horticulture, and [2] Department of Crop and Soil Science, Oregon State University, Corvallis, OR 97331, USA

[3] Department of Plant and Soil Sciences, Montana State University, Bozeman, MT 59751, USA

[4] Department of Botany and Plant Sciences, University of California, Riverside, CA, USA

[5] Martonvasar Research Institute, Martonvasar, Hungary

Abstract. A linkage map of barley was developed from 100 doubled haploid lines derived from a winter (Dicktoo) x spring (Morex) barley cross. Based on this map, quantitative trait locus (QTL) analyses were conducted to determine the chromosome location of genes controlling components of winter hardiness. We detected only one major interval on long arm of chromosome 7 accounting for significant QTL effects for winter hardiness. This region was found to be responsible for 39-79% of the phenotypic variation in field survival, 32% of the variation for LT_{50}, and 47% of the variation for growth habit. Two members of the barley dehydrin gene family, *Dhn1* and *Dhn2,* were also found within the region defining the winter hardiness QTL effects. Reasoning that genes controlling winter hardiness may be expressed during cold acclimation, we examined the expression pattern of six barley dehydrin gene family members in shoot tissue in response to cold temperature. *Dhn5*, located on chromosome 6, was the only dehydrin induced by low temperature under both controlled environments as well as under field conditions. Under these conditions, *Dhn1* and *Dhn2* were not induced to detectable levels by low temperature. The expression patterns of *Dhn5* in shoot tissue for Dicktoo and Morex were identical under the conditions studied, in spite of the known phenotypic differences in their winter hardiness. These results, together with the allelic structure of selected high and low survival doubled haploid lines, suggested that the Dicktoo alleles at the *Dhn1* and *Dhn2* are probably not the primary determinants of winter hardiness in barley.

Keywords. Barley, dehydrins, freezing tolerance, *Hordeum vulgare*, genome mapping, QTL, winter hardiness

1 Introduction

For winter cereals to survive winter conditions, they must acquire the ability to tolerate freezing temperatures in the fall. The ability of plants to harden in response to fall/winter conditions, i.e., cold acclimation or hardening, is considered the most important factor for winter survival. Cold acclimation is a complex genetic trait inducible by low temperature and results in a series of changes, leading to the final expression of the complex traits of winter hardiness. Winter hardiness of cereals is determined by complex interactions between genetic and environmental factors which are not well understood. Improvement of winter hardiness in winter cereals has been very slow. A major reason is thought to be due to the lack of intraspecific genetic variability (Gusta and Fowler , 1979).

The genetic nature of winter hardiness in winter cereals is most frequently interpreted as being genetically complex and quantitatively inherited (Limin and Fowler, 1983). Traditionally, the genetic basis of winter hardiness has been studied through quantitative genetic analysis. Genes controlling winter hardiness have been assigned to every chromosome in each of three genomes of wheat (reviewed in Thomashow, 1990). However, genetic analysis of complex traits, such as winter hardiness, is dependent on the interaction between environmental and genetic effects. Much of the complexity encountered in studying winter hardiness may be attributed to evaluation procedures, environmental effects, and the nature of the test population. Nevertheless, given the complexity of changes during cold acclimation, the components of winter hardiness in winter cereals may be under multi-genic control and thus quantitative phenotypic expression.

Many low temperature-responsive (*ltr*) genes have been isolated and characterized in barley (Hughes et al., 1994). However, the functional roles of these genes have not been determined yet. While the isolation and characterization of *ltr* genes are promising, the development of molecular marker technology and QTL mapping offers an alternative approach to resolve the complexity of interacting traits determining winter hardiness.

In 1970, Weiser proposed that plant cold acclimation involved the alteration of gene expression. Since then, evidence has been accumulating in supporting this notion. In all the systems studied, the induction of freezing tolerance is generally accompanied by the appearance of many new mRNAs and new polypeptides (see review by Chen et al., 1995). In the past few years, rapid progress has been made in the isolation and characterization of cDNA and genomic clones of cold-regulated (*cor*) or *ltr* genes (see review by Chen et al., 1995). There are few examples in which the putative functions of the *cor* gene products have been proposed (Chen et al., 1995). It is generally believed that

that some of these *cor* genes could be directly or indirectly responsible for the observed increases in freezing tolerance. However, a direct linkage between the expression of a specific gene and cold acclimation is still missing. In a few systems in which the over-expression of *cor* genes or expression of antisense sequence in transgenic plants have been achieved, these transgenic plants generally failed to show altered freezing tolerance. The negative results raise several important questions. For instance, (1) since there are many new mRNAs induced during cold acclimation and only few of them have been cloned and characterized, are we dealing with the right genes? (2), if cold hardiness is controlling by many genes as suggested by traditional inheritance studies, can we manipulate a single or a small number of genes to alter freezing tolerance?

By locating individual QTL with molecular markers, individual QTL genotypes can be determined with relative certainty. This allows for a dissection of complex traits, the measurement of individual gene effects, and the measurement of gene interactions.

2 Mapping QTL for Traits Associated with Winter Hardiness in Barley

A reference population of 100 F_1-derived doubled haploid (DH) lines from the cross of Dicktoo x Morex was used to map the components of winter hardiness. Dicktoo is a six-row winter feed barley of undetermined ancestry and mixed description. Morex is a spring habit 6-row malting barley. The average LT_{50} evaluated in 6 environments is -9.2°C and -1.3°C for Dicktoo and Morex, respectively (Kolar et al., 1991). This DH population has been repeatedly phenotyped and genotyped over a number of years. The most recent results were reported by Pan et al. (1994).

The winter survival data were collected from field plots at Bozeman, Montana and Corvallis, Oregon. In Montana, Morex and approximately half of the DH lines, experienced complete mortality. The mean survival of Dicktoo was 43% and the survival of the remaining DH lines ranged from 1 to 85%. Less-severe winter conditions at Corvallis, Oregon, led to complete survival of all DH lines, with a range of 1-100%. The survival of Dicktoo was 89% and that of Morex was 18%. The correlation between winter survival in the two field tests was 0.57.

QTL controlling traits associated with winter hardiness in barley (field survival, LT_{50}, growth habit, and crown fructan content) were mapped to chromosome 7 (Hayes et al., 1993; Pan et al., 1994). The largest QTL effects for all traits were detected in a 21% recombination interval (21 cM) on the long arm of chromosome 7, which is homologous with wheat chromosome 5 (Islam et al., 1981). Sutka and Snape (1989) and Roberts (1990) reported cold tolerance effects in wheat chromosome 5. QTL in this region accounted for 37-68% of the variation for three measures of freezing tolerance, 47% of the variation for growth habit, and 28% of the variation in crown fructan content (Hayes et al., 1993).

Multiple peak QTLs were detected in the three intervals defined by WG364B, mR, Dhn1/Dhn2, and BCD265B. The distance from WG364B to BCD265B is 48.3 cM. Confidence intervals for these QTL overlap and a common gene or gene cluster may be responsible.

3 Chromosomal Location and Cold-Induced Expression of Dehydrin Genes

During freezing stress, a lower water potential is established outside the cell. The difference in water potential between the intracellular liquid and the extracellular ice creates a driving force which drives the net movement of liquid water out of the cell across the plasma membrane. Thus, freezing stress is one type of drought stress, and both drought and freezing stress result in desiccation of the protoplasm of plant cells. In 1958, Levitt proposed that plants may use the general mechanism of resistance to desiccation, regardless of whether desiccation results from drought or freezing stress.

Dehydrins (Dhns) are a family of closely related proteins that accumulate in plants in response to a dehydration stress (Close et al., 1993). Numerous Dhns have been identified in many plant species in response to ABA or in response to various types of environmental stress or developmental changes involving dehydration (Close et al., 1993). Furthermore, Dhn-like proteins accumulate during cold acclimation (reviewed in van Zee et al., 1995). However, to date, a biochemical function has not been established for any of the Dhns.

As discussed above, we have identified a QTL on the long arm of barley chromosome 7, controlling winter hardiness traits. Two members of the barley *Dhn* gene family, *Dhn1* and *Dhn2*, mapped within the confidence interval of the winter hardiness QTL in the Dicktoo x Morex cross. The *Dhn1* and *Dhn2* loci are 10.1 cM distal to the mR locus and 13.3 cM proximal to the BCD265b locus (Pan et al., 1994). Before the *Dhn* loci were added to the linkage map, the QTL for freezing tolerance mapped between mR and BCD265b (Hayes et al., 1993). *Dhn3*, *Dhn4*, and *Dhn5* mapped to chromosome 6, *Dhn4* to chromosome 6 and 7, and *Dhn6* to chromosome 4. Thus, *Dhn3* through *Dhn6* loci have been mapped elsewhere in the barley genome where QTL effects for winter hardiness-related traits have not been detected.

We reasoned that (a) genes controlling winter hardiness may be located within the chromosome region controlling QTL for low temperature stress resistance; (b) one would expect complete co-segregation between phenotype and allele, if a particular locus is a primary determinant of trait expression; (c) winter hardiness genes should be induced by low temperature; and (d) those genes should be expressed in crown tissues, which are most critical tissues for overwintering of winter cereals. Since *Dhn1* and *Dhn2* loci are mapped within the interval of winter hardiness QTL, we asked whether the expression of *Dhn1* and *Dhn2* are induced by cold-temperature treatment in Dicktoo (a hardy genotype) and Morex (a non-hardy genotype) plants.

First, we analyzed soluble protein extract prepared from cold-treated shoot tissue of 3-week-old Dicktoo and Morex plants by immunoblotting. A single Dhn species approximately 86 kD was induced by cold treatment in both Dicktoo and Morex shoots (van Zee et al., 1995). In Dicktoo and Morex we have not observed significant differences in either the kinetic of Dhn5 protein accumulation or in the relative amounts of Dhn5 protein between these two freezing tolerance (Dicktoo) and freezing sensitive (Morex) genotypes.

The barley genome encodes approximately 10 *Dhns* (Close and Chandler, 1990). The nucleotide sequence for Himalaya barley *Dhns* and the corresponding map locations in the DH lines derived from a Dicktoo x Morex cross have been determined for *Dhn1* through *Dhn6* (Pan et al., 1994). To determine which member of *Dhn* gene family was induced by the cold temperature treatment, we probed blots of Dicktoo and Morex RNA with Himalaya barley *Dhn* cDNAs (*Dhn1* through *Dhn4*) and genomic clones (*Dhn5* and *Dhn6*). Northern analysis of total RNA prepared from cold-treated 3-week-old Dicktoo and Morex shoots using *Dhn1* as probe under low stringency conditions detect a primary transcript of approximately 2.1 kb. Under high-stringency conditions, only the *Dhn5* probe hybridized to the 2.1 kb transcript (van Zee et al., 1995). Under the same high-stringency conditions, *Dhn1, Dhn2, Dhn3, Dhn4* and *Dhn6* probe did not hybridize detectably to RNA prepared from cold-treated Dicktoo and Morex shoots.

We also collected crown samples from field grown barley plants and analyzed the expression of *Dhns* at protein and RNA levels. Again, *Dhn5* is the only *Dhn* family members that was expressed to a high level in barley plants hardened under natural field conditions.

To explore the possible direct associations of the *Dhn1* and *Dhn2* loci with low temperature tolerance, the most hardy and least hardy 20 DH lines were identified and their marker genotypes and field survivals were determined (van Zee et al., 1995). Within the 20 least hardy lines, there were 2 lines that had Dicktoo alleles at the *Dhn1* and *Dhn2* loci. Within the 20 most hardy lines, 1 line (DH-67) had Morex alleles at the *Dhn1* and *Dhn2* loci and the Dicktoo alleles at BCD265b, the distal locus. If *Dhn1* and/or *Dhn2* were the principal determinants of low temperature response in this population, complete co-segregation of marker genotype and phenotype would be expected.

Under our experimental conditions only *Dhn5* was specifically induced by low temperature, and yet no QTL effects for low temperature tolerance or other winter growth habit-related traits were detected in the vicinity of this locus. *Dhn1/Dhn2* were not detectably induced by low temperature under our conditions, and yet QTL effects for low temperature tolerance were detected in the vicinity of these loci. Furthermore, there was not a complete association of *Dhn1/Dhn2* allele type and field survival phenotype. Thus, these loci do not appear to be the primary determinants for the low temperature tolerance QTL observed on chromosome 7.

Acknowledgements. This research was supported, in part, by USDA competitive grant No. 89-37264-4756, and USAID/PSTC grant No. DHR-5542-G-SS-9030-00.

References

Chen THH, Burke MJ, Gusta LV (1995) Freezing tolerance in plants: an overview. *In* RE Lee, GJ Warren, LV Gusta, eds, Biological Ice Nucleation and its Applications. APS Press, pp 115-135

Close TJ, Chandler PM (1990) Cereal dehydrins: serology, gene mapping and potential functional roles. Aust J Plant Physiol **17**: 333-344

Close TJ, Fenton RD, Yang A, Ashgar R, DeMason DA, Crone DE, Meyer NC, Moonan F (1993) Dehydrins: the proteins. *In* TJ Close, EA Bray, eds, Responses of Plants to Cellular Dehydration during Environmental Stress. Amer Soc Plant Physiol, pp 104-118

Gusta LV, Fowler DB (1979) Cold resistant and injury in winter cereals. *In* H Mussell, RC Staples, eds, Stress Physiology in Crop Plants. Wiley-Interscience, New York, pp 159-178

Hayes PM, Blake TK, Chen THH, Tragoonrung S, Chen F, Pan A, Liu B (1993) Quantitative trait loci on barley (*Hordeum vulgare*) chromosome 7 associated with components of winterhardiness. Genome **36**: 66-71

Hughes MA, White AJ, Goddard NJ, Phillips JR, Dunn MA (1994) Molecular analysis of acclimation in cereals. *In* K Dorffling, B Brettschneider, H Tantau, K Pithan, eds, Crop Adaptation to Cool Climates. ECSP-EEC-EAEC, Brussels, pp 309-316

Islam AKMR, Shepherd KW, Sparrow DHB (1981) Isolation and characterization of wheat-barley chromosome addition lines. Heredity **46**: 161-174

Kolar SC, Hayes PM, Chen THH, Linderman RG (1991) Genotypic variation for cold tolerance in winter and facultative barley. Crop Sci **31**: 1149-1152

Levitt J (1958) Frost, drought, and heat resistance. *In* Protoplasmatologia VIII. Physiologie des Protoplasmas, Springer-Verlag, pp 1-87

Limin AE, Fowler DB (1983) Genetics of cold hardiness. *In* DB Fowler, LV Gusta, AE Slinkard, BA Hobin, eds, New Frontiers in Winter Wheat Production. Univ of Saskatchewan, pp 193-207

Pan A, Hayes PM, Chen F, Chen THH, Blake T, Wright S, Karsai I, Bedo Z (1994) Genetic analysis of the components of winterhardiness in barley (*Hordeum vulgare L.*). Theor Appl Genet **89**: 900-910

Roberts DWA (1990) Identification of loci on chromosome 5A of wheat involved in control of cold hardiness, vernalization, leaf length, rosette growth habit, and height of hardened plants. Genome **33**: 247-259

Sutka J, Snape JW (1989) Location of a gene for frost resistance on chromosome 5A of wheat. Euphytica **42**: 41-44

Thomashow MF (1990) Molecular genetics of cold acclimation in higher plants. Adv Genet **28**: 99-31

van Zee K, Chen F, Hayes PM, Close TJ, Chen THH (1995) Cold-specific induction of a dehydrin gene family member in barley. Plant Physiol **108**: 1233-1239

Weiser CJ (1970) Cold resistance and injury in woody plants. Science **169**: 1269-1278

Cold Acclimation in *Arabidopsis thaliana*: Function and Regulation of *COR* Genes

Michael F. Thomashow[1,2], Nancy N. Artus[1], Leonard Bloksberg[1], Sarah J. Gilmour[1], Eric J. Stockinger[1], Kathy Wilhelm[3], Matsuo Uemura[4], Murray Webb[4] and Peter L. Steponkus[4]

[1] Department of Crop and Soil Sciences, [2]Department of Microbiology, [3]Program in Genetics, Michigan State University, East Lansing, MI 48824, USA

[4] Department of Soil, Crop and Atmospheric Sciences, Cornell University, Ithaca, NY 14853 USA

Abstract. *Arabidospis thaliana* alters gene expression and increases in freezing tolerance in response to low non-freezing temperatures. One of our long range goals has been to determine whether certain *COR* (*co*ld-*r*egulated) genes that encode hydrophilic, "boiling-soluble" polypeptides have a role in freezing tolerance; another has been to determine the mechanism(s) by which *A. thaliana* senses low temperature and alters gene expression. Most of our work on *COR* gene function has focused on *COR15*. We have found that this gene encodes a 15 kD polypeptide that is targeted to the stromal compartment of chloroplasts. During import, the polypeptide is processed to a hydrophilic 9 kD polypeptide, designated COR15am, that shares a low degree of amino acid sequence similarity with group III LEA (*l*ate-*e*mbryogenesis-*a*bundant) proteins Transgenic *A. thaliana* plants that constitutively produce COR15am have been created and tested for freezing tolerance. The data indicate that COR15am enhances the freezing tolerance of chloroplasts in non-acclimated plants by abundant 2°C, nearly one-third of the increase that occurs upon cold acclimation of both transgenic and wild type plants. Constitutive expression of *COR15a* was also found to affect freezing tolerance at the cellular level: at freezing temperatures between -6 and -8 °C, the survival of protoplasts isolated from leaves of non-acclimated plants expressing *COR15a* was greater than that of protoplasts isolated from leaves of plants not expressing *COR15a*. The implications of these data regarding possible COR15a functions are discussed. In regard to gene regulation, the promoters of *COR15a* and *COR78* were found to be up-regulated in response to low temperature, drought and ABA. An examination of *COR* gene expression in the *A. thaliana abi* (*AB*A-*i*nsensitive) mutants indicated that low temperature and ABA regulation involve independent pathways. Gene fusion experiments established that *COR15a* has a functional DRE (*d*rought-*r*egulatory-*e*lement) that imparts cold- and drought-regulated gene expression in tobacco. Gel retardation studies indicated that nuclear extracts prepared from both cold-acclimated and non-acclimated *Arabidopsis* plants contain a protein(s) that binds to the *COR15a* DRE and that the avidity of the protein(s) for the element is not significantly altered with cold acclimation.

Keywords. ABA, *Arabidopsis thaliana*, cold acclimation, cold-regulated genes, freezing tolerance

1 Introduction

Many plants increase in freezing tolerance in response to low non-freezing temperatures, a phenomenon known as cold acclimation (Levitt, 1980). Over the years, considerable effort has been directed at understanding the molecular basis of this plant response. Knowledge in this area is not only of basic scientific interest, but has the potential to provide new strategies to improve the freezing tolerance of agronomically important plants. At present, the phenomenon is only partially understood; one cannot yet design from first principles a freezing tolerant tomato. Nevertheless, it is clear that cellular membranes are a primary site of damage during a freeze-thaw cycle and that the increase in freezing tolerance that occurs with cold acclimation involves changes in the cryobehavior and cryostability of cellular membranes (Steponkus et al., 1993). Moreover, it has been established that the changes in membrane cryobehavior associated with cold acclimation are brought about, largely, by changes in membrane lipid composition (Steponkus et al., 1993). A wide variety of other biochemical and physiological alterations have also been shown to occur with cold acclimation including changes in sugar composition, alterations in isozyme patterns and increases in the levels of soluble proteins and free amino acids such as proline (Guy, 1990). In these cases, however, it is unclear whether the changes contribute directly to freezing tolerance or constitute metabolic adjustments required for growth and development at low temperature.

In 1985, Guy et al. demonstrated that changes in gene expression occur during cold acclimation of spinach. Subsequent work by numerous investigators have demonstrated that alterations in gene expression are a central feature of plant cold acclimation (Thomashow, 1993). Two of the fundamental questions thus raised are: what are the functions of the genes induced during cold acclimation? and how do plants sense and alter gene expression in response to low temperature? The progress that we have made in addressing these questions studying *Arabidopsis thaliana COR* (*co*ld-*r*egulated) genes is summarized below.

2 *COR* Gene Function

The primary question that we have been addressing regarding *COR* gene function is whether any of these genes encode cryoprotective polypeptides. The initial stimulus to address this question was the work of Volger and Heber (1975). These investigators presented data indicating that cold-acclimated spinach and cabbage, but not non-acclimated plants, synthesize polypeptides that have potent cyroprotective activities *in vitro*; the proteins are highly active in protecting isolated thylakoid membranes against freeze-thaw damage. Subsequent studies by Hincha and colleagues (1989, 1990) have suggested that the putative cryoprotective polypeptides act by reducing membrane permeability during freezing and increasing membrane expandability during thawing. Unfortunately, the polypeptides have proven recalcitrant to purification, a situation that has hampered detailed characterization of the proteins. Indeed, specific proteins with cryoprotective activity have not been conclusively identified in cabbage and spinach; the genes encoding the putative cryoprotective polypeptides have not been isolated; and it remains to be established whether the cabbage and spinach polypeptides affect freezing tolerance *in vivo*.

As an exploratory experiment to determine whether *Arabidopsis* might produce polypeptides analogous to the putative cryoprotective polypeptides of cabbage and spinach, we asked whether transcripts encoding "boiling-soluble" polypeptides (polypeptides that remain soluble upon boiling) accumulated in *Arabidopsis* during cold acclimation (Lin et al., 1990). We asked this because the results of Volger and Heber (1975) indicated that at least some of the putative cabbage and spinach cryoprotective polypeptides did not coagulate upon heating to 95°C. What Lin et al. (1990) found was that transcripts encoding boiling-soluble polypeptides did indeed accumulate in *Arabidopsis* (as well as wheat) in response to low temperature. Subsequent efforts (see Thomashow et al., 1992) resulted in the identification of cDNA clones for four of the genes that encode boiling-soluble COR polypeptides: *COR78*, *COR47*, *COR15a* and *COR6.6*. Most of our work, described below, has been with *COR15a*.

2.1 Properties of the *COR15a*-Encoded Polypeptide

COR15a encodes a 15 kD polypeptide, COR15a, that is targeted to the stromal compartment of chloroplasts (Lin and Thomashow, 1992a). During import, COR15a is processed to a 9.4 kD polypeptide, designated COR15am. The COR15am polypeptide is hydrophilic, acidic (pI about 4.5), rich in alanine, lysine, glutamic acid and aspartic acid residues, and is predicted to form an amphipathic α-helix. The polypeptide is novel in that no homologs of known function have been described in other organisms. However, three closely related genes have been identified in *Brassica napus*, *BN115*, *BN26* and *BN19* (Weretilnyk et al., 1993), and one in *Arabidopsis*, *COR15b* (Wilhelm and Thomashow, 1993). The *Arabidopsis COR15b* gene is located immediately downstream of *COR15a* in the same relative 5' to 3' orientation. Thus, it would appear that the two genes arose from duplication of a progenitor gene. Interestingly, it has been shown by others that the same situation holds for three other *COR* genes that encode boiling-stable polypeptides: specifically *COR78* (also known as *LTI78* and *RDd29a*) (Nordin et al., 1993; Yamaguchi-Shinozaki and Shinozaki, 1993); *COR47* (Welin et al., 1994) and *COR6.6* (also known as *KIN2*) (Kurkela and Borg-Franck, 1992). In each of these cases, there are two closely related genes present in tandem array.

While no COR15am-related proteins of known function have been described, COR15am does share a low degree of amino acid sequence similarity (about 30%) with group III LEA (late-embryogenesis-abundant) proteins (Gilmour et al., 1992; Wilhelm and Thomashow, 1993). LEA protein are a diverse group of polypeptides synthesized late in embryogenesis, just prior to seed desiccation, and in seedlings in response to drought and ABA (Bray, 1993; Dure, 1993). While the function(s) of LEA proteins have not yet been established, the prevailing thought is that they probably help plant tissues tolerate the potentially damaging effects of dehydration associated with seed development and drought (Bray, 1993; Dure, 1993). This is of interest in regard to *COR* gene function as the cellular damage that occurs with freezing is primarily due to the round of dehydration-rehydration associated with the freeze-thaw cycle (Steponkus et al., 1993). Freezing tolerance must include tolerance to dehydration stress and thus, certain genes with roles in freezing tolerance might be expected to have roles in drought tolerance and *vice versa*. In this regard it is pertinent to note that *COR15a*, as well as *COR78*, *COR47*, and *COR6.6* are induced in response to drought and ABA (Hajela et al., 1990) and that certain *LEA* genes are induced in response to low temperature (Bray, 1993; Welin et al., 1994).

2.2 Evidence that *Arabidopsis COR15a* Encodes a Polypeptide Involved in Freezing Tolerance

To ascertain whether *COR15a* might have a role in freezing tolerance, we constructed transgenic plants that constitutively express the COR15am polypeptide under non-acclimating conditions at levels approximating those found in cold-acclimated wild type plants (Artus et al., 1994). We then determined the freezing tolerance of chloroplasts in non-acclimated transgenic and wild type plants; i.e., the freezing tolerance of chloroplasts that either contained, or did not contain, the COR15am polypeptide. This was done by freezing leaves to various subzero temperatures and after thawing, estimating damage to photosystem II by measuring chlorophyll fluorescence (F_v/F_m). The data indicated that the COR15am-containing chloroplasts were indeed more freezing tolerant than the chloroplasts that did not contain COR15am (Artus et al., 1994; Artus et al., in preparation). In particular, COR15am imparted about a 2°C increase in freezing tolerance. While this increase is small on an absolute scale, it is proportionally large; i.e., it is nearly one-third that which occurs upon cold acclimation of transgenic and wild type plants.

Additional experiments (Artus et al., in preparation) indicated that the effects of COR15am were not limited to the chloroplast organelle; they were also manifested at a cellular level. This was determined by protoplast freezing-tolerance assays. In particular, protoplasts isolated from leaves of non-acclimated plants that constitutively produced COR15am, and non-acclimated plants that did not produce COR15am, were frozen to various subzero temperatures, thawed, and the percent survival was determined by staining with fluorescein diacetate. In repeated tests, the freezing tolerance of COR15am-containing protoplasts was consistently greater than that of protoplasts without COR15am at freezing temperatures between -6 and -8°C.

2.3 Does COR15am Act Directly as a Cryoprotectant?

The transgenic plant experiments described above demonstrate that the COR15am polypeptide can enhance chloroplast and protoplast freezing tolerance in non-acclimated plants. However, the studies do not establish whether COR15am acts directly as a cryoprotectant, as Volger and Heber (1975) and Hincha and colleagues (1989, 1990) have suggested for the putative cryoprotective polypeptides of spinach and cabbage. That is, the data do not distinguish between a direct role as a cryoprotectant or an indirect role in freezing tolerance such as the polypeptide causing changes in, perhaps, sugar metabolism that result in enhanced freezing tolerance (a situation that might more easily explain the effects of COR15am on protoplast freezing tolerance).

As a test of the direct cryoprotectant hypothesis, we have asked whether COR15am has the ability to stabilize either membranes and/or proteins against the damaging effects of freezing *in vitro*. The membrane studies have been carried out in collaboration with Peter Steponkus and colleagues (Cornell University). In particular, the COR15am polypeptide, as well as COR6.6 (one of the other hydrophilic, "boiling-soluble" polypeptides that accumulates during cold acclimation), were expressed at high levels in *Escherichia coli*, purified to near homogeneity (Gilmour et al., unpublished) and then examined for their effects on the cryostability of lipid vesicles (Uemura et al., unpublished) and thylakoid membranes (Uemura et al., unpublished). In brief, it was found that COR6.6 and COR15am decreased the incidence of freeze-induced fusion (i.e., increased the cryostability) of phospholipid SUVs (small unilamellar vesicles) that were suspended and frozen in a Tris/Mes buffer (Uemura et al., unpublished). However, the addition of relatively low concentrations (10 mM) of solutes, such as sucrose or NaCl, also decreased the incidence of freeze-induced fusion, and in their presence, there was no apparent cryoprotective effect of the COR polypeptides. Furthermore, under no conditions tested did the COR polypeptides increase the cryostability of PC (phosphatidylcholine)-SUVs as measured by the incidence of freeze-induced leakage of intraliposomal fluorescent markers. In fact, when the PC-SUVs were frozen alone in buffer (i.e., the only condition that resulted in an apparent cryoprotective effect as measured by the incidence of freeze-induced fusion), addition of the COR polypeptides greatly *increased* the incidence of freeze-induced leakage after freezing to -10°C or lower. Parallel studies demonstrated that the COR polypeptides did not affect the dehydration-induced increase in the liquid crystalline-to-gel phase transition temperature of either DPPC (dipalmitoylphosphatidylcholine) or DOPC (dioleoylphosphatidylcholine) (Webb et al., unpublished) and had no effect on the occurrence of the lamellar-to-hexagonal II phase transition in 1:1 mixtures of DOPE (dioleoylphosphatidylethanolamine) and DOPC dehydrated over a range of osmotic pressures (Webb et al., unpublished). It was also found that COR15am had no cryoprotective effects on freeze-induced inactivation of either light-induced proton uptake or photophosphorylation in spinach thylakoids frozen *in vitro* (Uemura et al., in preparation). Thus, our studies to date indicate that neither COR15am nor COR6.6 have a direct cryoprotective effect on either lipid vesicles or thylakoid membranes that are frozen *in vitro*.

In regard to cryoprotection of proteins, Lin and Thomashow (1992b) reported that COR15a, the precursor of COR15am, is extremely effective in stabilizing lactate dehydrogenase (LDH) against freeze-thaw inactivation; on a concentration basis, it was found to be about 100-fold more effective than bovine serum albumin (BSA) in protecting LDH against freeze-inactivation. Thus, Lin and Thomashow (1992b) suggested that the *COR15a* gene might encode a cryoprotective polypeptide that helps stabilize proteins against freeze-thaw damage. However, recent results indicate that neither the COR15am nor COR6.6 polypeptides purified from *E. coli* have exceptional cryoprotective activity; initial experiments indicate that both proteins are about equal to BSA (on a concentration basis) in protecting LDH against freeze-thaw inactivation (Gilmour and Thomashow, unpublished). The reason for these differing results is not clear. However, the results are not necessarily contradictory as the two polypeptides are quite different: COR15a is a neutral polypeptide (pI of 7.0) with a hydrophobic N-terminus (part of the chloroplast targeting sequence) while COR15am is acidic (pI 4.5) and devoid of hydrophobic regions. Regardless, unlike the data with COR15a, the results with COR15am and COR6.6 do not suggest a direct cryoprotective role for these proteins in stabilizing proteins against freeze-thaw inactivation.

2.4 Summary and Future Directions

Do plants that cold acclimate have genes that encode cryoprotective polypeptides? This has been a controversial issue in plant cold acclimation research ever since the notion was suggested some twenty years ago by Volger and Heber (1975). Our results demonstrate that the cold-regulated *COR15a* gene encodes a protein that can enhance the freezing tolerance of chloroplasts and protoplasts. However, the results of the *in vitro* experiments conducted in collaboration with Peter Steponkus and colleagues argue against COR15am having a direct cryoprotective role in stabilizing membranes as Volger and Heber (1975) and Hincha and colleagues (1989, 1990) have suggested for the putative cryoprotective polypeptides of spinach and cabbage. In addition, there is no evidence at present indicating a role for COR15am in protecting proteins against freeze-thaw damage. Thus, in addition to conducting further tests for direct roles in cryoprotection, indirect roles such as the possibility that COR15am affects changes in cell metabolism that result in enhanced freezing tolerance must be considered. More detailed comparisons of the wild type and COR15am-producing transgenic plants in regard to differences in cryobehavior and biochemical composition may suggest additional testable hypotheses regarding the mode of COR15am action.

3 *COR* Gene Regulation

The overall pattern of gene expression in *Arabidopsis* seedlings is not dramatically altered with cold acclimation (Gilmour et al., 1988), a situation that is quite different from many other environmental "stresses" such as heat shock (Vierling, 1991). Specifically, two dimensional SDS-PAGE analysis of *in vivo* labeled polypeptides indicates that out of a total display of some 200 polypeptides, there are about a dozen differences in non-acclimated and cold-acclimated plants; i.e., appearance of new polypeptides or substantial increases or decreases in others. Of course, this experimental approach only detects changes in the levels of the most abundantly expressed genes. If the same degree of change occurs among all (approximately) 20,000 *Arabidopsis* genes, it would mean that expression of hundreds of genes is altered during cold acclimation.

Northern analysis has shown that the transcript levels for *COR78*, *COR47*, *COR15a* and *COR6.6* begin to accumulate within 4 h of transferring plants to low temperature (5°C) and that they remain elevated for as long as the plants are kept in the cold (for at least two weeks) (Hajela et al., 1990). Transfer of the plants back to normal growth temperatures (22°C) results in a rapid decrease in the levels of the transcripts; depending on the specific transcript, the levels fall to those found in non-acclimated plants within 4 to 8 h. The basic question that we wish to address is what mechanism(s) regulates the levels of these *COR* genes in response to low temperature?

3.1 Does ABA Mediate Cold-Regulated Gene Expression?

As noted above, the transcript levels for *COR78*, *COR47*, *COR15a* and *COR6.6* accumulate in response to ABA. This is intriguing since ABA levels are known to increase in certain plants, including *Arabidopsis* (Lång et al., 1994), in response to low temperature. Moreover, exogenous application of ABA has been shown to increase the freezing tolerance of certain plants (Chen and Gusta, 1983). Thus, as first suggested by Chen et al. (1983), the possibility is raised that low temperature gene expression is mediated through the action of ABA; i.e., ABA levels increase in response to low temperature which, in turn, affect changes in gene expression. To test this notion, we asked whether ABA regulation of *COR* gene expression was impaired in either the *abi1*, *abi2* or *abi3* ABA-insensitive mutants of *Arabidopsis* (Koornneef et al., 1984). The data indicated that it was: ABA-induced accumulation of *COR* transcripts was severely impaired in the *abi1* mutants (Gilmour and Thomashow, 1991). However, the *abi1* mutation had little, if any, effect on the accumulation of *COR* transcripts in response to low temperature. Thus, we concluded (Gilmour and Thomashow, 1991) that ABA and low temperature regulation of *COR* genes appeared to proceed through separate pathways. Nordin et al. (1991) reached the same conclusion in a similar study.

3.2 Characterization of the *COR15a* and *COR78* Promoters

Gene fusion studies have established that the promoters of both *COR15a* (Baker et al., 1994) and *COR78* (Horvath et al., 1993; Yamaguchi-Shinozaki and Shinozaki, 1993) are responsive to low temperature, ABA and drought. Interestingly, however, there are tissue specificity differences between the promoters. One regards constitutive expression. The *COR15a* promoter (-900 to +78) fused to the GUS reporter gene is expressed in anthers of non-acclimated plants (Baker et al., 1994) while the *COR78* promoter (-880 to +8) is not (Horvath et al., 1993). Moreover, there are differences in the low temperature tissue specificity of the promoters. For instance, in cold-acclimated plants, the *COR15a* promoter is expressed in all flower parts except the ovary (Baker et al., 1994) while the *COR78* promoter is only expressed in the sepals, petals and anther filaments (Horvath et al., 1993). This finding has interesting implications regarding potential *cis*-acting cold-regulatory elements. If there were only one element and it was solely responsible for cold-regulated gene expression, then one would not expect to see differences in tissue specificity. Thus, there must either be a family of *cis*-acting cold-regulatory elements, with members having different tissue specificity, and/or the *cis*-acting cold-regulatory element interacts with other tissue specific elements to give the cold regulation pattern of gene expression.

An inspection of the *COR15a* and *COR78* promoter sequences indicates that each has multiple potential ABREs (ABA-responsive elements), elements shown to be involved in both ABA- and drought-regulated expression of a number of genes (Bray, 1993). Presumably these elements contribute to the ABA- and drought-regulated expression of *COR15a* and *COR78*. In addition, there is a repeated sequence with a perfectly conserved 5 base core, CCGAC, that is present three times in the *COR15a* promoter and four times in the *COR78* promoter; we refer to this sequence as the C-Box. Promoter deletion experiments with *COR15a* were consistent with the C-Box being involved in cold-regulated gene expression (Baker et al., 1994). However, the first definitive evidence for this was provided by Yamaguchi-Shinozaki and Shinozaki (1994) who showed that the C-Box sequence from *RD29a* (*COR78*), which they refer to as a DRE (drought regulatory element), can impart drought and cold-regulated gene expression in *Arabidopsis*. Moreover, they found that the element does not impart ABA-regulated gene expression, a result that is consistent with the genetic studies indicating that low temperature and ABA can affect gene expression through independent pathways (Gilmour and Thomashow, 1991; Nordin et al., 1991). In accord with the results of Yamaguchi-Shinozaki and Shinozaki (1993), we have found that the C-Box from *COR15a* fused to the -90 CaMV promoter imparts both drought- and cold-regulated gene expression in tobacco (Bloksberg et al., in preparation). In addition, gel retardation assays indicate that there is a protein(s) in nuclear extracts of both cold-acclimated and non-acclimated *Arabidopsis* plants that binds to the C-Box element of *COR15a* and that the avidity of the protein(s) for the C-Box is not significantly altered with cold acclimation (Bloksberg et al., in preparation).

3.3 Summary and Future Directions

It is now established that the promoters of *COR15a* and *COR78* are activated in response to low temperature; that the C-Box element is involved in this regulation; and that there is a protein(s) in both non-acclimated and cold-acclimated plants that binds to the C-Box element. Major goals now include determining the identity of the C-Box binding protein(s) and nature of the low temperature signal transduction pathway responsible for activating it and/or associated factors. In addition, the relationship between cold- and drought-regulated gene expression that is mediated through the C-Box element must be determined. Do low temperature and drought produce a common stimulus that activates the same signal transduction pathway or do they activate separate signal transduction pathways that happen to act through the same *cis*-acting regulatory element? As a first step to address these questions, we are attempting to isolate cDNA clones for C-Box-binding proteins and to identify *Arabidopsis* mutants that either constitutively express *COR15a* at normal growth temperature or do not express it in response to low temperature.

Acknowledgements. Research conducted in the authors' laboratories was supported by grants from the National Science Foundation (IBN-9307348), the U.S.D.A National Research Initiative Competitive Grants Program (92-37100-7351) and the Michigan Agricultural Research Station to M.F.T.; and the U.S. Department of Energy (DE-FG01-84ER13214) and the U.S.D.A National Research Initiative Competitive Grants Program (93-37100-8835) to P.L.S.

References

Artus NN, Gilmour SJ, Lin C, Thomashow MF (1994) Use of transgenic *Arabidopsis* plants to evaluate the roles of cold-regulated genes in freezing tolerance. Plant Physiol **105**: S-169 (Abs #939)

Baker SS, Wilhelm KS, Thomashow MF (1994) The 5'-region of *Arabidopsis thaliana cor15a* has *cis*-acting elements that confer cold-, drought-, and ABA-regulated gene expression. Plant Mol Biol **24**: 701-713

Bray EA (1993) Molecular responses to water stress. Plant Physiol **103**: 1035-1040

Chen THH, Gusta LV (1983) Abscisic acid-induced freezing resistance in cultured plant cells. Plant Physiol **73**: 71-75

Chen H-H, Li PH, Brenner ML (1983) Involvement of abscisic acid in potato cold acclimation. Plant Physiol **71**: 362-365

Dure III L (1993) Structural motifs in LEA proteins. *In* TJ Close, EA Bray, eds, Plant Responses to Cellular Dehydration During Environmental Stress. American Society for Plant Physiologists, Rockville, pp 91-103

Gilmour SJ, Hajela RK, Thomashow MF (1988) Cold acclimation in *Arabidopsis thaliana*. Plant Physiol **87**: 745-750

Gilmour SJ, Thomashow MF (1991) Cold acclimation and cold-regulated gene expression in ABA mutants of *Arabidopsis thaliana*. Plant Mol Biol **17**: 1233-1240

Gilmour SJ, Artus NN, Thomashow MF (1992) cDNA sequence analysis and expression of two cold-regulated genes of *Arabidopsis thaliana*. Plant Mol Biol **18**: 13-21

Guy CL, Niemi KJ, Brambl R (1985) Altered gene expression during cold acclimation of spinach. Proc Natl Acad Sci USA. **82**: 3673-3677

Guy CL (1990) Cold acclimation and freezing stress tolerance: Role of protein metabolism. Annu Rev Plant Physiol Plant Mol Biol **41**: 187-223

Hajela RK, Horvath DP, Gilmour SJ, Thomashow MF (1990) Molecular cloning and expression of *cor* (*co*ld-*r*egulated) genes in *Arabidopsis thaliana*. Plant Physiol **93**: 1246-1252

Hincha DK, Heber U, Schmitt JM (1989) Freezing ruptures thylakoid membranes in leaves, and rupture can be prevented *in vitro* by cryoprotective proteins. Plant Physiol Biochem **27**: 795-801

Hincha DK, Heber U, Schmitt JM (1990) Proteins from frost-hardy leaves protect thylakoids against mechanical freeze-thaw damage *in vitro*. Planta **180**: 416-419

Horvath DP, McLarney BK, Thomashow MF (1993) Regulation of *Arabidopsis thaliana* L. (Heynh) *cor78* in response to low temperature. Plant Physiol **103**: 1047-1053

Koornneef M, Reuling G, Karssen CM (1984) The isolation and characterization of abscisic acid-insensitive mutants of *Arabidopsis thaliana*. Physiol Plant **61**: 377-383

Kurkela S, Borg-Franck M (1992) Structure and expression of *kin2*, one of two cold- and ABA-induced genes of *Arabidopsis thaliana*. Plant Mol Biol **19**: 689-692

Lång V, Mäntylä E, Welin B, Sundgerg B, Palva ET (1994) Alteration in water status, endogenous abscisic acid content, and expression of *rab18* gene during the development of freezing tolerance in *Arabidopsis thaliana*. Plant Physiol **104**: 1341-1349

Levitt J (1980) Responses of plants to environmental stress. Chilling, freezing, and high temperature stresses. Ed 2. Academic Press, New York

Lin C, Guo WW, Everson E, Thomashow MF (1990) Cold acclimation in *Arabidopsis* and wheat. A response associated with expression of related genes encoding 'boiling-stable' polypeptides. Plant Physiol **94**: 1078-1083

Lin C, Thomashow MF (1992a) DNA sequence analysis of a complementary DNA for cold-regulated *Arabidopsis* gene *cor15* and characterization of the COR15 polypeptide. Plant Physiol **99**: 519-525

Lin C, Thomashow MF (1992b) A cold-regulated *Arabidopsis* gene encodes a polypeptide having potent cryoprotective activity. Biochem Biophys Res Commun **183**: 1103-1108

Nordin K, Heino P, Palva ET (1991) Separate signal pathways regulate the expression of a low-temperature-induced gene in *Arabidopsis thaliana* (L.) Heynh. Plant Mol Biol **16**: 1061-1071

Nordin K, Vahala T, Palva ET (1993) Differential expression of two related, low-temperature-induced genes in *Arabidopsis thaliana* (L.) Heynh. Plant Mol Biol **21**: 641-653

Steponkus PL, Uemura M, Webb MS (1993) A contrast of the cryostability of the plasma membrane of winter rye and spring oat-two species that widely differ in their freezing tolerance and plasma membrane lipid composition. Adv in Low Temperature Biology **2**: 1-67

Thomashow MF, Gilmour SJ, Lin C (1992) Cold-regulated genes of *Arabidopsis thaliana*. *In* PH Li, L Christersson, eds, Advances in Plant Cold Hardiness. CRC Press, Boca Raton, pp 32-44

Thomashow MF (1993) Genes induced during cold acclimation in higher plants. Adv in Low Temperature Biology **2**: 183-210

Vierling E (1991) The roles of heat shock proteins in plants. Annu Rev Plant Phys Plant Mol Biol **42**: 579-620

Volger HG, Heber U (1975) Cryoprotective leaf proteins. Biochim Biophys Acta **412**: 335-349

Welin BV, Olson Å, Nylander M, Palva ET (1994) Characterization and differential expression of *dhn/lea/rab*-like genes during cold acclimation and drought stress in *Arabidopsis thaliana*. Plant Mol Biol **26**: 131-144

Weretilnyk E, Orr W, White TC, Iu B, Singh J (1993) Characterization of three related low-temperature-regulated cDNAs from winter *Brassica napus*. Plant Physiol **101**: 171-177

Wilhelm KS, Thomashow MF (1993) *Arabidopsis thaliana cor15b*, an apparent homologue of *cor15a*, is strongly responsive to cold and ABA, but not drought. Mol Gen Genet **236**: 331-340

Yamaguchi-Shinozaki K, Shinozaki K (1993) Characterization of the expression of a desiccation-responsive *rd29* gene of *Arabidopsis thaliana* and analysis of its promoter in transgenic plants. Mol Gen Genet **236**: 331-340

Yamaguchi-Shinozaki K, Shinozaki K (1994) A novel *cis*-acting element in an *Arabidopsis* gene is involved in responsiveness to drought, low-temperature, or high-salt stress. Plant Cell **6**: 251-264

Part 3

Salt Stress

In Search of a Function for Dehydrin TAS14

José Antonio Godoy, Rosa Luna, Maria de Mar Parra, Olga del Pozo and José Antonio Pintor-Toro

Instituto de Recursos Naturales y Agrobiología, C.S.I.C., E-41080 Sevilla, Spain

Abstract. A continuously growing number of genes induced by osmotic or salt stress have been described. A group of these genes defines the dehydrin family. Dehydrin genes were originally described as being expressed during late stages of embryogenesis, coinciding with the desiccation of the seed. A common characteristic among dehydrins is that they can be induced in vegetative tissue by both exogenous application of ABA and salt treatment. The are characterized by a conserved lysine-rich block (KIKEKLPG), that is predicted to form an amphipathic α-helix, and a stretch of serine residues. In spite of the extensive information about dehydrin proteins, their function(s) remain unknown, although several hypotheses have been considered. Our research is currently focused on unravelling the function of dehydrin TAS14. Toward this goal, the biochemical characterization of TAS14 protein is being carried out. TAS14 tissue distribution and subcellular localization will be shown as well as the analysis of *gus* reporter gene expression driven by TAS14 promoter. The possible interaction of TAS14 with other macromolecules will be also discussed.

Keywords. Abscisic acid, dehydrin, LEA proteins, *Lycopersicon esculentum*, osmotic stress, tomato

1 Introduction

An increasing number of stress-induced plant genes have been identified and characterized in the last decade. This continuing effort is contributing to the understanding of the responses of plants to stress and may ultimately provide molecular tools for the breeding of crops better adapted to extreme environments. Early work in our lab in the search for salt stress-induced genes led to the isolation of a cDNA we named TAS14. Sequence comparisons made clear that it was a member of a class of water stress-induced genes, that we are going to refer to as dehydrins (Close et al., 1989).

The expression of dehydrins is generally induced by environmental conditions that implicate some kind of water stress (drought, high salt, desiccation, cold) and by the exogenous application of abscisic acid (ABA). In addition, some dehydrins, like other dehydration-induced genes, are expressed during the dehydration of the seed at late stages of embryogenesis. The term *lea* (*l*ate *e*mbryogenesis *a*bundant), refers to these proteins and genes; dehydrins are those LEA proteins classified as group 2. Therefore, dehydrins as a general term include group 2 *lea* and some of the *rab* (responsive to ABA), *cor* (cold-regulated), *lti* (low temperature-induced) genes. Structurally, the different dehydrins share one main characteristic and some others are of general occurrence. The structural signature of the family is a highly conserved lysine-rich block which is present usually twice at the carboxy terminus present usually twice, but from once to many times within the protein. The usual dehydrin characteristics are: the presence of a stretch of serine residues, a second conserved domain with the consensus sequence DEYGN located at the N-terminus, a high content of glycine and threonine residues and the lack of cysteine and tryptophan residues. They are all highly hydrophilic and remain soluble after boiling. TAS14 protein presents all these structural characteristics, being thus a typical dehydrin.

The occurrence of dehydrins extend to gymnosperms and both mono- and dicotyledoneous species (Close et al., 1993a). Immunological evidence was found indicating that related proteins exist in cyanobacteria (Close and Lammers, 1993).

It seems clear that dehydrin gene expression is associated to cellular dehydration. The main point that remains to be established is to what extent they account for the observed desiccation tolerance of embryos, or they participate in the adaptive response of plants to drought. A second point that could help answering the first one and that has an interest in itself is: what are the molecular or biochemical properties of dehydrins?; could they be related to the tolerance to dehydration? Several possible functions for dehydrins have been discussed, mostly based on their predicted physiochemical properties (see Dure, 1993). These include ion or water traps, protein stabilizers and compatible solutes. From what we currently know about dehydrins an enzymatic activity seems unlikely; their wide range of sizes, their flexible structure predicted from their sequence and the high concentrations they seem to reach in the cell seem to exclude an enzymatic function for dehydrins.

Two aspects that might ultimately contribute to the understanding of dehydrins and their role in development and in the response to environmental stress will be discussed here. These are: the analysis of *tas14* expression and the analysis of its molecular properties.

2 The Expression of Dehydrins: Regulation and Tissue-Specificity

When and where in the plant are dehydrins expressed? What are the environmental cues that lead to dehydrin expression? What organs, tissues and cells accumulate dehydrins in response to stress? Is there any developmental regulation of dehydrin genes, apart from their known expression during embryogenesis? These are all the questions that we have tried to answer in many years of research focused on TAS14.

The regulation of dehydrin expression by different environmental factors has been extensively studied, mostly at the level of mRNA. Environmental factors that induce dehydrin mRNA accumulation include: drought, desiccation, salt stress and cold. Not all the dehydrins respond to all these factors, suggesting that there is some specialization between different dehydrins. For example, TAS14 mRNA accumulates under osmotic or salt stress, but not under low temperatures. An extensive analysis of the expression of *Arabidopsis* dehydrins revealed that the different genes analyzed varied in their relative inducibility by drought, cold or ABA (Welin et al., 1994). Developmentally regulated expression of dehydrins have been observed to occur in maturing seeds. In addition, the exogenous application of ABA can induce dehydrins expression. Since endogenous ABA levels increase both in response to water stress and during late stages of embryogenesis, ABA is supposed to be an endogenous regulator of dehydrin expression. Studies with ABA deficient or insensitive mutant indicate that this may be the case for their stress-induced expression in vegetative tissues, but not for their expression during embryogenesis (Vilardell et al., 1990, 1994); furthermore ABA is required for the stress induction of some, but not others, *Arabidopsis* dehydrins (Welin et al., 1994). Recently, the spectrum of differentially regulated dehydrins was further widened by the discovery of constitutive or "cognate" dehydrins that are expressed in vegetative tissues in the absence of stress (Robertson and Chandler, 1994; Welin et al., 1994).

We obtained antibodies against TAS14 protein and used them to follow the accumulation of TAS14 protein under different treatments and in different organs (Godoy et al., 1994). Several bands ranging from 19 to 21 kD induced by the treatments are recognized by anti-TAS14 antibodies. These bands correspond to different phosphorylated forms of a single polypeptide (Torres-Schumann et al., 1991). The pattern of accumulation of TAS14 protein mostly correlated with the previously observed accumulation of the transcript (Godoy at al., 1990). ABA, mannitol and salt treatments are all able to induce the accumulation of TAS14 protein in tomato seedlings. In salt-stressed plants TAS14 accumulates preferentially in stems and leaves, while only a slight and transient accumulation was detected in roots. These low protein levels in roots did not correspond with the fairly high mRNA levels, indicating a possible post-transcriptional regulation (Godoy et al., 1994).

Information about the organ, tissue and cell specificity of dehydrin expression is even scarcer. Maize RAB17 protein was detected by immunocytochemistry in all types of cells in mature embryos (Goday et al., 1994). Stress-induced expression in vegetative tissues has only been analyzed at this level for *Craterostigma plantagineum* dsp16, and was found to be homogeneous in desiccated roots and leaves (Schneider et al., 1993). We have used affinity-purified anti-TAS14 antibodies to study the tissue distribution of TAS14 protein in salt-stressed plants. A striking pattern was observed consisting of high staining of provascular tissue and outer cortex of stems and developing leaves, and an intense staining of adventitious root meristems (Godoy et al., 1994).

Additional information on the tissue-specific expression of dehydrins might be obtained by studying the pattern of GUS activity driven by *tas14* promoter in transgenic plants. Unfortunately, this information is also scarce for a number of reasons. Stress-induced expression of GUS was not observed, at least in mature plants, when rice *rab16* or maize *rab17* promoters were analyzed in transgenic tobacco or *Arabidopsis*; expression during seed development was observed, however, indicating that different regulatory pathways may be involved (Yamaguchi-Shinozaki et al., 1990; Vilardell et al., 1991, 1994). In contrast, *Craterostigma* cDeT6-19 promoter drives GUS expression in vegetative tissues in response to desiccation, as well as in seeds and pollen (Michel et al., 1994). A detailed histochemical analysis of stressed plants harbouring the cDeT6-19-GUS construct has not been reported, except for its desiccation- and ABA-induced expression in stomatal guard cells of leaves (Taylor et al., 1995). In an effort to fill this gap and complement the data obtained with the antibodies, we have analyzed the expression features conferred by *tas14* promoter in transgenic tobacco, using *tas14-gus* fusions. A fragment of *tas14* gene containing approximately 2500 bp of 5' upstream regulatory sequences and 150 bp of the 5' region of the transcript confers a dramatic inducibility by both ABA and osmotic stress to the reporter gene. The fused fragment is also able to drive developmentally regulated GUS expression in seeds. Histochemical analysis of salt-stressed transgenic tobacco plants reveals a pattern of GUS activity that is consistent with the immunohistochemistry results: GUS activity is detected associated to vascular system and to the outer cortex and is high in some, but not all, adventitious root primordia. In addition, stress-induced activity of TAS14 promoter was observed in leaf axils, and in stomatal guard cells and trichomes in the epidermis of leaves. High levels of GUS activity were detected associated to the vascular cylinder in roots (Parra et al., in preparation).

The analysis of the expression of dehydrins in non-stressed plants is most interesting since it could reveal a stress-independent regulation that could pinpoint possible unsuspected functions for dehydrins. In earlier studies no mRNA could be detected in roots, stems or leaves of non-stressed tomato plants (Godoy et al., 1990). Fluorometric assays of GUS activity of non-stressed tobacco plants harbouring the *tas14-gus* construct revealed a very low basal activity that was, however, 3 to 5-fold the activity detected in plants transformed with a promoterless ß-*glucuronidase* gene. This result indicated a basal level of expression of the transgene in the absence of stress. When analyzed histochemically, GUS activity in non-stressed plants could only be detected in the leaf axils. Expression of *tas14* in tomato seeds was paralleled by the GUS staining of transgenic tobacco seeds; interestingly, stain in seeds was restricted to the embryo, where it was fairly homogenous. Surprisingly, GUS activity was detected in flowers. All floral organs were stained, with quite distinct patterns. Stigma was stained from very early stages throughout flower development, while style and ovary were not stained. Anthers showed high levels of GUS activity at later stages of floral development, being highest just preceding anther dehiscence. Mature pollen grains were heavily stained. GUS activity in petals and sepals was restricted to stomatal guard cells and trichomes. Expression in floral organs was confirmed by northern blot of tomato flowers. A transcript hybridizing to *tas14* cDNA was detected that accumulated in all flower stages, from unopened floral buds to flowers 3-5 d post-anthesis. The transcript accumulates to different extents in all floral organs, correlating with the relative GUS activity detected in transgenic tobacco plants (Parra et al., in preparation).

We have just described the tissue-specific expression of *tas14* gene in stressed and non-stressed plants. The significance of these findings regarding TAS14 function is not clear. Several questions arise concerning the possible functional implications of the observed pattern and the regulation of *tas14* gene expression. Are those tissues that express *tas14* in response to stress simply the ones that are more dehydrated by the particular treatments used? Or are there further developmental restrictions superimposed to *tas14* environmental regulation? Could the expression of *tas14* gene during flower development be explained solely by developmentally regulated cellular dehydration? This could clearly be the case for the expression in anthers and pollen. Similarly, the expression in guard cells could be related to the osmotic imbalance associated with their function. However, it is not so evident that cellular dehydration may occur in stigma, trichomes or leaf axils of non-stressed plants. Assuming that the protein is expressed when and where needed, the expression of *tas14* in apparently non-stressed cells should have a role related to the physiology of these particular cell type. Which role this might be is so far unclear.

3 Localization of Dehydrins within the Cell

Another interesting point that could give hints on the possible dehydrin functions is the determination of the intracellular compartment where these proteins accumulate. Early fractionation experiments with rice RAB16 (Mundy and Chua, 1988) suggested that dehydrins were cytosolic proteins. We have analyzed the location of TAS14 protein in adventitious root meristem cells at the electron microscope level. Gold particles were found both in the cytoplasm and in the nucleus. No clear association to any structure in the cytoplasm was observed. In contrast, TAS14 protein in the nucleus is associated to transcriptionally active regions: euchromatin and nucleolus. Sections of untreated plants incubated in parallel with anti-TAS14 antibodies and sections of salt-stressed plants incubated with preimmune serum were all devoid of gold particles, discarding possible staining artefacts (Godoy et al., 1994). A nuclear-cytoplasmic location was also recently found for maize dehydrin in embryo and aleurone cells (Goday et al., 1994; Asghar et al., 1994), while an exclusively cytoplasmic localization has been reported for *Craterostigma* dsp16 (Schneider et al., 1993). Even within the maize embryo, the relative distribution between nucleus and cytosol seems to vary from one cell type to another (Asghar et al., 1994). This variety of results may indicate that dehydrin distribution within the cell may be regulated by cell type, treatment or other unknown factors. It is possible that this distribution may be affected by phosphorylation as it happens with other nuclear proteins.

What are the functional implications of this dual nuclear-cytosolic localization? TAS14 protein could stabilize cellular components that are common to nucleus and cytosol. However, its association to transcriptionally active chromatin indicates a more specific target, i.e. the transcriptional machinery. Dehydrins could also be implicated in nuclear transport. In fact, it has pointed out the structural similarities existing between dehydrins and Nopp140, a protein that binds nuclear localization signals (NLS) and participates in nucleolar trafficking (Asghar et al., 1994, Goday et al., 1994). This point has been investigated by Dr. Pages's group, having demonstrated the ability of maize RAB17 protein to bind NLS peptides *in vitro* (Goday et al., 1994).

4 Molecular and Biochemical Characteristics of Dehydrin

Little is known about the actual structural characteristics and biochemical properties of dehydrins, although some predictions can be made from their amino acid sequence (see Dure, 1993). Interestingly, the most strictly conserved part of dehydrins sequence, the lysine-rich block, is the only part of the molecule that is predicted to form a defined structure: an amphiphilic alpha-helix, in which hydrophobic, acidic and basic residues are grouped in different faces of the helix (Godoy et al., 1990). Helices of this type have been shown to mediate interactions with other macromolecules, namely DNA and proteins. We have analyzed both possibilities.

4.1 Analyzing Possible Interactions of TAS14 Protein with DNA

The observation that TAS14 accumulates in the nucleus opened the possibility that it might interact with DNA. In order to test this hypothesis we performed polyacrylamide gel mobility shift assays using a complex mixture of oligonucleotides with random sequences and recombinant TAS14 protein from different sources. Bacterially expressed TAS14, non-phosphorylated or phosphorylated *in vitro* with casein kinase II, and *tas14* cDNA *in vitro* transcription/translation product were all used in these assays. Consistent evidence for an interaction of TAS14 and the oligonucleotides could not be found in any case. Our results indicate that TAS14 does not interact directly with naked DNA. Interaction mediated by other proteins cannot be ruled out.

After the negative results with DNA we went on to check whether TAS14 protein interacts with proteins.

4.2 Analyzing Possible Interactions between TAS14 Molecules

The possibility that TAS14 interacts with itself to form homo-oligomers was investigated through three different approaches: 1) determination of its molecular weight under non-denaturing conditions by gel filtration chromatography, 2) chemical crosslinking, and 3) two hybrid assay.

In order to estimate a mol wt for native TAS14 protein, 250 μg of recombinant TAS14, either non phosphorylated or *in vitro* phosphorylated with casein kinase II, were analyzed by chromatography through a Sephacryl-S200 column equilibrated in 100 mM acetate buffer, pH 5.5, 100 mM KCl. TAS14 elution volumes were estimated by determining the protein content of the different fractions and confirmed by western blotting. Mol wts estimated from the elution volumes for phosphorylated and non-phosphorylated recombinant TAS14 are 27.5 and 30.9, respectively. Similar results were obtained with native maize dehydrin purified from embryos (Close et al., 1993b). These values are 2 to 3 fold higher than the TAS14 mol wt determined from its sequence, indicating possible self-aggregation of TAS14 polypeptide. However, contradictory evidence is described below, indicating that this may not be the case. An alternative explanation is that the mol wt calculated from gel filtration chromatography are overestimated due to anomalous migration. In fact, TAS14 and other dehydrins show a mobility lower than that expected in polyacrylamide gel electrophoresis (Godoy et al., 1990, 1994; Mundy and Chua, 1988; Vilardell et al., 1990).

We tried to covalently bind interacting TAS14 molecules with crosslinking agents and then analyze the complex by SDS-PAGE and western blot. Two crosslinking agents with different specificity were used: 1) dimethyl-pimelimidate-2 HCl (DMP), that forms covalent bonds between amine groups, and 2) 1-ethyl-3-(3-dimethylaminopropyl) carbodiimide hydrochloride (EDC), that binds closely located amine and carboxylic acid groups. Only one band of the size of denatured TAS14 was observed after incubation with DMP, even with the highest amounts of TAS14 protein used. On the other hand, a second band, with a molecular size 2 kD higher than that of TAS14, was observed that increased with the time of incubation in EDC. Due to the small size increase and to the small distance range in which EDC can crosslink, we interpret this size change as due to intermolecular bonds affecting TAS14 protein mobility.

The two-hybrid system is a yeast-based genetic assay to detect specific protein-protein interactions *in vivo* by functional restoration of a eukaryotic transcriptional activator (Bartel et al., 1993). TAS14 cDNA was cloned into two separate vectors: one that generated a hybrid protein with the GAL4 DNA-binding domain (pGBT) and one that generated a hybrid protein with the GAL4 activation domain (pGAD). Both domains are required in close proximity for normal functioning of the GAL4 activator. The two hybrid plasmids were then cotransformed into a yeast host strain that has a *lacZ* and a *HIS3* reporter genes under the control of upstream GAL4 binding sites. Expression of the reporter genes would indicate that the two hybrid proteins interact. Yeast cells cotransformed with both plasmids, pGBT-TAS and pGAD-TAS, did not grow on medium lacking histidine and were white in a ß-galactosidase filter assay, indicating that TAS14 does not interact with itself.

4.3 Analyzing Possible Interactions of TAS14 with other Proteins

We have also investigated the possible interactions of TAS14 protein with other cellular proteins by two different experimental approaches: affinity chromatography and ligand blot assays. Additionally, the analysis of TAS14-containing protein extracts should also determine whether TAS14 protein interacts with itself.

In the affinity chromatography assays total labeled protein extracts are incubated with immobilized TAS14 protein and bound proteins are eluted and analyzed by SDS electrophoresis. *In vivo* labeled proteins from 4-day-old tomato seedlings were sequentially incubated with Glutathione Sepharose beads with no protein bound, same beads saturated with glutathione-S-transferase and beads saturated with GST-TAS14 fusion protein. After washing, proteins bound to each type of beads were eluted and analyzed by SDS-PAGE and fluorography. Two bands of 40 and 42 kD were observed among the elution products of the GST-TAS14 beads that were not present in eluates from GST or glutathione beads, indicating that they correspond to proteins specifically retained by TAS14. These proteins are detected in all extracts tested, including extracts from seedlings, untreated or treated with NaCl, mannitol or ABA, and extracts from leaves, flowers and fruits. These results suggests that TAS14-bound proteins are constitutively expressed. Retention of these proteins by TAS14 was not affected by the phosphorylation state of TAS14 (Luna et al., in preparation).

Interaction of TAS14 with other proteins was also investigated using a ligand blot assay. Whole protein extracts from tomato seedlings were analyzed by SDS-PAGE. Resolved proteins were transferred to PVDF filters that were then incubated with TAS14 protein phosphorylated *in vitro* with [γ-^{32}P]ATP and casein kinase II. After incubation for several hours, the filter was washed and autoradiographed. Several bands were observed after autoradiography in lanes corresponding to both untreated and salt-treated seedlings, potentially corresponding to proteins that interact with TAS14 (Luna et al., in preparation).

It must be noted that TAS14 protein was not detected in the affinity chromatography or ligand blot experiments, providing an additional evidence that interaction between TAS14 molecules does not occur.

We have discussed the information we have gathered so far regarding the expression pattern and molecular properties of dehydrins. Our results indicate that the dehydrin *tas14* gene might be expressed in non-desiccating cells and that TAS14 protein interacts with other cellular proteins. The identification of these proteins will be a major step towards the understanding of dehydrin function and to determine which role, if any, they play in the adaptive response of plant cells to dehydration.

References

Asghar R, Fenton RD, DeMason DA, Close TJ (1994) Nuclear and cytoplasmic localization of maize embryo and aleurone dehydrin. Protoplasma **77**: 87-94

Bartel PL, Chien C-T, Sternglanz R, Fields S (1993) Using the two-hybrid system to detect protein-protein interactions. *In* DA Hartley, ed, Cellular Interactions in Development: a Practical Approach. Oxford University Press, Oxford, pp 153-179

Close TJ, Kort AA, Chandler PM (1989) A cDNA-based comparison of dehydration-induced proteins (dehydrins) in barley and corn. Plant Mol Biol **13**: 95-108

Close TJ, Fenton RD, Moonan F (1993a) A view of plant dehydrins using antibodies specific to the carboxy terminal peptide. Plant Mol Biol **23**: 279-286

Close TJ, Fenton RD, Yang A, Asghar R, DeMason DA, Crone DE, Meyer NC, Moonan F (1993b) Dehydrin: the protein. *In* TJ Close, EA Bray, eds, Plant Responses to Cellular Dehydration During Environmental Stress. American Society of Plant Physiologists, Rockville, MD, pp 104-118

Close TJ, Lammers PJ (1993) An osmotic stress protein of cyanobacteria is immunologically related to plant dehydrins. Plant Physiol **101**: 773-779

Dure L III (1993) Structural motifs in Lea proteins. *In* TJ Close, EA Bray, eds, Plant Responses to Cellular Dehydration During Environmental Stress. American Society of Plant Physiologists, Rockville, MD, pp 91-103

Goday A, Jensen AB, Culiáñez-Macià FA, Albà MM, Figueras M, Serratosa J, Torrent M, Pagès M (1994) The maize abscisic acid-responsive protein Rab17 is localized in the nucleus and cytoplasm and interacts with nuclear localization signals. Plant Cell **6**: 351-360

Godoy JA, Pardo JM, Pintor-Toro JA (1990) A tomato cDNA inducible by salt stress and abscisic acid: nucleotide sequence and expression pattern. Plant Mol Biol **15**: 695-705

Godoy JA, Luna R, Torres-Schumann S, Moreno J, Rodrigo RM, Pintor-Toro, JA (1994) Expression, tissue distribution and subcellular localization of dehydrin TAS14 in salt-stressed tomato plants. Plant Mol Biol **26**: 1921-1934

Michel D, Furini A, Salamini F, Bartels D (1994) Structure and regulation of an ABA- and desiccation -responsive gene from the resurrection plant *Craterostigma plantagineum*. Plant Mol Biol **24**: 549-560

Mundy J, Chua NH (1988) Abscisic acid and water stress induce the expression of a novel rice gene. EMBO J **7**: 2279-2286

Robertson M, Chandler PM (1994) A dhydrin cognate protein from pea (*Pisum sativum* L.) with an atypical pattern of expression. Plant Mol Biol **26**: 805-816

Schneider K, Wells B, Schmelzer E, Salamini F, Bartels D (1993) Desiccation leads to the rapid accumulation of both cytosolic and chloroplastic proteins in the resurrection plant *Craterostigma plantagineum* Hothchst. Planta **189**: 120-131

Taylor JE, Renwick KF, Webb AAR, McAinsh MR, Furini A, Bartels D, Quatrano RS, Marcotte WR, Hetherington AM (1995) ABA-regulated promoter activity in stomatal guard cells. Plant J **7**: 129-134

Torres-Schumann S, Godoy JA, del Pozo O, Pintor-Toro JA (1991) Salt-induced TAS14 protein is highly phosphorylated *in vivo*. J Plant Physiol **139**:115-118

Vilardell J, Goday A, Freire MA, Torrent M, Martinez MC, Torne JM, Pagés M (1990) Gene sequence, developmental expression and protein phosphorylation of RAB-17 in maize. Plant Mol Biol **14**: 423-432

Vilardell J, Mundy J, Stilling B, Leroux B, Pla M, Freyssinet G, Pagés M (1991) Regulation of the maize rab17 gene promoter in transgenic heterologous systems. Plant Mol Biol **17**: 985-993

Vilardell J, Martínez-Zapater JM, Goday A, Arenas C, Pagès M (1994) Regulation of the rab17 gene promoter in transgenic *Arabidopsis* wild-type, ABA-deficient and ABA-insensitive mutants. Plant Mol Biol **24**: 561-569

Welin BV, Olson A, Nylander M, Palva ET (1994) Characterization and differential expression of dhn/lea/rab-like genes during cold acclimation and drought stress in *Arabidopsis thaliana*. Plant Mol Biol **26**: 131-144

Yamaguchi-Shinozaki K, Mino M, Mundy J, Chua N-H (1990) Analysis of an ABA-responsive rice gene promoter in transgenic tobacco. Plant Mol Biol **15**: 905-912

Crucial Reactions for Salt Tolerance in Yeast

Ramón Serrano, Rashid Ali, Francisco A. Culiañez-Maciá, Ana Espinosa, Alejandro Ferrando, María J. García, Roberto Gaxiola, Heinz-U. Gläser, José A. Márquez, Vicente J. Miralles, Consuelo Montesinos, Françoise Montrichard, José R. Murguía, Gabino Ríos, Pedro L. Rodriguez and Carlos Romero

Instituto de Biologia Molecular y Celular de Plantas, Universidad Politécnica de Valencia-C.S.I.C., E-46022 Valencia, Spain

Abstract. Random over-expression of genes in multicopy plasmids and gene disruptions have uncovered rate-limiting steps for salt tolerance in the model organism *Saccharomyces cerevisiae* (baker's yeast). In cells growing in glucose media intracellular sodium toxicity is the major component of salt stress. In minimal synthetic media sodium toxicity primarily affects methionine biosynthesis. This is due to sodium inhibition of the 3',5'-bisphosphate nucleotidase encoded by the *HAL2/MET22* gene. The sodium-extrusion ATPase encoded by the *ENA1/PMR2* gene is the major determinant of sodium homeostasis, together with its complex regulatory system, which includes the calcineurin and *HAL3* (*HAL1*) pathways.

Keywords. Calcineurin, *Saccharomyces cerevisiae*, salt stress, sodium toxicity, sodium transport, yeast

1 The Identification of Rate-Limiting Steps in Salt Tolerance

The progressive salinization of irrigated land compromises the future of agriculture in the most productive areas of our planet (Ashraf, 1994). Classical genetic improvement by breeding requires no basic knowledge about mechanisms. However, in the case of modern genetic engineering a basic understanding is essential. Before genetic engineering of salt tolerance could be attempted, rate-limiting factors should be identified. This means the most sensitive cellular phenomena and the most important defense reactions. Genes corresponding to these crucial reactions should be cloned and over- expressed in transgenic plants.

Salinity has two possible toxic effects on cells: the non-specific osmotic stress and the specific toxicity of the sodium and chloride ions to defined cellular systems. As far as with salt stress, organisms develop defenses for their salt-sensitive metabolic reactions based on osmotic adjustment (osmolyte synthesis), ion transport systems (both at the plasma membrane and at the vacuole) and stress-defensive proteins. In addition to the enzyme systems directly involved in these phenomena, regulatory pathways controlling these systems may also be determinants of salt tolerance. To add complexity to this view, responses at the level of whole plant, such as salt exclusion from the leaves, CAM metabolism and salt glands, make important contributions to plant salinity tolerance (Serrano, 1995).

This situation is clearly more complicated than the one faced by scientists who started the genetic engineering of hormone levels with biosynthetic enzymes, of insect resistance with toxic proteins, of virus resistance with coat proteins, of herbicide resistance with relevant enzymes etc. In the case of salinity it is not clear how to select the most promising genes for genetic engineering. In addition, genes controlling the most important phenomena relevant for salt tolerance have not yet been isolated. One example is our ignorance about the molecular basis of ion transport in plant vacuoles. Compartmentalization of NaCl within leaf vacuoles is an outstanding feature of halophytic plants. But as the genes encoding the relevant transport systems are not available, it cannot be tested if crop plants develop improved salt tolerance by engineering enhanced vacuolar ion accumulation.

Two different transgenic tobacco plants have established a correlation between osmolyte synthesis and salt tolerance. Plants expressing the bacterial enzymes mannitol-1P 5-dehydrogenase (Tarczynski et al., 1993) and levansucrase (Pilon-Smits et al., 1995) accumulate moderate levels (0.2-0.6 mg/g fresh weight) of mannitol and fructan, respectively, and are significatively protected from salt stress. The mechanism of protection is unclear because the contribution of the engineered osmolytes to overall osmotic adjustment is small. Therefore, either mannitol or fructan accumulate in particularly sensitive tissues (meristems?) or they have unexpected regulatory effects.

These results with osmolytes do not detract importance to ion transport and, in order to improve salt tolerance of crop plants, both osmotic adjustment and ion transport will probably need to be simultaneously manipulated.

Facing with all these uncertainties, we decided some time ago to systematically explore crucial reactions for salt tolerance in a model organism, the baker's yeast *Saccharomyces cerevisiae*. The results of this analysis could illuminate studies with higher plants, where many of the molecular genetics approaches available for yeast are not easy to be implemented. For example, our basic approach has been to randomly over-express yeast genes in multicopy plasmids. This involved transformation of yeast cells with yeast genomic DNA libraries constructed in plasmids capable of multicopy replication in yeast. From the thousands of transgenic yeasts so generated, those exhibiting improved growth under salt stress were selected and the responsible genes in the plasmids analyzed. They should correspond to rate-limiting steps for growth under salt stress. This approach has been complemented by the characterization of salt-sensitive phenotypes produced by mutations in specific genes. As described below, the important result is that both approaches pointed to the same rate-limiting steps in yeast salt tolerance.

2 Sodium Extrusion is an Important Determinant of Salt Tolerance

Two genes obtained by the random over-expression approach described above, *HAL1* (Gaxiola et al., 1992) and *HAL3* (Ferrando et al., 1995), have similar mechanisms of salt tolerance: over-expression of these genes increases the expression of the *ENA1/PMR2* gene, encoding the major sodium extrusion pump of *S. cerevisiae*. A null mutation of *ENA1/PMR2* dramatically decreases salt tolerance (Haro et al., 1991). The products of *HAL1* and *HAL3* genes are cytoplasmic proteins without homologies in data banks. They probably define novel signal transduction pathways. By increasing the amount of the sodium pump, over-expression of *HAL1* or *HAL3* decreases intracellular sodium levels and improves salt tolerance. These genes are important determinants of *ENA1/PMR2* expression because null mutations of either genes decrease expression of the sodium pump and reduce salt tolerance. Nevertheless, they are not the only regulators of sodium transport. By a different approach, based on isolation of salt sensitive mutants, the calcium-dependent protein phosphatase calcineurin has also been found to be implicated in *ENA1/PMR2* expression (Nakamura et al., 1993; Mendoza et al., 1994). Calcineurin and *HAL3* seem to have additive effects on *ENA1/PMR2* expression, suggesting two regulatory pathways converging on the expression of this crucial gene (Ferrando et al., 1995). Other important regulatory pathways controlling *ENA1/PMR2* expression are those corresponding to the *HOG1* MAPKinase and to protein kinase A (José A. Marquez and R. Serrano, in preparation).

That sodium extrusion at the plasma membrane is a determinant of salt tolerance was not unexpected. The importance of the above described work is that this presumption was rigorously demonstrated by gain and loss of function through gene manipulation.

Extrapolation of these results to higher plants, however, may be dangerous. In *S. cerevisiae* cells growing on glucose media, osmotic adjustment is very efficient and proceeds by rapid glycerol synthesis. On the other hand, osmolyte synthesis in higher plants may be limiting and NaCl accumulation within the vacuole results in osmotic adjustment for most of the cellular volume without need of osmolyte synthesis. Excessive sodium extrusion at the plasma membrane may be detrimental under these circumstances and the optimal situation for the plant cell may be provided by a balance between salt uptake at both plasmatic and vacuolar membranes. (Serrano, 1995). Nevertheless, under conditions of low salinity, when ion toxicity may be more important than osmotic adjustment, sodium extrusion at the plasma membrane may be beneficial for plant cells. Genetic engineering of this trait could be attempted with known genes for sodium extrusion from either yeast or bacteria (Serrano, 1995). It must be emphasized that the cloning of Na^+ and Cl^- transport systems from plant plasmatic and vacuolar membranes should be a priority in salinity studies. They could provide the best tools for effective manipulation of salt tolerance in transgenic plants.

3 A Sodium-Sensitive 3',5'-bisphosphate Nucleotidase as Determinant of Salt Tolerance

Many cellular enzymes are inhibited by high NaCl concentrations (greater than 0.4 M) but yeast growth is inhibited when intracellular sodium concentrations are only 50-100 mM. An important target under these conditions seems to be the 3',5'-bisphosphate nucleotidase encoded by the *HAL2* gene (Murguía et al., 1995).

The route of sulfate assimilation starts with the formation of activated sulfate in the form of adenosine-5'-phosphosulfate (APS), a reaction catalyzed by ATP sulfurylase:

$$ATP + SO_4^{-2} = APS + 2\ P_i$$

This reaction is thermodynamically unfavourable (DG = + 12 kJ/mol) and a universal strategy for displacing the equilibrium is to phosphorylate APS to make 3'-phosphoadenosine-5'-phosphosulfate (PAPS). This reaction, catalyzed by APS kinase, drives the accumulation of activated sulfate (phosphosulfate), which is used for either reduction to sulfite or transfer to lipids, proteins and carbohydrates (Serrano, 1995). A by-product of these reactions is 3'-phosphoadenosine-5'-phosphate (PAP), which is hydrolyzed to AMP by a specific nucleotidase. In *S. cerevisiae* this enzyme is very sensitive to sodium inhibition and it is encoded by *HAL2*, a gene which by over-expression improves salt tolerance. This suggests that inhibition of the *HAL2* nucleotidase during salt stress determines the inhibition of yeast growth. On the other hand, this gene is required for methionine biosynthesis and it has been also indicated as *MET22* (Gläser et al., 1993). Apparently, incorporation into methionine is the major fate of sulfate in *S. cerevisiae*. A defect in *HAL2/MET22* would result in PAP accumulation, product inhibition of PAPS reductase (the enzyme generating sulfite) and blockage of methionine biosynthesis. During salt stress, partial blockage would occur and, accordingly, methionine supplementation may improve salt tolerance. In addition to providing the essential amino acid, methionine inhibits sulfate uptake and therefore no deleterious PAP accumulation occurs in methionine-containing media.

This mechanism of salt toxicity may be relevant for plants because enzymes homologues to the *HAL2* nucleotidase exist in plant tissues and sulfate activation *via* the PAPS route is operative in higher plants. Interestingly, sulfate reduction in plant chloroplasts occurs probably through glutathione thiosulfonate as activated form of the sulfate, while PAPS is mainly used in sulfolipid biosynthesis (Anderson and Beardall, 1991). The inhibition of this route in salt-stressed plants may also determine growth inhibition because of perturbation of chloroplast membranes.

4 Perspectives for the Biotechnology of Salt Tolerance

Transgenic plants over-expressing specific genes would be important in assessing rate-limiting steps in salt tolerance. Step by step, several rate limiting processes could be improved and a substantial increase in the tolerance of crop plants to salinity could be gradually obtained.

First progress in this direction has been made with the osmotic component of salt stress. Preliminary results with transgenic tobacco plants expressing mannitol and fructan biosynthetic enzymes have highlighted the crucial role of small molecules of compatible osmotica (osmolytes) in salt tolerance. This work emerged from physiological and biochemical studies which identified the function and biosynthesis of osmolytes and it has opened a very promising approach to the biotechnology of salt tolerance. Some questions remain about the fitness of different osmolytes for different organisms and stresses and about their mechanism of action at the low concentrations obtained in transgenic plants.

The results obtained with yeast have provided additional genes which may be of biotechnological relevance. They cover another aspect of salt stress: ion transport and toxicity. The *HAL2* nucleotidase and the *HAL1* and *HAL3* regulatory genes are being tested in transgenic plants. The salt-sensitive nucleotidase has good chances of improving salt tolerance because it corresponds to an enzyme conserved between yeast and plants. The yeast regulatory genes *HAL1* and *HAL3* have a mechanism with less chances of being conserved in plants: they control the expression of the sodium pump. Nevertheless, some signal transduction pathways activated by stress exhibit striking conservation in different organisms (for example, those involving the transcription factors HSF and AP1) and there are chances that the salt stress pathway could also be conserved.

In retrospective, it seems that the sodium stress mechanisms mediated by *HAL1*, *HAL2*, *HAL3*, *ENA1/PMR2* and calcineurin could not be easily demonstrated without a convenient model system such as *S. cerevisiae*. The plant homologues of these yeast genes could provide important clues for salt tolerance in higher plants. However, in order to uncover the peculiarities of salt stress in plants, it would be important to implement random over-expression and mutagenesis approaches in model plant systems. The experimental system developed by the group of R. Walden at Cologne (random integration of T-DNA containing a strong promoter in tobacco protoplasts; Hayashi et al., 1992) could provide the required methodology for the first approach. *Arabidopsis* genetics (Meyerowitz and Somerville, 1994) would be essential for a mutational analysis of salt tolerance. It is expected that genes corresponding to crucial reactions such as sodium and chloride transporters, sodium and chloride sensitive enzymes and the corresponding regulatory systems could be identified by a combination of both approaches.

It is obvious that our view of salinity tolerance as a succession of bottle-necks which can be progressively enlarged constitutes an oversimplification of the problem. However, we are comforted by a sentence of Roger Bacon: "Truth emerges more readily from error than from confusion".

Acknowledgements. This work was supported by the European Molecular Biology Laboratory (Heidelberg), by a grant of the Spanish programme of Biotechnology (BIO91-1313, Madrid) and by grants of the European Union (BIO2-CT-93-0400 and AIR3-CT-94-1508, Brussels).

References

Anderson JW, Beardall J (1991) Molecular Activities of Plant Cells. An Introduction to Plant Biochemistry. Blackwell, Oxford, pp 230-232

Ashraf M (1994) Breeding for salinity tolerance in plants. Crit Rev Plant Sci **13**: 17-42

Ferrando A, Kron SJ, Rios G, Fink GR, Serrano R (1995) Regulation of cation transport in *Saccharomyces cerevisiae* by the salt tolerance gene *HAL3*. Mol Cell Biol **15**: 5470-5481

Gaxiola R, Larrinoa IF, Villalba JM, Serrano R (1992) A novel and conserved salt-induced protein is an important determinant of salt tolerance in yeast. EMBO J **11**: 3157-3164

Gläser HU, Thomas D, Gaxiola R, Montrichard F, Surdin-kerjan Y, Serrano R (1993) Salt tolerance and methionine biosynthesis in *Saccharomyces cerevisiae* involve a putative phosphatase gene. EMBO J **12**: 3105-3110

Haro R, Garciadeblas B, Rodriguez-Navarro A (1991) A novel P-type ATPase from yeast involved in sodium transport. FEBS Lett **291**: 189-191

Hayashi H, Czaja I, Lubenow H, Schell J, Walden R (1992) Activation of a plant gene by T-DNA tagging: auxin-independent growth *in vitro*. Science **258**: 1350-1353

Mendoza I, Rubio F, Rodriguez-Navarro A, Pardo JM (1994) The protein phosphatase calcineurin is essential for NaCl tolerance of *Saccharomyces cerevisiae*. J Biol Chem **269**: 8792-8796

Meyerowitz EM, Somerville CR (1994) *Arabidopsis*. Cold Spring Harbor Laboratory Press

Murguía JR, Belles JM, Serrano R (1995) A salt-sensitive 3'(2'),5'-bisphosphate nucleotidase involved in sulfate activation. Science **267**: 232-234

Nakamura T, Liu Y, Hirata D, Namba H, Harada S, Hirokawa T, Miyakawa T (1993) Protein phosphatase type 2B (calcineurin)-mediated, FK506-sensitive regulation of intracellular ions in yeast is an important determinant for adaptation to high salt stress conditions. EMBO J **12**: 4063-4071

Pilon-Smits EAH, Ebskamp MJM, Paul MJ, Jeuken MJW, Weisbeek PJ, Smeekens SCM (1995) Improved performance of transgenic fructan-accumulating tobacco under drought stress. Plant Physiol **107**: 125-130

Serrano R (1995) Salt tolerance in plants and microorganisms: toxicity targets and defence responses. Int Rev Cytol (in press)

Tarczynski MC, Jensen RG, Bohnert HJ (1993) Stress protection of transgenic tobacco by production of the osmolyte mannitol. Science **259**: 508-510

Cloning Salt Tolerance Genes by Insertional Mutagenesis Tagging and Gene Complementation

José M. Pardo, Imelda Mendoza, Rafael Prieto, Francisco J. Quintero, Alonso Rodriquez-Navarro, Shuli Yang, Dae-Jin Yun, Iraida Amaya, Satomi Takeda, Albino Maggio, Matilde Paino D'Urzo, Paul M. Hasegawa and Ray A. Bressan

Department of Horticulture, Purdue University, West Lafayette, IN 47907-1165, USA

Abstract. The plasmid pMN24 carrying a functional copy of the nitrate reductase gene *Nit1* was used for insertional mutagenesis of nit$^-$ *Chlamydomonas reinhardtii*. Over 6800 nit$^+$ transformants were screened for mutations causing NaCl sensitivity. Four mutants with salt sensitivity that cosegregated with nit$^+$ were isolated. Salt sensitivity of these mutants also cosegregated with the presence of a DNA fragment hybridizing to the *Nit1* plasmid pMN24. Over 8000 EMS mutagenized yeast colonies were screened for salt sensitivity and eleven sensitive lines were isolated. One of these, SS2, was complemented for salt tolerance with a genomic library from wild type yeast. A salt tolerant transformant was found to harbor a DNA fragment encoding the calcineurin regulatory subunit (*CNB1*) gene. Gene disruption and segregation analysis indicated the *SS2* and *CNB1* loci were allelic. Molecular modification of calcineurin catalytic subunit (CNA) to remove the calmodulin (CAM)-binding and auto-inhibitory domains was performed. The subsequent truncated *CNA* (*CNAtr1*) gene, along with a *CNB1* gene were over-expressed in yeast. This resulted in a 5 fold increase in salt tolerance. The *CNAtr1* and *CNB1* genes were also transformed into tobacco and several tobacco lines expressing yeast CNAtr1 and CNB1 transcripts were isolated. Preliminary experiments indicate that *CNAtr1/CNB1* transgenic tobacco plants have altered growth on salt.

Keywords. Calcineurin, *Chlamydomonas reinhardtii*, gene complementation, *Saccharomyces cerevisiae*, salt tolerance, yeast

1 Introduction

It has long been hypothesized that stress adaptation involves the co-ordinate regulation of many genes probably at several levels but at least at the transcriptional level. This regulation would involve a primary signal reception, a signal transduction cascade, and a set of signal-activated genes (master genes) encoding regulatory transcription factors that control the expression of several other genes (slave genes) that participate in the manifestation of the tolerance phenotype (Fig. 1).

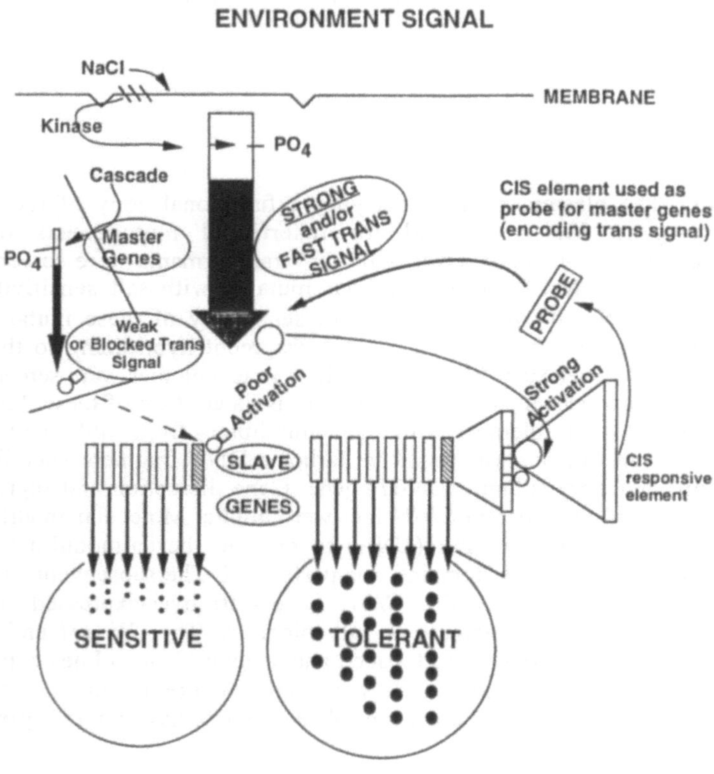

Fig. 1. Transduction of environmental stress signal to the induction of genes that alter phenotype and allow adaptation. Various screening procedures can be used to identify highly expressed genes in tolerant state vs. low expression in sensitive state. Responsive elements on the promoters of these "slave" genes can be used to identify "master" regulatory genes and signaling systems that perceive environmental changes

Thus, an often used approach to identify stress tolerance genes has been to look for genes whose expression is induced by the condition of stress. Indeed many genes have been identified this way and some appear likely to participate in stress adaptation (Bohnert et al., 1995). However, many of these genes appear to be involved in stress in a very general way and may not function to alleviate substantially the inhibitions imposed by any particular stress (Kononowicz et al., 1993).

A much more direct and in many ways more promising approach is the functional identification of stress tolerance genes by genetic complementation of stress sensitive mutants (Mendoza et al., 1994; Serrano and Gaxiola, 1994). Here we describe our efforts to identify salt sensitive mutants in both yeast and the unicellular alga *Chlamydomonas reinhardtii* and to clone subsequently the genes that are controlling salt sensitivity/tolerance.

Invoking the above-mentioned scenario of co-ordinate regulation, one might anticipate that genes identified directly through a functional test would likely encode regulatory proteins because they are more likely to limit the ability of the cell to adjust. However, genes encoding proteins that play a major direct role in adjustment to salinity might also be found in this way such as the *ENA1* gene encoding a Na^+ pump (Haro et al., 1991).

2 Insertional Mutagenesis of *Chlamydomonas* and Isolation of Salt-Sensitive Mutants

About 6800 nit^+ clones isolated following transformation of the 305CW15 (*nit1 cw15 mt*$^+$) *Chlamydomonas* mutant strain with pMN24 (the transformation plasmid pUC119 that carries a functional copy of the *Nit1* gene) were analyzed for their ability to grow on minimal media supplemented with either 150 mM NaCl or 30 mM LiCl. Four isolates (S4, S18, S46, S66) showed slow growth on the Na^+ and Li^+ containing media (sod^- phenotype) (Fig. 2) and another (S33) on the Li^+ medium only containing (lit^- phenotype).

3 Genetic Analyses Indicate that Four Different Recessive Genes Mediate Na^+ and Li^+ Tolerance in the Mutants

Genetic crosses between the strain 305d (*nit1 km1 mt*$^-$) and either the parental strain 305CW15 or sod^- or lit^- mutants were carried out to determine the number of mutations involved in these phenotypes and whether or not they co-segregate with the nit^+ phenotype. If the sod^-/lit^- phenotypes were caused by an insertion of pMN24, then in backcrosses all of the progeny that displayed Na^+ and/or Li^+ sensitivity should be able to grow on minimal medium containing nitrate as the sole nitrogen source (nit^+). The results are presented in Table 1. No sod^- or lit^-

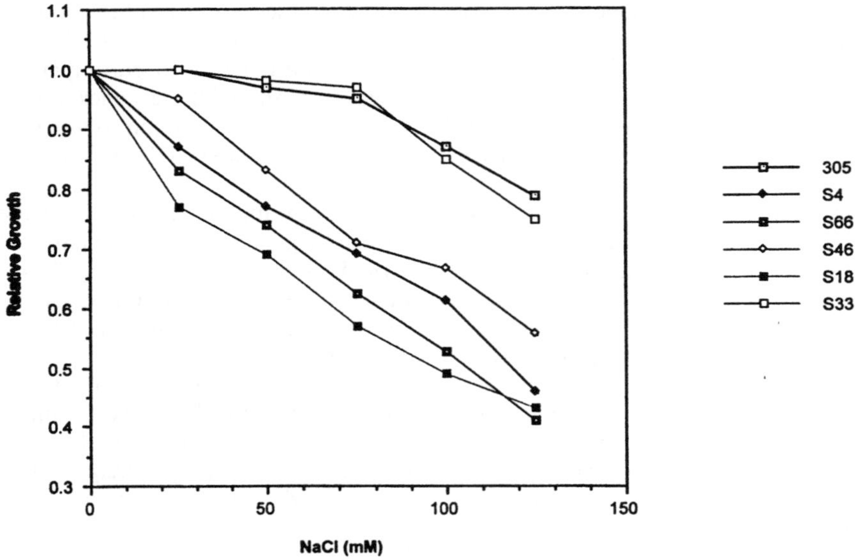

Fig. 2. Sensitivity in wild type (305) and mutant strains (S4, S18, S33, S46, S66) of *Chlamydomonas*. Liquid cultures of the different strains growing at exponential phase were harvested and the cells transferred to media containing different salt concentrations (0, 25, 50, 75, 100, 125 mM). Relative growth was calculated from the growth curves of the above cultures

segregants emerged from crosses between 305d and 305CW15, which indicates that the sod$^-$ and lit$^-$ phenotype only appeared as a result of mutagenic events and not from differences in genetic background. In the cross of 305d and either S4, S18, S33, S46 or S66, segregation ratios close to 1:1 for sodium and/or lithium tolerance indicate that a single locus is responsible for the sod$^-$ and lit$^-$ phenotypes. The cosegregation between Na^+ and Li^+ sensitivity in sod$^-$ mutants indicates that they are caused by either the same mutation or two closely linked mutations. In the crosses 305d x S18 and 305d x S46 recombinants of either nit$^+$ sod$^+$ or nit$^-$ sod$^-$ were found, which indicates that the sod$^-$ phenotype is not segregating with an insertion of a functional *Nit1* gene (Table 1). The most likely explanation for a mutant phenotype that does not cosegregate with a *Nit* insertion (nit$^+$) is the loss of the original mutagenic *Nit1* insertion from the genome. The segregation of the cross 305d x S33, similar to the case of S4 and S66, is consistent with the presence of a locus for lit$^-$ that is linked to one of the functional *Nit1* insertions.

Genetic crosses between the different mutant strains were performed to determine if these mutations were allelic. The cross S4 x S18 gave no sod$^+$/lit$^+$ segregants, indicating that these two mutants are defective in the same locus or two closely linked loci. This gene was named *sod1*. In the rest of the crosses, the segregation ratios for sod$^-$/lit$^-$ were close to 3:1, denoting that S46, S66 and

Table 1. Genetic crosses between wild type and mutant strains of *Chlamydomonas*. Segregation ratio is given on parenthesis. P, parental line

	Segregation		Recombinants			
	$nit^+ : nit^-$	$sod^+ lit^+ : sod^- lit^-$	$nit^+ sod^-/lit^-$	$nit^- sod^+/lit^+$	$nit^+ sod^+/lit^+$	$nit^- sod^-/lit^-$
305CW15 x 305d	0 : 189 (0 : 1)	189 : 0 (1 : 0)	0	189 (P)	0	0
S4 x 305d	134 : 124 (1 : 1)	136 : 122 (1 : 1)	122 (P)	124 (P)	12	0
S18 x 305d	80 : 67 (1 : 1)	73 : 74 (1 : 1)	65 (P)	58 (P)	15	9
S33 x 305d	91 : 81 (1 : 1)	86 : 86 (1 : 1)	86 (P)	81 (P)	5	0
S46 x 305d	70 : 73 (1 : 1)	78 : 65 (1 : 1)	64 (P)	72 (P)	6	1
S66 x 305d	108 : 96 (1 : 1)	101 : 103 (1 : 1)	103 (P)	96 (P)	5	0

S33 are defective in three unlinked loci that we have named *sod2*, *sod3* and *lit1* respectively.

Heterozygous diploids carrying both wild type and mutant loci for either sod or lit were obtained from crosses between the strain 305d (nit$^-$ km^R) and each of the mutants S4, S18, S46, S66 (nit$^+$ sod$^-$) and S33 (nit$^+$ lit$^-$). All diploids isolated showed sod$^+$/lit$^+$ phenotype, which indicates that all of the mutations for sod$^-$ and lit$^-$ are recessive.

4 Molecular Analyses of the Mutants

To determine whether these mutants contained sequences derived from the pMN24 plasmid that are integrated into the genome, genomic DNA from S4, S46, S66 and S33 was digested with a variety of restriction enzymes and analyzed by Southern blot using pUC119 as the labeled hybridization probe. Three of the mutants (S4, S33 and S66) presented two bands and the fourth mutant (S46) just one band (data not shown). These patterns indicated that these transformants contained only one or a few copies of the integrated plasmid.

Cosegregation of the sod$^-$ and lit$^-$ phenotypes with the integrated plasmid was tested by analyzing on Southern blot PstI digested genomic DNA of five progeny nit$^+$ sod$^-$/lit$^-$ and one nit$^+$ sod$^+$/lit$^+$ from backcrosses of the four different mutants (Fig. 3). In the case of S46, the plasmid was detected in all the progeny indicating that the sod$^-$ phenotype is not linked with the insertion of *Nit1*. On the other hand, S4 and S33 mutants exhibited two PstI bands in the five sod$^-$/lit$^-$ segregants but no bands in the sod$^+$/lit$^+$, indicating that the sod$^-$/lit$^-$ phenotype is linked to these bands. In the case of S66 mutant, one of the bands (3.7 kb) is missing in the segregant sod$^+$ which indicates cosegregation between the sod$^-$ phenotype and this specific band.

Therefore, the results from both genetic and molecular analyses are consistent with the conclusion that insertion of pMN24 into the nuclear genome disrupted a gene involved in Na^+ and/or Li^+ tolerance in three out of the five mutants analyzed.

5 Isolation and Characterization of NaCl Sensitive Mutants in Yeast

In order to identify yeast genes required for cellular adaptation to high salinity stress, a mutational analysis was initiated in the yeast *Saccharomyces cerevisiae*. Wild type cells were mutagenized by EMS and screened for inability to grow in solid media supplemented with 0.7 M NaCl (a moderate NaCl shock for yeast). Among 8100 EMS survivors screened, eleven NaCl sensitive mutants were isolated. Because high salt combines ion toxicity and osmotic imbalance, these mutants were arranged as specifically ion-sensitive or osmotic-sensitive after scoring their growth in media supplemented with increasing amounts of LiCl, KCl or PEG (Table 2). Li^+ can be used as an analog of Na^+ because Na^+ and Li^+ fluxes are mediated by the same transport systems in yeast (Haro et al., 1993). However, since internal Li^+ is more toxic than Na^+, effective growth inhibition is obtained at mM concentrations. The mutant strain SS2 was selected for further characterization because of its high sensitivity to NaCl (IC_{50} for mutant SS2 was 0.1 M NaCl, about 5 fold lower than the value for the parental strain, IC_{50}=0.47 M NaCl), and to LiCl (IC_{50} were 7.7 mM and 29 mM, respectively). Since SS2 had identical growth rates to the wild type strain in KCl or PEG, the mutant SS2 was specifically sensitive to Na^+ and Li^+ but not osmo-sensitive.

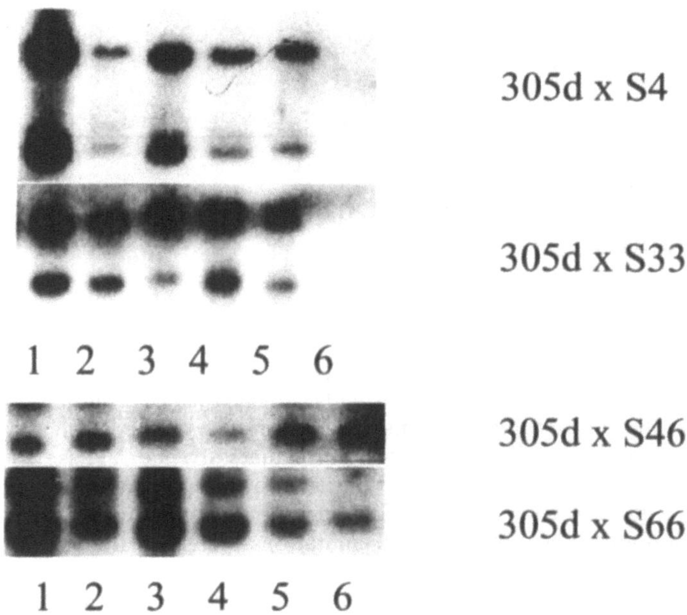

Fig. 3. DNA hybridization analysis of different segregants for nit^+ and either sensitive or resistant to NaCl and/or LiCl. DNA from five segregants nit^+ sod^-/lit^- and one segregant nit^+ sod^+/lit^+ isolated from the crosses between wild type (305d) and the mutant strains S4, S33, S46 and S66 were digested with PstI, fractionated on agarose gels and analyzed by Southern hybridization using the labeled plasmid pUC119 as probe. Lane 1-5: segregants nit^+ sod^-/lit^-. Lane 6: segregant nit^+ sod^+/lit^+

Table 2. NaCl sensitive mutants isolated from 8100 survivors to EMS mutagenesis

Strain	Growth in solid media			IC_{50} in liquid cultures	
	0.7 M NaCl	1.5 M KCl	2 M E.G.	NaCl [M]	LiCl [mM]
w.t.	+	+	+	0.47	29
SS 2	-	+	+	0.11	7.7
SS10	-	+	+	0.10	15
SS13	-	+	+	0.20	20
SS19	-	+	+	0.20	8.7
SS20	-	+	+	0.27	nd
SS21	-	+	+	0.20	5.5
SS 1	-	-	-	0.33	nd
SS11	-	-	-	0.33	nd
SS 3	-	-	-	0.19	nd
SS 4	-	-	-	0.20	nd
SS15	-	+	-	0.27	nd

6 The SS2 Mutant Strain is Defective in Na^+ Homeostasis

To further characterize the SS2 mutation, we compared the time course of Li^+ accumulation in SS2 to that of the parental strain after incubation in medium containing 30 mM Li^+ and 1 mM K^+ (data not shown). SS2 cells clearly accumulated more Li^+ than control cells, suggesting that mutant SS2 had a defect in the cellular mechanisms controlling the homeostasis of alkaline cations. Na^+ and Li^+ effluxes in yeast are mediated by the ATPase encoded by *ENA1*, and *ENA1* mutants are very sensitive to these ions (Haro et al., 1991). However, a genetic cross between SS2 and an *ena1Δ* strain resulted in a NaCl tolerant diploid, indicating that both mutations were not allelic. Because the SS2 mutation mimicked the phenotype of an *ena1* mutation, we measured Li^+ efflux in SS2 cells pre-loaded with Li^+. Mutant cells showed a diminished net Li^+ efflux when compared to wild type cells (0.53 *versus* 0.97 nmol mg^{-1} min^{-1}, respectively) (Fig. 4). Since the *ENA1* gene is inducible by NaCl, the *ENA1* promoter activity in SS2 cells was determined using a *ENA1-lacZ* fusion. Upon NaCl induction, the *ENA1-lacZ* fusion expression was reduced by two fold in SS2 cells, indicating that the reduced Li^+ efflux might be due to a defect in the signaling pathway controlling the induction of *ENA1* by salt stress.

In yeast cells uptake of K^+, Na^+ and Li^+ is through the same transport system. The K_ms for Na^+, Li^+ and K^+ influx vary depending on the growth conditions. Under Na^+ stress, the transport system changes to a new state with high affinity for K^+ while the affinity for Na^+ remains essentially unchanged, thereby allowing a better discrimination between K^+ and Na^+ (Haro et al., 1993). To test the effect of the SS2 mutation on K^+/Na^+ selectivity, wild type and SS2 cells were grown in medium supplemented with NaCl concentrations that produced similar Na^+ loading and an intracellular Na^+/K^+ ratio close to 1 in both strains.

In these conditions, wild type cells exhibited the high affinity mode of K^+ uptake (measured as Rb^+ uptake) with a K_m for Rb^+ of 80 μM ± 10 μM (the K_i of K^+ was 10 μM ± 2 μM), whereas the SS2 mutant cells showed low-affinity Rb^+ influx kinetics (K_m for Rb^+ was 1.6 mM ± 0.2 mM, and K_i of K^+ was 0.2 mM ± 0.02 mM). Therefore, SS2 cells also had a defective transition of the K^+ transport system from the low-affinity to the high-affinity state in response to Na^+ stress.

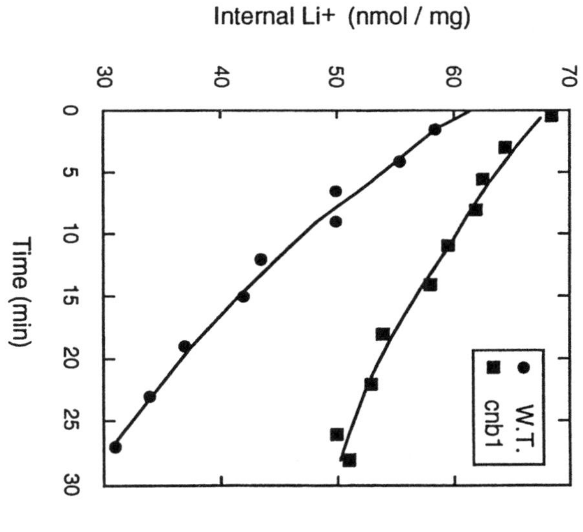

Fig. 4. Wild type and *cnb1* cells were loaded with Li^+ for 30 min in AP medium containing 0.2 mM KCl and 50 mM LiCl, diluted with four volumes of 62.5 mM KCl in AP medium, and the Li^+ content of the cells was determined at intervals to measure net Li^+ efflux

7 Isolation of the *CNB1* Gene by Genetic Complementation of SS2

Since the Na^+ sensitive phenotype of strain SS2 segregated consistently as a recessive mutation in a single nuclear locus (not shown), we undertook the molecular cloning of the gene restoring the capacity of SS2 to grow in 0.7 M NaCl. SS2 cells were electroporated in the presence of a *S. cerevisiae* genomic DNA plasmid library. One NaCl tolerant clone was isolated out of 7870 primary transformants. The functional region was mapped to a DNA fragment containing the gene *CNB1* which encodes the activatory subunit of the Ca^{2+}/calmodulin-dependent protein phosphatase PP2B, also known as calcineurin. Gene disruption of the *CNB1* locus and segregation analysis demonstrated that the mutation SS2 was allelic to *CNB1*. A genetic cross between a *cnb1Δ* disruptant and the mutant SS2 rendered a NaCl sensitive diploid that, after sporulation, produced 100% NaCl sensitive haploid segregants.

8 Structure of Calcineurin

Calcineurin (CaN) is a Ca^{2+}- and calmodulin-dependent protein phosphatase consisting of a catalytic subunit (CNA) and a regulatory [activatory] subunit (CNB) (Fig. 5).

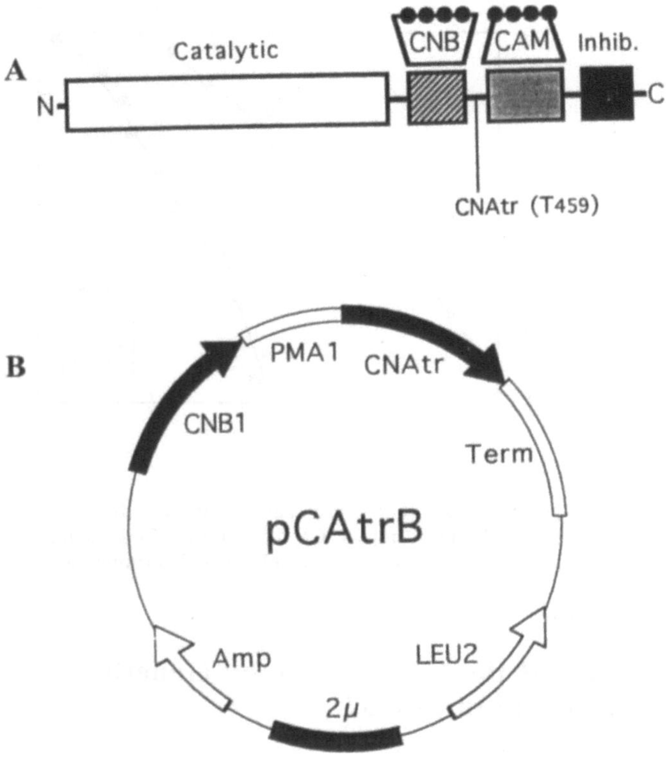

Fig. 5. A truncated *CNA2* allele lacking the carboxy-terminal Ca^{2+}/calmodulin binding and the auto-inhibitory domains was produced by PCR (A) and inserted between the promoter and transcriptional terminator regions of the yeast plasma membrane (H^+)ATPase *PMA1* gene (B). CNB1p was expressed using its own gene promoter. The vector backbone of pCAtrB was the multicopy plasmid YEp351

CNA is inactive in the absence of CNB. The phosphatase activity of CaN is modulated by intracellular Ca^{2+} in a complex way. The regulatory CNB subunit is able to bind up to four molecules of free Ca^{2+} and to stimulate the phosphatase activity of the CNA/CNB complex. Similarly to calmodulin (CAM), CNB contains four "EF-hand" motifs for Ca^{2+} binding, but these proteins are not functionally equivalent. CaN can also bind and be stimulated by Ca^{2+}/CAM complexes through the catalytic CNA subunit, thus yielding the fully active trimeric enzyme (Kincaid, 1993). Based on biochemical evidence, four functional domains have been distinguished in the catalytic CNA subunit of calcineurin: a catalytic core with high similarity to protein phosphatases PP2A, a CNB-binding domain, a calmodulin-binding site and an auto-inhibitory domain (Fig. 5). Removal of the CAM-binding and the auto-inhibitory domains of the CNA subunit yields an activated core CNA/CNB calcineurin that does not require CAM/Ca^{2+} binding (Hubbard and Klee, 1989; Parsons et al., 1994). We anticipated that the expression of a truncated catalytic subunit of calcineurin might increase the Na^+ and Li^+ tolerance of yeast cells.

9 Effects of Constitutive Activation of Calcineurin on Ion Homeostasis

To test the effect of constitutive activation of yeast calcineurin on ion homeostasis, a truncated CNA subunit (CNAtrl), missing both the CAM-binding and the auto-inhibitory domains, was constructed by PCR. The recombinant allele *CNAtrl* encoded a catalytic subunit truncated after threonine at position 459. To allow expression, *CNAtrl* was cloned between the promoter and the transcriptional termination signal of the plasma membrane (H^+) ATPase gene, *PMA1*. The native *CNB1* gene was also cloned in the same plasmid for the concomitant expression of the regulatory subunit (plasmid pCAtrB, Fig. 5). As anticipated, the simultaneous over-expression of the truncated CNAtrl and the CNB subunits increased the Li^+ tolerance of wild type cells by almost 5-fold (Fig. 6). On the other hand, co-expression of the full length CNA2p along with the regulatory CNB subunit did not increase substantially the Li^+ tolerance, suggesting that there may be other factor(s) limiting the activity of the native calcineurin.

The high tolerance to LiCl (and NaCl) of cells expressing the truncated calcineurin might result from an increased activity of the efflux mechanism (the *ENA* ATPases) or to a substantially reduced rate of Li^+ influx (a *TRK1*-dependent faculty). Consistent with this contention, the expression of *CNAtrl/CNB* in *Δena1-4 Δtrk1* cells failed to increase their low Li^+ tolerance (not shown). The expression of the *ENA1* gene was addressed in a yeast strain containing an integrated copy of a *ENA1-lacZ* fusion. These cells were transformed with the plasmid pCAtrB and the ß-galactosidase activity was determined after induction with NaCl. NaCl-stressed control cells increased the transcriptional activity of the *ENA1* promoter by 5 fold. Over-expression of CNAtrl/CNB resulted in a constitutive 3-fold increase in *ENA1* promoter activity in the absence of stress, and a ß-galactosidase activity 8 fold higher when cells were challenged with NaCl. Immunodetection of ENA1p by western blot showed protein levels in agreement with the transcriptional activities measured. Furthermore, cells transformed with pCAtrB and preloaded with Li^+ showed a rate of Li^+ efflux substantially higher than that of control cells transformed with the vector, in agreement with a higher activity of the ENA1p.

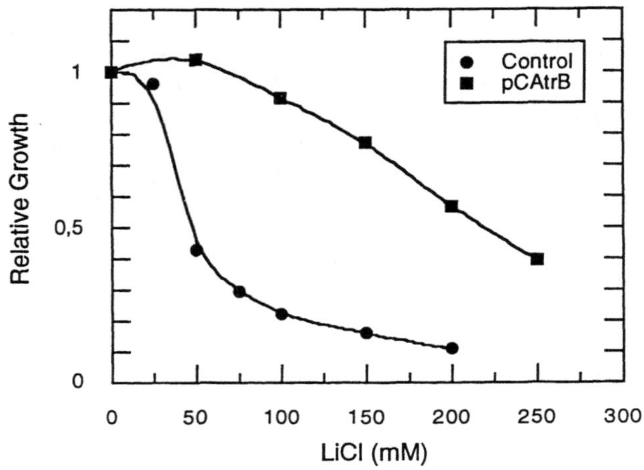

Fig. 6. The growth rate of YP4-7a cells (MATa ura3 leu2) transformed with either the plasmid pCAtrB or the vector as a control was measured in YNB medium containing increasing concentrations of LiCl. The growth rate in each medium was normalized to the value obtained in the absence of LiCl

The effect of calcineurin truncation on the uptake system shared by K^+, Na^+ and Li^+ was also surveyed. Cells with a *Δena1-4* deletion lacking significant Li^+ efflux were used for Li^+ uptake measurements to avoid interference by the constitutively expressed *ENA1*-ATPase in cells expressing the truncated calcineurin. A double-reciprocal plot of initial rates of Li^+ influx at several external concentrations of Li^+ demonstrated that cells containing the plasmid pCAtrB had a reduced affinity for Li^+ uptake. In the presence of 1mM K^+, the apparent Km for Li^+ was 50 mM in control cells and 125 mM in cells transformed with pCAtrB. Taken together, these results indicate that continued activation of calcineurin results in a permanent state of adaptation to ion stress, where the K^+ transport system is in the high affinity mode and the ENA1 protein is expressed in a constitutive fashion. Furthermore, the responsiveness of the *ENA1* gene is enhanced in response to ion stress. A consequence of this precocious and enhanced adaptation is a dramatic increase in survival when the cells over-expressing *CNAtr1/CNB* are challenged with otherwise lethal concentrations of LiCl and NaCl. Thus, over 90% of yeast cells transformed with pCAtrB survived after plating in 0.3 M LiCl or 1.2 M NaCl, conditions where control cells failed to develop colonies. Therefore, calcineurin activity is a rate-limiting component of the signaling pathway that mediates adaptation and tolerance to Na^+ stress.

We also transformed tobacco plants with *CNAtr1/CNB* and have recovered plants that are over-expressing both CNAtrl and CNB transcripts. We do not know yet the extent of expression at the protein level or the tissue specificity, if any, of the expression. However, in preliminary tests of the ability of these plants to grow on salt we have observed that *CNAtr1/CNB* transformed plants grow more rapidly in the presence of NaCl.

References

Bohnert HJ, Nelson DE, Jensen RG (1995) Adaptations to environmental stresses. Plant Cell **7**:1099-1111

Haro R, Garciadeblas B, Rodriguez-Navarro A (1991) A novel P-type ATPase from yeast involved in sodium transport. FEBS Letters **291**: 189-191

Haro R, Bañuelos MA, Quintero FJ, Rubio F, Rodriguez-Navarro A (1993) Genetic basis of sodium exclusion and sodium tolerance in yeast. A model for plants. Physiol Plant **89**: 868-874

Hubbard MJ, Klee CB (1989) Functional domain structure of calcineurin A: mapping by limited proteolysis. Biochemistry **28**: 1868-1874

Kincaid R (1993) Calmodulin-dependent protein phosphatases from microorganisms to man. A study in structural conservation and biological diversity. *In* S Shenolikar and AC Nairn, eds, Advances in Second Messenger and Phosphoprotein Researh. Vol 27. Raven Press, Ltd., New York, pp 1-23

Kononowicz AK, Raghothama KG, Casas AM, Reuveni M, Watad A-EA, Liu D, Bressan RA, Hasegawa PM (1993) Osmotin: Regulation of gene expression and function. *In* TJ Close, EA Bray, eds, Plant Responses to Cellular Dehydration During Environmental Stress. Amer Soc Plant Physiol, Rockville, MD, pp 144-158

Mendoza I, Rubio F, Rodriguez-Navarro A, Pardo JM (1994) The Protein Phosphatase Calcineurin Is Essential for NaCl Tolerance of Saccharomyces Cerevisiae. J Biol Chem **269**:8792-8796

Parsons Jn, Wiederrecht GJ, Salowe S, Burbaum JJ, Rokosz LL, Kincaid RL, O'Keefe SJ (1994) Regulation of calcineurin phosphatase activity and interaction with the FK506/FK-506 binding protein complex. J Biol Chem **269**: 19610-19616.

Serrano R, Gaxiola R (1994) Microbial models and salt stress tolerance in plants. Critical Rev in Plant Sci **13**: 121-138

Part 4

Drought Stress

Gene Expression during Dehydration Stress in the Resurrection Plant *Craterostigma plantagineum*

Dorothea Bartels, Antonella Furini, Christine Bockel, Wolfgang Frank and Francesco Salamini

Max-Planck-Institut für Züchtungsforschung, D-50829 Köln, Germany

Abstract. The resurrection plant *Craterostigma plantagineum* is being used as an experimental model system to dissect pathways leading to desiccation tolerance. This plant can recover from severe desiccation within 24 h of contact with water. During drying or ABA treatment novel gene products accumulate rapidly in leaves and other tissues. It is proposed that these gene products contribute to the protection of cellular structures during dehydration. It has been a major goal to isolate genes and determine parameters which regulate the expression of the responsive genes. One approach was to characterize promoters. Three promoters from different *lea* type genes and three promoters from genes encoding inducible enzymes of the carbohydrate metabolism have been analyzed for their responsiveness to drought and ABA in transient assays and in transgenic plants. The promoter activities in transgenic plants suggest differences in the mode of the regulatory pathways and point to at least two transduction pathways. Evidence has been accumulated that the activation of at least one promoter in vegetative tissues depends on a gene product related to the *Arabidopsis abi3*-gene. In gel shift mobility assays it was shown that nuclear proteins from ABA treated callus and leaves specifically bind sequence elements in the promoter region. Other environmental factors such as light interact with desiccation stress and modulate the expression level of the proteins involving post-transcriptional and post-translational mechanisms. To dissect events in this signal transduction chain, two other experimental approaches have been initiated. A T-DNA based tagging experiment was carried out in order to obtain positive mutations over- expressing gene products downstream of the ABA signal. One such mutant has been obtained and has been molecularly characterized. A subtractive and differential screening is being used to isolate cDNA clones which encode transcripts expressed during the very early events of dehydration.

Keywords. ABA, *Craterostigma plantagineum*, drought stress, promoter, resurrection plants, signal transduction mutants

1 Introduction

In many breeding programmes classical genetic selection has provided an enormous improvement of drought tolerance in crop plants. Many traits have been identified which are potentially important parameters in conferring drought tolerance. In breeding programmes these desired traits have been transferred from close relatives into crop plants. Most of the traits are complex, and the molecular basis of them is not understood but their understanding is essential in assessing biotechnological applications and in a targeted manipulation of crop plants. Molecular studies of drought stress in plants have initially been descriptive and yielded a large number of genes with a potential role in the cellular responses of plants to water deficit (Skriver and Mundy, 1990; Bray, 1993; Ingram and Bartels, 1996). A second focus of drought stress research has been revealing mechanisms which control gene expression during drought. All these studies have been carried out using a variety of different plants subjected to water deficit ranging from mild to severe stress. The molecular studies of dehydration stress have so far focused on experimental systems which express desiccation tolerance such as developing embryos (Quatrano, 1987) or "resurrection plants" (Gaff, 1971; Bartels et al., 1990), on genetic model plants like *Arabidopsis thaliana* (Schnall and Quatrano, 1992; Yamaguchi-Shinozaki and Shinozaki, 1993; Kiyosue et al., 1994) or on agricultural important plants like tomato, pea or sunflower to study responses to water stress at the genetic and molecular level (Bray, 1990; Guerrero et al., 1990; Almoguera et al., 1993).

2 *C. Plantagineum* as a Model System

Over the passed years we have extensively studied molecular mechanisms underlying the desiccation tolerance in the resurrection plant *Craterostigma plantagineum*. *C. plantagineum* is being used as an experimental model system to dissect pathways leading to desiccation tolerance. This plant can recover from severe desiccation within 24 h of contact with water. During drying or ABA treatment novel gene products accumulate rapidly in leaves and other tissues (Bartels et al., 1990). Two approaches have been used to isolate genes with altered expression levels during dehydration. Differentially expressed genes were either isolated from leaf tissues completely dehydrated or from leaf tissue dehydrated for 1 h. An overview of the genes characterized is given in Table 1.

With many of the isolated genes, RNA northern blot experiments were performed and three categories of transcript accumulations have been observed:

1. transcripts which accumulate abundantly upon dehydration and persist during dehydration, but disappear upon rehydration.
2. transcripts which accumulate very early during dehydration but only transiently.
3. transcripts which are down-regulated during dehydration. Examples for these accumulation patterns are shown in Fig. 1.

Table 1. Genes upregulated during drought stress

Pathways	**Functions**
Protection factors	LEA proteins[1] Elip homologues[2]
Synthesis of sugars	Sucrose synthase Sucrose-phosphate synthase
Transport	Putative water channel proteins
Signalling molecules	*cDT1* gene (callus)

[1] Piatkowski et al., 1990
[2] Bartels et al., 1992

Northern experiments like those shown in Fig. 1 and additional studies with ABA and protein synthesis inhibitors provided evidence that at least three different signal transduction pathways exist; a possible scheme is shown in Fig. 2.

A major goal of the research has been to isolate genes and determine parameters which regulate the expression of the response genes. For this we are using the following three experimental approaches:

- to dissect promoters of ABA-responsive genes
- to isolate mutants in the signal transduction pathway
- to isolate early drought ABA-responsive genes

Three promoters from different *lea* type genes and two promoters from genes encoding inducible enzymes of the carbohydrate metabolism (sucrose synthase and sucrose-phosphate-synthase) have been analyzed for their responsiveness to drought and ABA in transient assays to dehydration stress or ABA, a dot matrix sequence comparison did not reveal significant motif homologies. The best characterized promoter is the *cDeT 27-45* promoter (Michel et al., 1993). The structural gene *cDeT 27-45* encodes a *lea* type protein. Transient expression assay of this promoter revealed 5' sequences which confer ABA responsiveness. It was shown for this promoter that nuclear proteins from ABA treated callus and leaves specifically bind sequence elements in the promoter region (Nelson et al., 1994). By electrophoretic mobility shift assays ABA independent and ABA dependent binding activities could be distinguished. The ABA inducible binding activity is correlated with mRNA and protein accumulation as well as acquisition of desiccation tolerance. It was not possible to clone genes encoding the DNA binding proteins by Southern-Western type screening of a cDNA expression library. Therefore some of these binding proteins have been purified biochemically using specific DNA affinity chromatography.

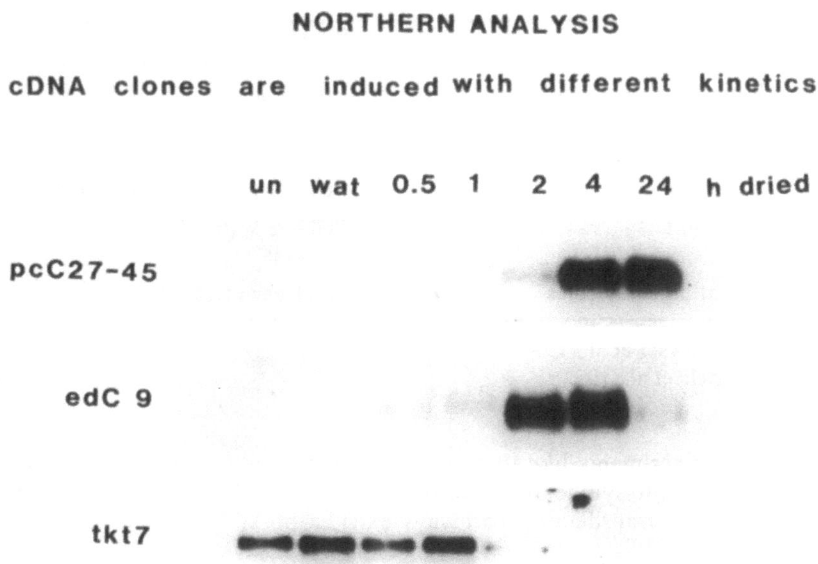

Fig. 1. Poly $(A)^+$RNA was isolated from the following samples: fully hydrated leaves (un), detached leaves incubated in water (wat), detached leaves dried for 0.5, 1, 2, 4 or 24 h. Equal amounts of poly $(A)^+$RNA were size fractionated in denaturing formaldehyde gels and transferred to nylon filters. Identical filters were probed with ^{32}P-labeled inserts of the following *Craterostigma* cDNA clones: *pcC27-45* (Bartels et al., 1990), *edc9* (Bockel and Bartels, unpublished), *tkt7* corresponding to transketolase 7 (Bernacchia et al., 1995)

3 Regulation of Gene Expression

The observation that callus of *Craterostigma plantagineum* becomes tolerant to desiccation when cultured in the presence of the plant hormone ABA has been used in the approach to isolate mutants. This treatment induces the same set of transcripts which are expressed upon drying in the whole plant (Bartels et al., 1990). This observation is being used in a search for genes regulating the ABA response. For this objective a T-DNA activation tagging experiment was set up. The rationale was to transform *Craterostigma* with a T-DNA-tagging vector (made available by R. Walden, MPI Köln; Walden et al., 1991) and transgenic callus was selected capable of surviving a severe dehydration treatment in the absence of exogenously supplied ABA. This approach allowed us to select one transgenic callus which was able to withstand the dehydration process without ABA incubation (Furini, 1995). Analysis of RNA transcripts in this callus showed that genes which are constitutively expressed in this callus which in non-transformed tissues can only be induced by ABA treatment or desiccation. This indicates the constitutive activation of a gene downstream of ABA. The

ABA. The tagged gene (*cDT-1*) and corresponding cDNA clones were isolated. *cDT-1* is constitutively expressed in the transgenic callus line and by ABA treatment in the untransformed callus or by drying in leaves. The *cDT-1* gene is present in multiple copies in the *Craterostigma* genome. DNA sequence analysis of cDNA and genomic clones revealed some unusual features of the *cDT-1* gene such as the absence of a plausible open reading frame and the presence of a stretch of adenosines. The mechanism of cDT-1 action is not yet understood and will be the subject of further research.

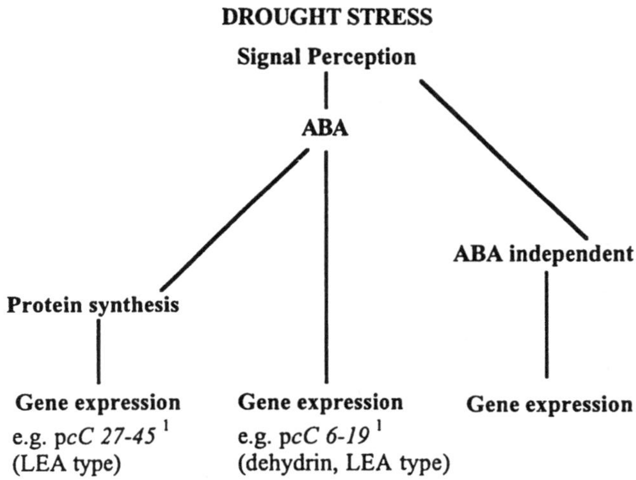

[1] for references see Piatkowski at al.,1990, Michel et al., 1993, Michel et al., 1994

Fig. 2. Possible signal transduction schemes during drought stress

Acknowledgements. This work was supported in part by the EU PTP programme and by an EU International Scientific Cooperation Programme.

References

Almoguera C, Coca MA, Jordans J (1993) Tissue-specific expression of sunflower heat shock proteins in response to water stress. Plant J **4**: 947-958

Bartels D, Schneider K, Terstappen G, Piatkowski D, Salamini F (1990) Molecular cloning of abscisic acid modulated genes which are induced during desiccation of the resurrection plant *Craterostigma plantagineum*. Planta **181**: 27-34

Bartels D, Hanke C, Schneider K, Michel D, Salamini F (1992) A desiccation-related *Elip*-like gene from the resurrection plant *Craterostigma plantagineum* is regulated by light and ABA. EMBO J **11**: 2771-2778

Bernacchia G, Schwall G, Lottspeich F, Salamini F, Bartels D (1995) The transketolase gene family of the resurrection plant *Craterostigma plantagineum*: differential expression during rehydration phase. EMBO J **14**: 610-618

Bray EA (1990) Drought-stress-induced polypeptide accumulation in tomato leaves. Plant Cell Environ **13**: 531-538

Bray EA (1993) Molecular responses to water deficit. Plant Physiol **103**: 1035-1040

Furini A (1995) T-DNA tagging and analysis of desiccation- and ABA-induced genes of the resurrection plant *Craterostigma plantagineum* (Hochst.) by *Agrobacterium*-mediated transformation. Thesis, Univ. of Köln

Gaff DF (1971) Desiccation-tolerant flowering plants in Southern Africa. Science **174**: 1033-1034

Guerrero FD, Jones JT, Mullet JE (1990) Turgor-responsive gene transcription and RNA levels increase rapidly when pea shoots are wilted. Sequence and expression of three inducible genes. Plant Mol Biol **15**: 11-26

Ingram J, Bartels D (1996) The molecular basis of dehydration tolerance in plants. Annu Rev Plant Physiol Plant Mol Biol **47**: 377-403

Kiyosue T, Yamaguchi-Shinozaki K, Shinozaki K (1994) Characterization of two cDNAs (ERD10 and ERD14) corresponding to genes that respond rapidly to dehydration stress in *Arabidopsis thaliana*. Plant Cell Physiol **35**: 225-231

Michel D, Salamini F, Bartels D, Dale P, Baga M, Szalay A (1993) Analysis of a desiccation and ABA-responsive promoter isolated from the resurrection plant *Craterostigma plantagineum*. Plant J **4**: 29-40

Michel D, Furini A, Salamini F, Bartels D (1994) Structure and regulation of an ABA- and desiccation-responsive gene from the resurrection plant *Craterostigma plantagineum*. Plant Mol Biol **24**: 549-560

Nelson D, Salamini F, Bartels D (1994) Abscisic acid promotes novel DNA-binding activity to a desiccation-related promoter of *Craterostigma plantagineum*. Plant J **5**: 451-458

Piatkowski D, Schneider K, Salamini F, Bartels D (1990) Characterization of five abscisic acid-responsive cDNA clones isolated from the desiccation-tolerant plant *Craterostigma plantagineum* and their relationship to other water-stress genes. Plant Physiol **94**: 1682-1688

Quatrano RS (1987) The role of hormones during seed development. *In* RD Davies, ed, Plant Hormones and their Role in Plant Growth and Development. Martinus Nijhoff Publishers, Dordrecht, pp 494-514

Schnall JA, Quatrano RS (1992) Abscisic acid elicits the water stress response in root hairs of *Arabidopsis thaliana*. Plant Physiol **100**: 216-218

Skriver K, Mundy J (1990) Gene expression in response to abscisic acid and osmotic stress. Plant Cell **2**: 502-512

Walden R, Hayashi H, Schell J (1991) T-DNA as a gene tag. Plant J **1**: 281-288

Yamaguchi-Shinozaki K, Shinozaki K (1993) Characterization of the expression of a desiccation-responsive *rd29* gene of *Arabidopsis thaliana* and analysis of its promoter in transgenic plants. Mol Gen Genet **236**: 331-340

Expression of the Maize *rab17* Gene in Response to Abscisic Acid and Osmotic Stress

Anders B. Jensen[1,2], John Mundy[2] and Montserrat Pagès[1]

[1] Departament de Genética Molecular CID, C.S.I.C., E-08034 Barcelona, Spain.
[2] Department of Plant Physiology, University of Copenhagen, DK-1353 Copenhagen K, Denmark

Abstract. Expression of the maize abscisic acid (ABA)-responsive gene *rab17*, was studied in seeds and vegetative tissues. RNA gel blot analysis of *rab17* mRNA accumulation in response to desiccation and exogenous applied ABA, detected expression in all tissues tested. NaCl, in concentrations higher than 50 mM activated accumulation of *rab17* mRNA expression in leaf tissue of maize, whereas a weak accumulation was observed in root tissue. Barley aleurone layers also responded to NaCl by an increase in *rab* mRNA levels. Moreover, an inhibitor of ABA biosynthesis, fluridone, completely prevented *rab* gene induction in response to 250 mM NaCl in aleurone layers. In the corresponding protoplasts, which can not synthesize ABA, no accumulation of endogenous gene products was detected and transiently transformed homologous rice and maize gene promoters did not respond to increased osmotic levels. These results suggest that accumulation of *rab17* mRNA in response to desiccation and high salt concentrations are dependent on increased levels of ABA.

Keywords. Abscisic acid, barley aleurone layer, gene expression, maize, NaCl, osmotic stress, *Zea mays*

1 Introduction

Previous studies on the regulation of gene expression during embryogenesis in *Zea mays* L. have shown that a group of stage-specific mRNAs and proteins which accumulate in late embryogenesis are induced precociously in immature embryos when ABA is supplied to the culture medium (Goday et al., 1988). ABA is supposed to mediate a large number of physiological and developmental processes in plants such as embryo maturation, seed dormancy and the response of plant to osmotic stress (Zeevaart and Creelman, 1988).

The response of plant cells to ABA include modulation of specific mRNAs and proteins in embryo and vegetative tissues (Skriver and Mundy, 1990). The maize ABA responsive *rab* genes are induced during late embryogenesis when ABA levels are high, and they are also ABA and water stress inducible in vegetative tissues (Gomez et al., 1988; Vilardell et al., 1990).

Many ABA responsive genes are controlled at the transcriptional level. Functional dissection of *rab* promoters using transient expression assays in protoplasts has revealed the existence of a specific sequence element (ABRE) which confers ABA responsiveness to the reporter gene (Skriver et al., 1991). Moreover, leucine zipper-type proteins binding to the ABRE have been identified (Guiltinan et al., 1990; Oeda et al., 1991). However, recent studies have shown that the expression of *rab* genes are also controlled by a variety of other processes and that changes in ABA levels are not always correlated with changes in gene expression (Pla et al., 1991). For example, studies of *rab* gene expression in *viviparous* mutants which are deficient or insensitive to ABA (Robertson, 1955; Robichoud et al., 1980; Neil et al., 1986) have shown that both *rab17* and *rab28* mRNA accumulated to normal levels during embryogenesis of the *vp2* mutant. However, in seedlings of *vp2* subjected to water stress, no induction of *rab* mRNAs was detected whereas in the wild type plants high expression levels were observed (Pla et al., 1989, 1991). Although the *vp2* mutant has reduced but significant levels of ABA in embryo (Neil et al., 1986), these results indicate that expression of *rab* mRNAs is dependent upon an increase in ABA that occurs in response to dehydration in wild type vegetative tissues.

To further confirm this hypothesis an expression study of *rab* genes in different tissues in response to desiccation and treatments with ABA or osmotic compounds such as sodium chloride (NaCl), polyethylenglycol (PEG) and mannitol was undertaken.

2 Materials and Methods

2.1 Plant Material

Seeds of *Zea mays* W64A (pure inbred line) were sterilized in ethanol for 5 min followed by 10 min in 5% calcium hypochlorite and washed in distilled water, then placed on wetted paper over a layer of vermiculite and incubated at 24°C with a 16 h photoperiod for 6 d. The seedlings were transferred to 0.5 x liquid MS medium and grown hydroponically for the last 48 h and the last 24 h in the presence of ABA, NaCl, polyethyleneglycol or mannitol as stated in the respective figure text. Aleurone layers and protoplasts were isolated from *Hordeum vulgare* cv. Himalaya as described by Skriver et al., (1991). Briefly, barley seeds were separated from the embryos, cut in two parts and incubated in water for 3 d. The aleurone layers were isolated from the endosperm and incubated in protoplast incubation medium in the presence of ABA and/or

NaCl. Fluridone was added from a stock to a final concentration of 10 μg/ml containing 5% ethanol.

2.2 RNA and Blot Analysis

Total RNA was prepared according to Leah et al., (1991) RNAs (10 μg/lane) were separated on 1% formaldehyde agarose gel electrophoresis and blotted onto Hybond-N nylon membranes (Amersham). Equal loading of RNA was confirmed by ethidium bromide staining. Hybridization was performed under high stringency conditions (Sambrook et al., 1989) using a random primed ^{32}P labeled 0.85 kb EcoRI-XhoI fragment containing the full length cDNA of *rab17*. Approximately 10^6 protoplasts (corresponding to 50 seeds) or 50 seeds were used for RNA extraction of barley aleurone protoplasts and layers. Abscisic acid (Sigma) was dissolved in ethanol (100 mM stock solution) and diluted to 100 μM in water. Mannitol and PEG 8000 (Sigma) were dissolved in H_2O. Fluridone (1-methyl-3-phenyl-5-[3-(trifluoromethyl)phenyl]-4-(1H)-pyridone) was purchased from Elenco and added from a 4 mg/ml stock solution in ethanol.

2.3 Osmotic Potential Measurements

The osmotic potentials of growth media containing various solutes were determined using a Wescor 5100 vapour pressure osmometer. The osmometer was calibrated with a NaCl solution.

3 Results and Discussion

3.1 Accumulation of *rab17* in Vegetative Tissues

To determine the tissue specificity of *rab17* mRNA accumulation in response to desiccation, 8-day-old seedlings were removed from the growth medium and subjected to fast severe stress for 4 h. RNA was extracted from dissected roots, stem and leaves. Northern blot analysis showed that *rab17* mRNA accumulates in all parts of the young seedling, although a slightly lower level was detected in roots.

To investigate whether the *rab17* mRNA accumulates in the same tissue in response to NaCl and ABA, similarly grown seedlings were subjected to different concentrations of ABA or NaCl. Accumulation of *rab17* mRNA was detected in both leaves and roots when 20 μM ABA was supplied to the growth medium. Leaves also responded to high NaCl concentrations by increased levels of *rab17* mRNA, but surprisingly, no induction of *rab17* mRNA was observed in roots in response to NaCl concentrations from 50 mM to 500 mM. However, a long exposure of the blot revealed a weak band corresponding to *rab17*.

To test whether other osmotic compounds such as mannitol and PEG (mol wt 8000) could induce *rab17* gene expression in leaves, we measured the osmotic potential of the growth medium containing 250 mM NaCl. Concentrations of 5% (w/vol) PEG and 10% (w/vol) mannitol were found to give similar osmotic potentials as the presence of 250 mM NaCl added to the growth medium (0.98 MPa). Northern blot analysis showed that neither the presence of 10% mannitol nor 5% PEG activated *rab17* gene expression.

Although, we have not measured the ABA concentration in response to these treatments, it has been reported that solutions of osmotica including mannitol can increase ABA levels (Creelman and Zeevaart, 1985; Zeevaart and Creelman, 1988). However, it could be possible that higher concentrations or a longer exposure to PEG and mannitol might activate *rab17* gene expression.

3.2 Transient Transformation of *rab* Promoters in Protoplasts

Isolated protoplasts are unable to produce ABA in response to a lowering in the osmotic potential of the incubation medium (Weiler et al., 1982; Loveys and Robinson, 1987). Therefore, protoplasts might be a good system to test whether the responsiveness of the *rab17* to osmotic stress is due to an increase in ABA concentration. Since aleurone protoplasts have been successfully applied to delineate regulatory elements in promoters of several plant genes (Jacobsen and Close, 1991; Skriver and Mundy, 1991), and because *rab17* homologous genes have been isolated from barley seeds (Close et al., 1989) we used that system.

First, Northern blot analysis was performed of barley aleurone layers to test whether the endogenous barley *rab* genes respond to high NaCl concentrations. As shown in Fig. 1A, no accumulation of *rab17* homologous mRNA was detected in the untreated control or layers treated with 5% ethanol (Fig. 1A lanes 9 and 3, respectively). Furthermore, *rab* gene induction was only observed in response to 250 mM NaCl but not to concentrations of 50 mM NaCl (lanes 1 and 2). Likewise, accumulation of *rab* mRNA was observed when 20 μM ABA was added to the culture medium. However, no additive effect was detected for ABA and 250 mM NaCl added together as compared to the level of expression seen for ABA alone (lanes 4 and 7). Moreover, fluridone, an inhibitor of ABA biosynthesis, blocked the induction of gene expression in response to NaCl, suggesting that ABA is important for activation of *rab* gene expression in barley aleurone layers.

In contrast, RNA gel blot analysis of barley aleurone protoplasts did not detect *rab* mRNA accumulation in response to 50 mM or 250 mM NaCl (Fig. 1B, lane 1 and 2). As expected, an activation of gene expression was observed both for ABA applied alone or in combination with NaCl. Surprisingly, in the presence of ABA and NaCl together, bands more intense than those induced by ABA alone were detected (compare lane 3 and 4 with lane 7). This is probably due to a slightly unequal loading of the RNA.

Transformation experiments of barley aleurone protoplasts, with various reporter gene constructs of *rab17* (Vilardell et al., 1990) or *rab16* (Mundy and Chua, 1988) fused to the chloramphenicol acetyltransferase (CAT) coding region were performed. In these experiments, we were unable to detect any activation of CAT activities over background levels, using a wide range of NaCl

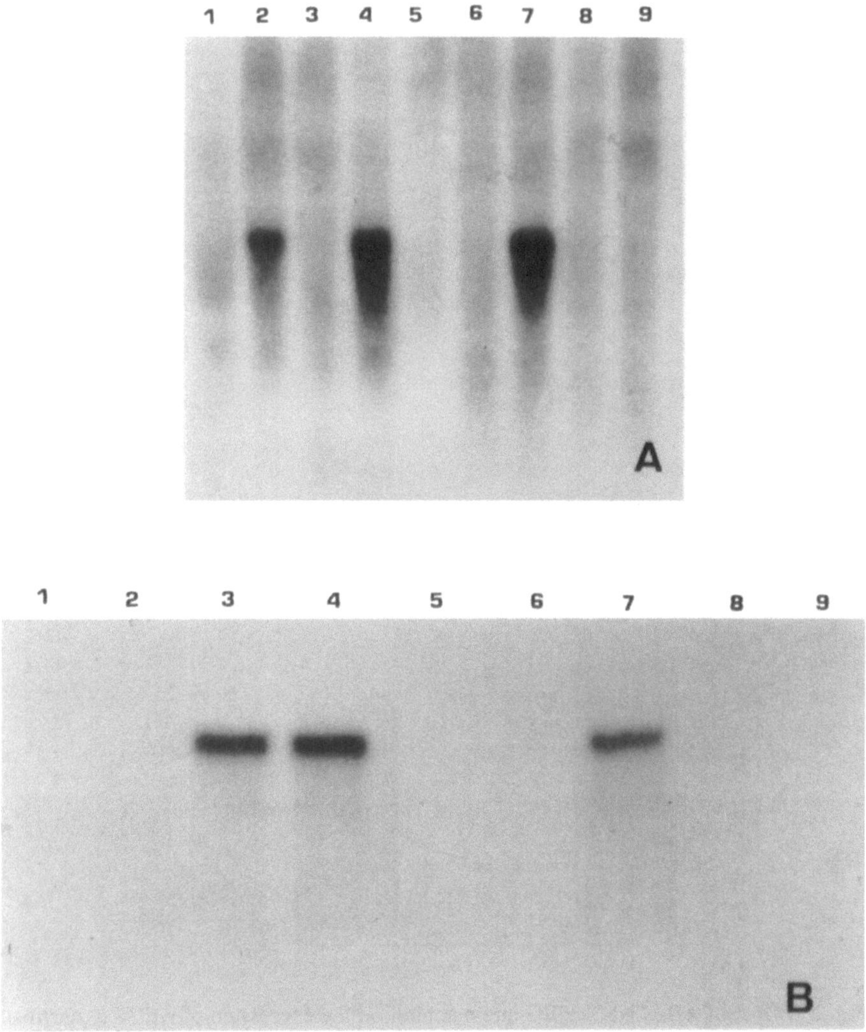

Fig. 1. Northern blot analysis of *rab* genes in barley aleurone layers and protoplasts. (A) Accumulation of *rab* mRNA in aleurone layers in response to 20µM ABA and/or various NaCl concentrations. Total RNAs (10µg) were loaded on each lane. 1) 50 mM NaCl, 2) 250 mM NaCl, 3) 5% ethanol, 4) 250 mM NaCl+20µM ABA, 5) 50 mM+10µg/ml fluridone, 6) 250 mM NaCl+10µg/ml fluridone, 7) 20µM ABA, 8) 10 µg/ml fluridone, 9) H20. (B) Expression of *rab* genes in aleurone protoplasts using the same conditions as above. Total RNAs were loaded on each lane. 1) 50 mM NaCl, 2) 250 mM NaCl, 3) 50 mM NaCl+20µM ABA, 4) 250 mM NaCl+20 µM ABA, 5) 50 mM+10 µg/ml fluridone, 6) 250 mM NaCl+10 µg/ml fluridone, 7) 20 µM ABA, 8) 10 µg/ml fluridone, 9) 5% ethanol. Northern blots were hybridized with a 0.85 kb fragment of the *rab17* cDNA

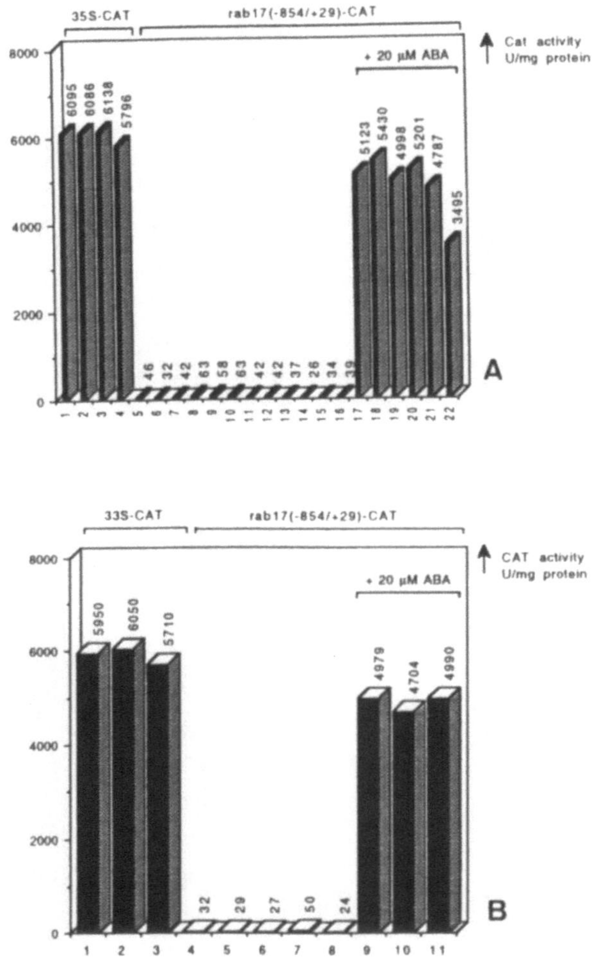

Fig. 2. Effect of ABA, NaCl, PEG and mannitol on the expression of *rab17* 5' upstream sequences. A fragment from positions -854 to +29 of *rab17* was transcriptionally fused to the CAT reporter gene. Data on the top of the columns are from three independent transient expression assays. (A) Columns 1-4, 35S(-941/+8)-CAT incubated in the presence of 0, 10, 50 and 250 mM NaCl. Columns 5 to 16, *rab17*-CAT incubated in the presence of 0, 0.01, 0.05, 0.25, 0.5, 1, 5, 10, 25, 50, 150 and 250 mM NaCl. Columns 17-22, *rab17*-CAT incubated in the presence 20 µM ABA and 0, 1, 10, 50, 150 and 250 mM NaCl. (B) Columns 1-3, 35S(-941/+8)-CAT incubated in the presence 0.5% PEG, 10% mannitol. Columns 4-8, *rab17*-CAT incubated in the presence of 0; 2.5% PEG; 5% PEG; 5% mannitol; 10% mannitol. Columns 9-11, *rab17*-CAT incubated in the presence 20 µM ABA and 0; 5% PEG; 10% mannitol

concentrations (Fig. 2). Viability tests showed that NaCl concentrations up to 250 mM did not significantly affect aleurone protoplasts.

In conclusion, expression of the *rab17* gene was detected in all maize tissues in response to desiccation and exogenous ABA. Sodium chloride, in concentration higher than 50 mM activate accumulation of *rab17* mRNA expression in leaf tissue of maize and in barley aleurone layers. Moreover, an inhibitor of ABA biosynthesis, fluridone, completely prevented *rab* gene induction in response to 250 mM NaCl. In the corresponding protoplasts neither accumulation of endogenous gene mRNAs nor activation of transiently transformed homologous rice and maize gene promoters responded to high salt levels, PEG or mannitol. One explanation of the lack of induction might be that PEG and mannitol do not stimulate ABA biosynthesis in the amounts used. Surprisingly, very low levels of *rab17* mRNA accumulation was detected in roots of seedlings grown in the presence of high NaCl concentrations. The effect of salt on plant cells has two components; osmotic stress and ion toxicity (Serrano and Gaxiola, 1994). It could be possible that these two components influence *rab17* mRNA expression differently in roots and leaves of young seedlings and we have not measured the concentrations of ABA in seedlings subjected to NaCl, mannitol or PEG. Similar results have been described by Mundy and Chua (1988) who have reported that the effect of NaCl on *rab* mRNA levels in rice suspension cultures is mediated entirely through ABA, since an additive effect of combined NaCl and ABA treatment was not observed.

Acknowledgements. This work was funded in part by the European Communities BIOTECH Programme, as part of the Project of Technological Priority 1993-1996, and in part by grant BIO-94-0750 from Plan Nacional de Investigación Científica y Desarrollo Tecnológico.

References

Close TJ, Kortt AA, Chandler PM (1989) A cDNA-based comparison of dehydration-induced proteins (dehydrins) in barley and corn. Plant Mol Biol **13**: 95-108

Creelman RA, Zeevaart JAD (1985) Abscisic acid accumulation in spinach leaf slices in the presence of penetrating and non-penetrating solutes. Plant Physiol **77**: 25-28

Goday A, Sanchez-Martinez D, Gomez J, Puigdomenech P, Pages M (1988) Gene expression in developing *Zea mays* embryos: regulation by abscisic acid of a highly phosphorylated 23-25 group of proteins. Plant Physiol **88**: 564-569

Gomez J, Sanchez-Martinez D, Stifel V, Rigau J, Puigdomenech P, Pages M (1988) A gene induced by the plant hormone abscisic acid in response to water stress encodes a glycine-rich protein. Nature **334**: 262-264

Guiltinan MJ, Marcotte WR, Quatrano RS (1990) A plant leucine zipper protein that recognizes an abscisic acid response element. Science **250**: 267-271

Jacobsen JV, Close CJ (1991) Control of transient expression of chimaeric genes by gibberellic acid and abscisic acid in protoplasts prepared from mature barley aleurone layers. Plant Mol Biol **16**: 713-724

Leah R, Tommerup H, Svendsen I, Mundy J (1991) Biochemical and molecular charaterization of three barley seed proteins with antifungal properties. J Biol Chem **266**: 1464-1473

Loveys BR, Robinson SP (1987) Abscisic acid synthesis and metabolism in barley leaves and protoplasts. Plant Sci **49**: 23-30

Mundy J, Chua N-H (1988) Abscisic acid and water-stress induce the expression of a novel rice gene. EMBO J **7**: 2279-2286

Neil SJ, Horgan R, Parry AD (1986) The carotenoid and abscisic acid content of viviparous kernels and seedlings of *Zea mays* L. Planta **169**: 87-96

Oeda K, Salinas J, Chua N-H (1991) A tobacco b-Zip transcription activator (TAF-1) binds to a G-box like motif conserved in plant genes. EMBO J **10**: 1793-1802

Pla M, Goday A, Vilardell J, Gomez J, Pages M (1989) Differential regulation of ABA-induced 23-25 kDa proteins in embryo and vegetative tissues of the viviparous mutants of maize. Plant Mol Biol **13**: 385-394

Pla M, Gomez J, Goday A, Pages M (1991) Regulation of the abscisic acid-responsive gene rab28 in maize viviparous mutants. Mol Gen Genet **230**: 394-400

Robertson DS (1955) The genetic vivipary in maize. Genetics **40**: 745-760

Robichaud CS, Wong J, Sussex IM (1980) Control of *in vitro* growth of viviparous embryo mutants of maize by abscisic acid. Dev Genet **1**: 325-330

Sambrook J, Fritsch EF, Maniatis T (1989) Molecular Cloning: A Laboratory Manual. Cold Spring Harbor Laboratory Press, Cold Spring Harbor, NY

Serrano R, Gaxiola R (1994) Microbial models and salt stress tolerance in plants. Crit Rev Plant Sci **13** 121-138

Skriver K, Mundy J (1990) Gene expression in response to abscisic acid and osmotic stress. Plant Cell **2**: 503-512

Skriver K, Olsen FL, Rogers JC, Mundy J (1991) Cis-acting elements responsive to gibberellin and its antagonist abscisic acid. Proc Natl Acad Sci USA **88**: 7266-7270

Vilardell J, Goday A, Freire MA, Torrent M, Martinez C, Torne JM, Pages M (1990) Gene sequence, developmental expression, and protein phosphorylation of RAB-17 in maize. Plant Mol Biol **14**: 423-432

Weiler EW, Schnabl H, Hornberg C (1982) Stress-related levels of abscisic acid in guard cell protoplasts of *Vicia faba* L. Planta **154**: 24-28

Zeevaart JAD, Creelman RA (1988) Metabolism and physiology of abscisic acid. Ann Rev Plant Physiol Plant Mol Biol **39**: 439-473.

The Role of Abscisic Acid in the Regulation of Gene Expression during Drought Stress

Elizabeth A. Bray[1], Meena S. Moses[1], Eunsook Chung[1] and Ryozo Imai[2]

1 Department of Botany and Plant Sciences, University of California, Riverside, CA 92521, USA
2 Laboratory of Plant Hormone Function, RIKEN, 2-1 Hirosawa, Wako-shi, Saitama 351-01, Japan

Abstract. A subset of drought-induced genes are regulated by the plant hormone abscisic acid (ABA). The currently characterized ABA-induced genes fall into a number of different categories including hydrophilic genes, or genes that are similar to late embryogenesis abundant (*lea*) genes originally identified in cotton, and those which may be involved in prevention of secondary stresses. An example of the first category is the tomato gene, *le25*. The deduced gene product maintains a conserved distribution of negatively- and positively-charged amino acids as demonstrated using helical wheel plots. It is proposed that this *lea* gene product plays a role in the control of solute concentration in the cytoplasm. The tomato gene, *le16*, is an example of a gene which may be involved in the protection of the plant against secondary stresses. LE16 is similar to non-specific lipid transfer proteins, and is one of a few that is induced in response to osmotic stresses. It is predicted that this gene product is important in the control of pathogens during stress. It is clear that many additional studies, and new methods are needed, to evaluate the role of drought-induced gene products.

Keywords. ABA, abscisic acid, drought stress, *Lycopersicon esculentum*, tomato, water deficit

1 Introduction

During many abiotic stresses, there is an increase in the concentration of the plant hormone ABA. ABA is involved in the ability of plants to adapt to stress through many different plant processes including alterations in gene expression, metabolism, and stomatal closure. However, a thorough understanding of ABA's role in plant adaptation has not been achieved.

Genes induced by ABA have been identified and isolated using mutants that are deficient in ABA biosynthesis from *Arabidopsis*, maize and tomato (Pla et al., 1989; Cohen and Bray, 1990; Lång and Palva, 1992; Gosti et al., 1995; Grillo et al., 1995). The genes that are known to be regulated by ABA fall into several different classes. Although there are now many genes isolated which are expressed during drought stress, still little can be confidently stated about the function of drought-induced gene products. Therefore, this is a major goal in agriculture.

2 Possible Roles of Drought-Induced Genes

Many different genes, which can be divided into several different classes based on sequence identities, predicted function and expression characteristics, have been studied with respect to drought stress (Bray, 1994). A number of these drought-induced genes are similar to late embryogenesis abundant genes, and as such are proposed to play an important role in the ability of vegetative organs to survive stress. Other genes may be involved in adaptation to secondary stresses, accumulation of osmolytes, or regulation of important cellular functions.

2.1 Late Embryogenesis Abundant Genes

Several families of proteins were originally identified in cotton which are expressed during seed development (Dure et al., 1981). A number of these genes are expressed during the maturation phase of seed development in which the embryo is preparing for survival of desiccation and entering into dormancy. Some of these genes are also thought to be induced in response to changes in ABA concentration during embryogenesis. Many of the deduced gene products fall into an interesting category of proteins which are extremely hydrophilic, indicating that the genes are cytoplasmically located. Proposals have ensued that the genes are involved in the protection of cellular function during desiccation by, for example, replacing water, and protecting membranes and essential proteins.

In tomato, we identified two genes that are expressed during stress of vegetative organs which are related to genes that are expressed during seed development. One of these, LE4, is a dehydrin (Close et al., 1989) and another, LE25, is similar to the cotton protein D-113 and other recently cloned genes (Fig. 1). Both of these proteins are extremely hydrophilic, yet, each have other unique predicted characteristics.

2.1.1 Structural Characteristics of LE25

There are currently 7 other plant sequences in the gene banks that have an interesting pattern of amino acid sequence similarity to LE25 (Fig. 1; Dure, 1993). All of these proteins are predicted to have an α-helical domain in the N-terminus ranging from 55 to 80 amino acids, resulting in approx. 15 to 22 turns of the helix. In addition, the helices are amphipathic with the negatively-charged amino acids on one face of the helix and 2 domains on either side with the majority of the positively-charged amino acids (Fig. 2). The C-termini differ in length and have considerably reduced homology (Fig. 1; Dure, 1993). The length of the C-terminal domain ranges from 16 in LE25 to 100 in D-113 from cotton. These regions are predicted to form random coils or beta sheets and in a few instances additional α-helices. Since the C-termini are not conserved, this region is not predicted to be involved in a common function.

```
 Le                  MQTGKDAAS  AAKAGMEKTK  ANVQEKAERM  TTRDPLKKEM
 Gh                   **SM****A  S*******A*  *SM***VDQ*  K****NE***
 Gm   MQGGKKAGES  IKE*ATNIGA  S*********  *T********  *A***VQ*EL
 Ha   MQSGKNAAAS  AKE*AANV*A  S*********  *SL***G*K*  *AH**MQ***
At1          MQS  MKE*ASNI*A  S**S**D***  *TLE****K*  K****VQ*Q*
At2          MQS  AK*KIS*M**  T**EK*VICQ  *KAD*****A  MA*TKEE**I
 Pm   MQPVKNAAAA  AKEKMSN**A  TT*EKLDK**  *SAE**MHKS  *AHSKGER*I
 Lm                                        MA*****K*  RAHT*ED*GI

 Le   **EKKEDRVA  AAEMGKRDAK  AQHAAEKQGA  ATTGTGTTGY  GATGNHTTF
 Gh   *R*R**E*QE  D**LR*QE*R  HHN*TAGHVG  GGGIG**GYT  TAGY*IGDTG
 Gm   **Q***AKMN  Q**LD*QA*R  QHNT*A**S*  T*A*HMGH*H  HT**TG*GTA
 Ha   *R****E*KH  E**YE*QA**  EHN**Q**TT  GIGTGTHSYT  TTNVTGHRTG
At1   **QV***KIN  Q***Q**ETR  HTN**M*EA*  GAGTGLGL*.  ..
At2   *HQRRKAKE*  D*N*DMHM**  *A**ED*LM*  KQSHYHV*DH  *PHVSQQASV
 Pm   *HQE**AKE*  E***R*HS**  *ENKIQREEE  KHLHHRAGAG  IMNPT**ATH
 Lm   *A*R*QQMME  Q**LN*QQ**  EEN**Q*EQ*  RAATVPG*RR  RRGTAAWRTP

 Gh   GYGGTGGHDN  RGYPTAGSGY  DTGRQDDLSS  MGFGGDTEGA  YGTTGNQDFP
 Gm   TYSTTGEYGQ  PMGAHQTSAM  PGHGTGQPTG  HVTEGVVGSH  PIGTNRGPGG
 Ha   TGGI
At2   PSPAPVMAHG  ...
 Pm   TTTTTAPPPG  KYI
 Lm   APSARPPAPG  GPHGAQPAER  WDPDWWSTPG  GGYT

 Gh   NAASNNAGTR  RNTRGGTQDD  PYYRSY
 Gm   TATAHNTRAG  GKPNDYGYGT  GGT
```

Fig. 1. Alignment of predicted polypeptides with similarities to LE25. *Le* (*Lycopersicon esculentum*; Q00747), *Gh* (*Gossypium hirsutum*; P09441), *Gm* (*Glycine max*; Q01417), *Ha* (*Helianthus annuus*; S23528) *At1* (*Arabidopsis thaliana* No. 1; Z27256, note: an expressed sequence tag that is not full length), *At2* (*Arabidopsis thaliana* No. 2; S38453, note: an expressed sequence tag that is not full length), *Pm* (*Psuedotsuga menziesii*; Z49711), and *Lm* (*Lemna minor*; JQ2283). The * indicates identity with LE25. Similarity is greatest in the N-terminal domain of this gene family. Alignment was done using the PILEUP program in the Genetic Computer Group Package (Devereux et al., 1984). The BLAST server at NIH was used to identify and retrieve sequences

2.1.2 Expression of *le25*

To begin to determine potential functions and thoroughly study the mechanism of regulation, the promoter from *le25* was fused with GUS and the pattern of GUS expression was studied (Imai et al., 1995). It was found that the *le25* promoter is active in all tissues in response to drought stress and ABA treatment. The level of GUS activity was controlled by the tissue of expression. These results indicate that LE25 can function in all cells.

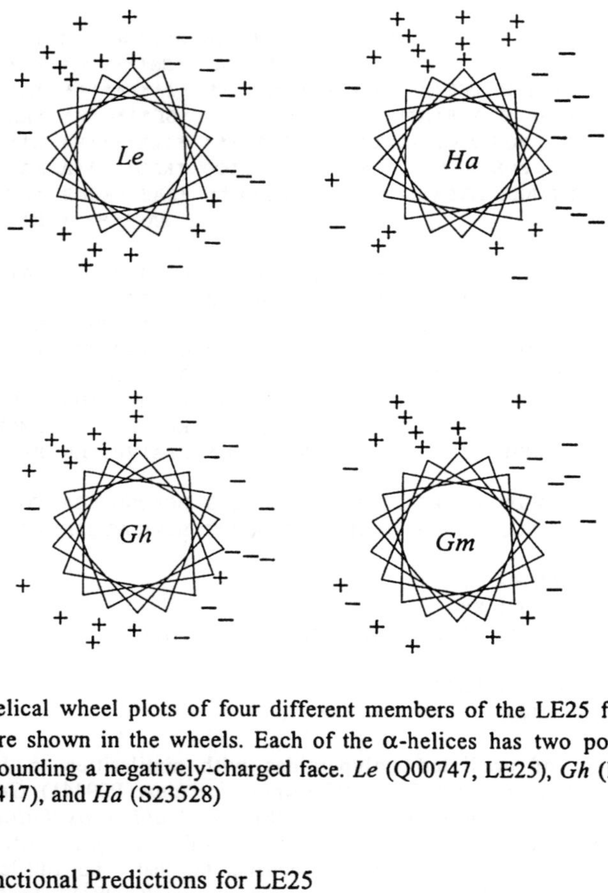

Fig. 2. Helical wheel plots of four different members of the LE25 family. Only the charges are shown in the wheels. Each of the α-helices has two positively-charged faces surrounding a negatively-charged face. *Le* (Q00747, LE25), *Gh* (P09441, D-113), *Gm* (Q01417), and *Ha* (S23528)

2.1.3 Functional Predictions for LE25

The extended amphipathic α-helical regions in this group of LEA proteins can be used to support an argument that these proteins are involved in the binding of ions that accumulate during drought stress and seed desiccation. During cotton seed embryogenesis, there are as many as 1.3 x 10^9 molecules of D-113 per cell, with a concentration as high as 283 μM (Roberts et al., 1993). It has been predicted that these proteins are present as water binding proteins. We make an additional prediction, based on the conservation of the helical domain but not the random coil region, that this family of proteins plays a role in the control of solute concentration in the cytoplasm.

2.2 Prevention of Secondary Stresses

A new category of stress-induced genes has recently been recognized. These genes are not predicted to play an important role in adaptation to the primary stress, water deficit, but to protect the plant from subsequent stress from pathogens. Osmotin has been the most carefully studied in this class of genes and has been shown to protect plants against pathogen attack (Kononowicz et al., 1993). Other possible examples of this type of gene include lipid transfer proteins, which some members of this class of genes have been shown to inhibit pathogenic activity (Terras et al., 1992; Molina and Garcia-Olmedo, 1993), and protease inhibitors such as BND*22* (Lee Downing et al., 1992).

2.2.1 Lipid Transfer Proteins

A family of proteins have been identified which have been named non-specific lipid transfer proteins. This name has arisen because some members of the family have been shown to transfer phospholipids in an *in vitro* assay (Kader et al., 1984). However, the gene and expression characteristics are not consistent with a general role in intracellular lipid transport. The proteins are not cytoplasmic proteins; the deduced proteins all contain a signal sequence indicating that the proteins are synthesized as precursor proteins that are inserted into the endomembrane system. Experimental evidence indicates that the proteins are secreted (Sterk et al., 1991) and are located in the cell wall (Thoma et al., 1993). The proteins have been found in the epidermis (Sossountzov et al., 1991; Sterk et al., 1991; Thoma et al., 1993) or in the aleurone layer at the interface between the endosperm and the maternal tissues of the seed coat (Kalla et al., 1994). At least three of the genes are expressed during stress: two genes from tomato, *Tsw12* (Torres-Schumann et al., 1992) and *le16*, and one from barley, *blt4* (Hughes et al., 1992). Recently, similar proteins have been shown to have antipathogenic activity (Terras et al., 1992; Molina et al., 1993). In addition, in barley the *Ltp* genes are induced by the fungal pathogen *Erysiphe graminis*. We have not yet identified a pathogen that can induce *ltp* (*le16*) expression in tomato. Other possible functions can be proposed for lipid transfer proteins during stress: inhibition of fungal or other pathogen infection, and formation (build up) of the cuticle.

There are now DNA sequences encoding lipid transfer proteins available from 17 different species (Fig. 3). Hallmark features of the predicted gene product include, 8 conserved Cys. A consensus amino acid sequence is shown in Fig. 3. Several of the proteins have been shown to be substrates for CDPK (Neumann et al., 1995). However, the amino acids that have been shown to be phosphorylated are not well conserved among all of the known LTPs.

```
con                        signal sequence        L CG V S      L PCL YL
Bn     ....MAGLMK LACLVLACMI VAGPITSNAA *S**T*-*G- -Y*A***G**
Bo     ....MAGLVK LSCLVLACMI VAGPIATNAA *S**T*-*G- -N*A***G**
At     .......... ..MLALGLHD CGRSNTSNAA *S**S*N*S- -N*A***G**
Hv3    ....MARAQV LLMAAALVLM LTAAPRAAVA *N**Q*D*K- --*K***T**
Ta     .......AQV MLMAVALVLM LAAVPRAAVA *D**H*D*L- --*R***S**
Os     .......... .......... .......... *T**Q*N*A- --*G***T**
Sv     MARSMKLAVA IAVVAAAAAV VLAATTSEAA *T**Q*S*A- --*G***S**
Zm     ..MARTQQLA VVATAVVALV LLAAATSEAA *S**Q*A*A- --*A***S**
Hv     ....MARAAA TQLVLVAMVA AMLLVATDAA *S**Q*S*A- --*S***S**
Hv2    .......... .......... .........A *T**Q*S*A- --*G***A**
Le     ....MEMFGK IACFVVFCMV VV-APHAES- *S**E*T*G- --*A***P**
Nt     ....MEMVGK IACFVVLCMV VV-APHAEA- *S**Q*Q*G- --*A***P**
Le2    ....MEMVSK IACFVLLCMV VV-APHAEA- *T**Q*TAG- --*A***P**
Nt2    ....MEIAGK IACFVVLCMV VA-APCAEA- *T**Q*T*N- --*A***A**
So     ..MASSAVIK LACAVLLCIV VA-APYAEAG *T**M*S*K- --*A***G**
Gh     ...MASMVMN VLCVAVACMV FS-ASYADA- *S**Q*T*G- --*V***G**
Dc     ...MGVLRSS FVAMMVMYMV LATTPNAEAV *T**Q*TGA- --*A***G**
Hv4    ..MARTAATK LALVPLVAAM LLVAADAH-H *RPGELCLG- -P*RR----*
Le3    ....MTSRTL VMLILVITYN SIAVKGSNGH PCSSTFF*AL IQ*I**RAS*


con         GP        CC GVK L     A  T DK    AC CLK  A        LD
Bn     TQN--**LPR G**T**TN*N NM*RTTP**Q Q**R**VGA* NSF-PT**AA
Bo     TQN--**LPR G**T**TN*N NM*RT*P**Q Q**R**VGA* NSF-PT**AA
At     LQG--*VIPP A**S***N*N SI*KT*P**Q Q**N**QGA* RALGSG**AG
Hv3    QG-GP**-SG E**N***D*H NQ*QSSG**Q TV*N***GI* RGI-HN**LN
Ta     QG-GP**-SG Q**D***N*H NQ*RSQS**Q S**N***GI* RGI-HN**ED
Os     RG-GA**-SA A**S***S*K AA*ST*A**R T**N***NA* RGI-KG**AG
Sv     RGQGS**-SA G**S***S*N SA*RT*A**R A**N***NA* RGI-RG**VG
Zm     RGQGS**-SA G**S***S*N NA*RT*A**R A**N***NA* AGV-SG**AG
Hv     RGNGAK*-PV A**S***R*A GA*QS*A**Q A**R***SL* TSI-KG**MG
Hv2    KGSGTS*-SA G**S***R*A GL*RS*A**Q AT*R***SV* G----AY*AG
Le     EGR--**L-G G**G***G*L GA*KTPE**K T**T***SA* NSI-KG**TG
Nt     QGR--**L-G S**G***G*L GA*KSLS**K T**T***SA* NAI-KG**MG
Le2    QGR--**L-G G**G***N*L GS*KT*A**K T**T***SA* NAI-KG**LN
Nt2    RNT--**L-G R**G***A*V NS*RT*E**Q I**T***SA* GAI-SG**LG
So     KGG--PLG-G G**G***A*N AA*AT*P**K T**N***SA* NAI-KG**YG
Gh     AAG--**VPP A**N***G*N NA*KT*P**Q T**G***GIL AAN-TR**LN
Dc     RSQVNV*VPL T**NV**G*N NA*RT*L**R T**G***QT* NAV-TG**LN
Hv4    RKRQRH*PSA G**S***R*A GL*RS*A**Q AT*R***SVP ARTTPA*PQA
Le3    VPFSSV*PSE A**ASI*A*G Q--------- -P*L*VLING PPI-SGV*RN
```

Fig. 3 - cont.

con	AA LP C	GVNIPY	IS C	V
Bn	R**G**KA*-	----******	K**K*TN*NS	*R
Bo	R**G**KA*-	----******	K**K*TN*NS	*R
At	R**G**KA*-	----******	K**T*TN*KT	*R
Hv3	N**S**SK*-	----N*****	T**PDID*SR	*Y
Ta	N*RS**PK*-	----***L**	T**L*ID*SR	*-
Os	N**S**SK*-	----******	T**A*ID*SR	*S
Sv	K**S**SK*-	----******	T**T*TD*SR	*S
Zm	N**S**SK*-	----******	T**T*TD*SR	*N
Hv	KVSG**GK*-	----******	P**M*TD*NK	*H
Hv2	R**G**SR*-	----******	T**A*VD*SK	*H
Le	K**G**GV*-	----******	K**P*TD*ST	*Q
Nt	K**G**GA*-	----******	K**P*TD*SK	*Q
Le2	K**G**SV*-	----K*****	K**P*TD*ST	*Q
Nt2	K**G**ST*-	----******	K**P*TD*SK	*Q
So	K**G**GM*-	----**H***	A**P*TN*NA	*H
Gh	N*NS**GK*-	----****G*	K**PNID*SK	*H
Dc	A**G**AR*-	----****P*	K**PTTD*NR	*V
Hv4	SPPGAASASP	TRSAP*STAL	RS*DRTRAPH	*SSDRRLVG
Le3	M*VQ**EK*T	ANFEQCEFGK		

Fig. 3. Alignment of lipid transfer proteins from 14 different species. *Bn* (*Brassica napus*, U22174), *Bo* (*Brassica oleracea*, wax9b, L33905), *At* (*Arabidopsis thaliana*, LTP1, S73825), *Hv3* (*Hordeum vulgare*, P07597), *Ta* (*Triticum aestivum*, P24296), *Os* (*Oryza sativa*, P23096), *Sv* (*Sorghum bicolor*, S33461), *Zm* (*Zea mays*, PLTP, PS00597), *Hv* (*Hordeum vulgare*, CW-19, S49198), *Hv2* (*Hordeum vulgare*, CW18, S28871), *Le* (*Lycopersicon esculentum*, LE16), *Nt* (*Nicotiana tabacum*, Q03461), *Le2* (*Lycopersicon esculentum*, TSW12, P27056), *Nt2* (*Nicotiana tabacum*, S22168), *So* (*Spinacea oleracea*, P10976), *Gh* (*Gerbera hybrida*, S44100), *Dc* (*Daucus carota*, EP2, P27631), *Hv4* (*Hordeum vulgare*, P25307), and *Le3* (*Lycopersicon esculentum*, 5B, S39552). Similar or identical amino acids are indicated with a *. Amino acid sequences were aligned using the PILEUP Program in the Genetics Computer Group Package (Devereux et al., 1984). The BLAST server at NIH was used to identify similar protein sequences

Expression characteristics and the *in vitro* function of proteins with a similar sequence are the only pieces of information that we can use presently to predict a role for LTPs during stress. Since the stress-induced LTP is expressed in trichomes (unpublished results), we favor a role in prevention of secondary stresses from plant pathogens. To test this hypothesis, plants over- and under-expressing *le16*, and the *in vitro* antipathogenic activity are being tested.

3 Conclusions

In the past 10 years, there has been significant progress in the identification of stress-induced genes. In the next 10 years, we look forward to a significant improvement in our understanding of the role of plant hormones in stress adaptation and understanding the role of stress-induced gene products. New methods are needed in order to test the function of ABA-induced and other stress-induced genes. Since the identification of a large number of genes can now be easily achieved, and a potential gene function can be determined through similarity to other genes, it is essential that methods be devised to conclusively determine the roles of stress-induced gene products.

References

Bray EA (1994) Alterations in gene expression in response to water deficit. *In* AS Basra, ed, Stress-Induced Gene Expression in Plants. Harwood Academic Publishers, pp 1-23

Close TJ, Kortt AA, Chandler PM (1989) A cDNA-based comparison of dehydration-induced proteins (dehydrins) in barley and corn. Plant Mol Biol **13**: 95-108

Cohen A, Bray EA (1990) Characterization of three mRNAs that accumulate in wilted tomato leaves in response to elevated levels of endogenous abscisic acid. Planta **182**: 27-33

Devereux J, Haeberli P, Smithies O (1984) A comprehensive set of sequence analysis programs for the VAX. Nucleic Acid Res **12**: 387-395

Dure III L, Greenway SC, Galau GA (1981) Developmental biochemistry of cotton seed embryogenesis and germination. XIV. Changing mRNA populations as shown by *in vitro* and *in vivo* protein synthesis. Biochemistry **20**: 4162-4168

Dure III L (1993) Structural motifs in Lea proteins. *In* TJ Close, EA Bray, eds, Plant Responses to Cellular Dehydration During Environmental Stress. Current Topics in Plant Physiology 10, pp 91-103

Gosti F, Bertauche N, Vartanian N, Giraudat J (1995) Abscisic acid-dependent and -independent regulation of gene expression by progressive drought in *Arabidopsis thaliana*. Mol Gen Genet **246**: 10-18

Grillo S, Leone A, Xu Y, Tucci M, Francione R, Hasegawa PM, Monti L, Bressan RA (1995) Control of osmotin gene expression by ABA and osmotic stress in vegetative tissues of wild-type and ABA-deficient mutants of tomato. Physiol Plant **93**: 498-504

Hughes MA, Dunn MA, Pearce RS, White AJ, Zhang L (1992) An abscisic acid-responsive, low temperature barley gene has homology with a maize phospholipid transfer protein. Plant, Cell and Environ **15**: 861-865

Imai R, Moses MS , Bray EA (1995) Expression of an ABA-induced gene of tomato in transgenic tobacco during periods of water deficit. J Exp Bot **46**: 1077-1084

Kader J-C, Julienne M, Vergnolle C (1984) Purification and characterization of a spinach-leaf protein capable of transferring phospholipids to mitochondria or chloroplasts. Eur J Biochem **139**: 411-416

Kalla R, Shimamoto K, Potter R, Stein Nielsen P, Linnestad C, Olsen O-A (1994) The promoter of the barley aleurone-specific gene encoding a putative 7 kDa lipid transfer protein confers aleurone cell-specific expression in transgenic rice. Plant J **6**: 849-860

Kononowicz AK, Raghothama KG, Casas AM, Reuveni M, Watad A-EA, Liu D, Bressan R, Hasegawa PM (1993) Osmotin: regulation of gene expression and function. *In* TJ Close, EA Bray, eds, Plant Responses to Cellular Dehydration During Environmental Stress. Current Topics in Plant Physiology 10, pp 144-158

Lång V, Palva ET (1992) The expression of a *rab*-related gene, *rab18*, is induced by abscisic acid during the cold acclimation process of *Arabidopsis thaliana* (L.) Heynh. Plant Mol Biol **20**: 951-962

Lee Downing W, Mauxion F, Fauvarque M-O, Reviron M-P, de Vienne D, Vartanian N, Giraudat J (1992) A *Brassica napus* transcript encoding a protein related to the Kunitz protease inhibitor family accumulates upon water stress in leaves, not in seeds. Plant J **2**: 685-693

Molina A, Garcia-Olmedo F (1993) Developmental and pathogen-induced expression of three barley genes encoding lipid transfer proteins. Plant J **4**: 983-991

Neumann GM, Condron R, Thomas I, Polya GM (1995) Purification, characterization and sequencing of a family of *Petunia* petal lipid transfer proteins phosphorylated by plant calcium-dependent protein kinase. Plant Sci **107**: 129-145

Pla M, Goday A, Vilardell J, Pagès M (1989) Differential regulation of ABA-induced 23-25 kDa proteins in embryo and vegetative tissues of the *viviparous* mutants of maize. Plant Mol Biol **13**: 385-394

Roberts JK, DeSimone NA, Lingle WL, Dure III L (1993) Cellular concentration and uniformity of cell-type accumulation of two Lea proteins in cotton embryos. Plant Cell **5**: 769-790

Sossountzov L, Ruiz-Avila L, Vignolis F, Jolliot A, Arondel V, Tchang F, Grosbois M, Guerbette F (1991) Spatial and temporal expression of a maize lipid transfer protein gene. Plant Cell **3**: 923-933

Sterk P, Booij H, Schellekens GA, VanKammen A, De Vries SC (1991) Cell-specific expression of the carrot EP2 lipid transfer protein gene. Plant Cell **3**: 907-921

Terras FR, Goderis IJ, Van Leuven F, Vanderleyden J, Cammue BPA, Broekaert WF (1992) *In vitro* antifungal activity of a radish (*Raphanus sativus* L.) seed protein homologous to nonspecific lipid transfer proteins. Plant Physiol **100**: 1055-1058

Thoma S, Kaneko Y, Somerville C (1993) A non-specific lipid transfer protein from *Arabidopsis* is a cell wall protein. Plant J **3**: 427-436

QTL Analysis of Stress Responses as a Method to Study the Importance of Stress-Induced Genes

Steve Quarrie[1], Alexander Heyl[1], Andrew Steed[1], Claude Lebreton[1] and Vesna Lazić-Jančić[2]

[1] John Innes Centre, Norwich Research Park, Colney, Norwich NR4 7UH, UK
[2] Maize Research Institute, Zemun Polje, 11080 Belgrade-Zemun, Serbia

Abstract. Many genes are known to be activated in response to abiotic stresses. However, it is still not clear whether stress-activated gene products have a significant effect on improving the adaptation of crop plants to abiotic stresses. While transformation studies with stress-induced genes have shown some significant effects on specific physiological processes, there is no evidence yet that the agronomic performance of crop plants can be improved using this approach. Molecular marker technologies, using restriction fragment length polymorphisms (RFLPs) and markers derived with polymerase chain reactions (PCR) allow high density genetic maps to be prepared. These can be used to identify loci regulating expression of quantitative traits (QTL). The candidate gene approach was used to study the possible roles of several stress-induced gene products. Coincidence of a QTL for a trait and a locus for a stress-induced gene could indicate a functional relationship between the two. We mapped cDNAs for the maize *rab*17 and *rab*28 and rice *rab*16 genes in an F_2 maize population and identified QTL for a wide range of responses to drought stress. However, no coincidence was found between the stress-induced cDNAs and QTL for stress responses or yield under drought. Using the bulked segregants C_0 and C_8 from the Tuxpeño Sequia population which differ markedly in yield under severe drought, several RFLP markers revealed changes in allele frequencies as a result of recurrent selection for yield under drought. Some of the RFLP probes showing changes in allele frequencies had homologies with genes likely to be involved in plant stress responses.

Keywords. Bulked segregant analysis, drought-induced genes, maize, mapping, QTL analysis, *Zea mays*

1 Introduction

Many genes are known to be activated in response to abiotic stresses (e.g. Dubcovsky et al., 1995; Gosti et al., 1995; Mäntylä et al., 1995; van Zee et al., 1995, and see Close et al., 1993; Nagao et al., 1990; Skriver and Mundy, 1990 for, reviews). However, it is still not clear whether stress-activated gene products have a significant effect on improving the adaptation of crop plants to abiotic stresses. While transformation studies with stress-induced genes have shown some significant effects on specific physiological processes, there is no evidence yet that the agronomic performance of crop plants can be improved using this approach. Molecular marker technologies, using restriction fragment length polymorphisms (RFLPs) and markers derived with polymerase chain reactions (PCR) allow high density genetic maps to be prepared. These can be used to identify loci regulating expression of quantitative traits (QTL). With cDNAs and sequence data information now available for many stress-induced genes, the candidate gene approach (as illustrated by Causse et al., 1995) can be used to study the possible roles of these stress-induced gene products. Coincidence of a QTL for a trait and a locus for a stress-induced gene could indicate a functional relationship between the two.

An alternative approach using molecular markers can be carried out using population bulks selected for differences in trait expression. The method is explained in detail by Tanksley et al. (1995). Differences in allele frequencies between the two bulks, identified by Southern blotting and hybridization, should indicate those genomic regions that have a role in determining the consequences of the selection pressure. By using cDNAs to stress-specific genes as hybridization probes to test populations selected to differ in stress responses, the likelihood of those genes being important in determining stress responses can be assessed.

We have used these two methods to test the likely importance of several drought-induced genes in determining the physiological and yield responses of maize plants to drought stress.

2 QTL Analysis of Candidate Genes

Two maize inbred lines differing markedly in drought resistance (yield under drought) and in expression of many other drought-related traits (Polj17, drought resistant, and F-2, drought susceptible) were crossed and a population of 81 F_2 plants mapped with RFLP markers (Lebreton et al., 1995). These plants were also scored for a large number of drought-related traits in a soil glasshouse experiment. After three weeks without water, leaf water status, stomatal conductance, leaf and xylem abscisic acid (ABA) concentrations, and leaf fluorescence characteristics were measured. At maturity, root pulling force (a measure of the number and distribution of roots) and nodal root number were measured.

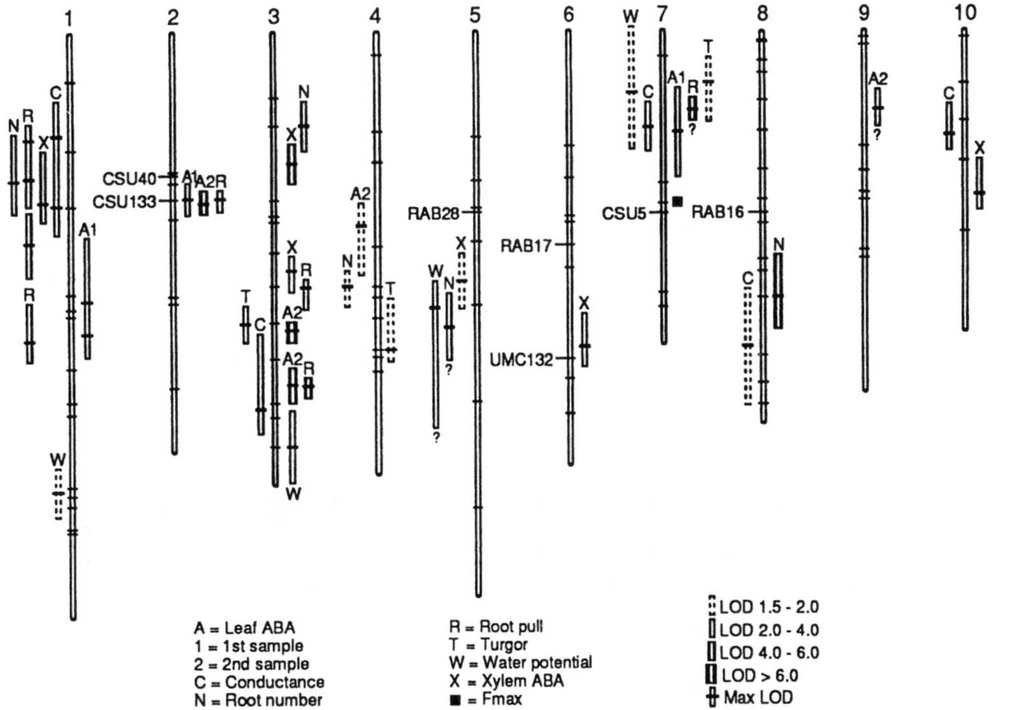

Fig. 1. Maize genetic map showing the locations of QTL for stomatal conductance, ABA concentrations, leaf water relations and fluorescence maximum, root pulling force and nodal root number. Bars and symbols to the left or right of chromosomes indicate the allele having the greater mean value: F-2 on the left and Polj17 on the right. The ends of the bars indicate the positions where location likelihoods for the QTL are only 20% those at the maximum likelihood position (indicated by the thick bar). Key markers referred to in the text are indicated

The F_2 population was mapped with 103 RFLP loci, using mainly CSU, UMC, BNL and NPI probes. These probes included several cDNAs either that are known to be genes induced by drought (*rab17* and *rab28* from maize, and *rab*16 from rice) or that have homology with genes involved in drought responses, namely *csu5* (a thiol-protease, Keith et al., 1993), *csu40* (a glutaredoxin, identified in the MaizeDB), and *csu133* (a pathogenesis-related protein similar in structure to osmotin, identified in the MaizeDB). A genetic map of the ten chromosomes was constructed using MAPMAKER (Lander at al.,1987) modified for the Macintosh, V 2.0 NDP (JL Proctor, A Rafalski, R Hubner, S Tingey, Du Pont and WIBR).

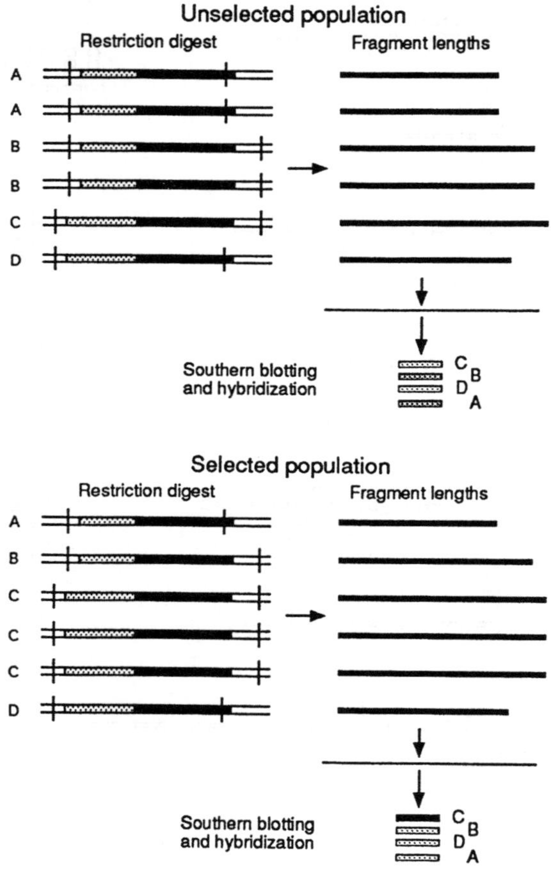

Fig. 2. A schematic representation of allele frequencies for a gene responding to selection pressure in the C_0 and C_8 populations revealed by Southern blotting and hybridization with its cDNA. Four alleles (A-D) are indicated, with modifications either to the coding sequence (solid bars) or the upstream regulatory sequences (shaded bars)

The location of QTL for each of the traits was assessed using MAPMAKER/ QTL, an interval mapping software package (Lander and Botstein, 1989). The locations of QTL for the traits were dispersed amongst the ten chromosomes (Fig. 1), though chromosomes 1, 3 and 7 in particular contained QTL for a wide range of drought-related traits. None of the *rab* genes (chromosomes 5, 6 and 8) was located at any significant QTL for drought-related traits. Although in subsequent trials of F_3 families from this maize cross (Quarrie et al., unpublished data), *rab16* was significantly associated with QTL for yield, the QTL effects were present across all water availability regimes. *Rab16* is therefore unlikely to be associated with yield maintenance specifically under droughted conditions. Of the possible stress-related *csu* markers, only *csu133* (synonym *umc371*) on chromosome 2 was associated with any stress-related QTL (leaf ABA content and root pulling force).

3 Bulked Segregant Analysis with C_0 and C_8 Populations

The Tuxpeño Sequia maize population (C_0) from CIMMYT) has been selected through eight generations of recurrent full-sib mating (C_8) on the basis of grain yield under severe drought conditions (Fischer et al., 1989). A comparison of the two populations under a range of soil water availabilities showed that the C_8 population consistently outyielded the C_0 population by about 0.8 Mg ha^{-1}, equivalent to a yield increase of about 50% in severely droughted environments (Bolaños and Edmeades, 1993). Changes in the allele frequencies between the two populations were examined using RFLP probes. A schematic representation of the distribution of alleles between the C_0 and C_8 populations and use of the Bulked Segregant Analysis (BSA) method are shown in Fig. 2. Alleles of genes conferring a selective advantage in maintaining yield under droughted conditions will tend to increase in frequency during each of the cycles of selection.

Seedlings of the two populations were grown in a controlled environment and extracted for DNA (equal weights of 100 seedlings pooled from the C_0 and C_8 populations). The DNA was cut with restriction enzymes, separated according to size and Southern blotted onto membranes using standard procedures. DNA of Polj17 and F-2 was also blotted onto the membranes to help in identifying the chromosomal location of multilocus probes. Membranes were probed with 115 RFLP markers from the CSU, UMC, BNL and NPI collections, and cDNAs to stress-related genes from various sources, including *rab16* and *rab17*. The chromosomal location of alleles of those markers that had been mapped in the Polj17 x F-2 F_2 population was deduced by comparison with the tracks of Polj17 and F-2 on the C_0, C_8 filters. The chromosomal location of other markers was inferred by reference to other published maps with those markers.

The majority of the probes showed no visible differences in the allele frequencies between C_0 and C_8 with any of the restriction enzymes used (Fig. 3).

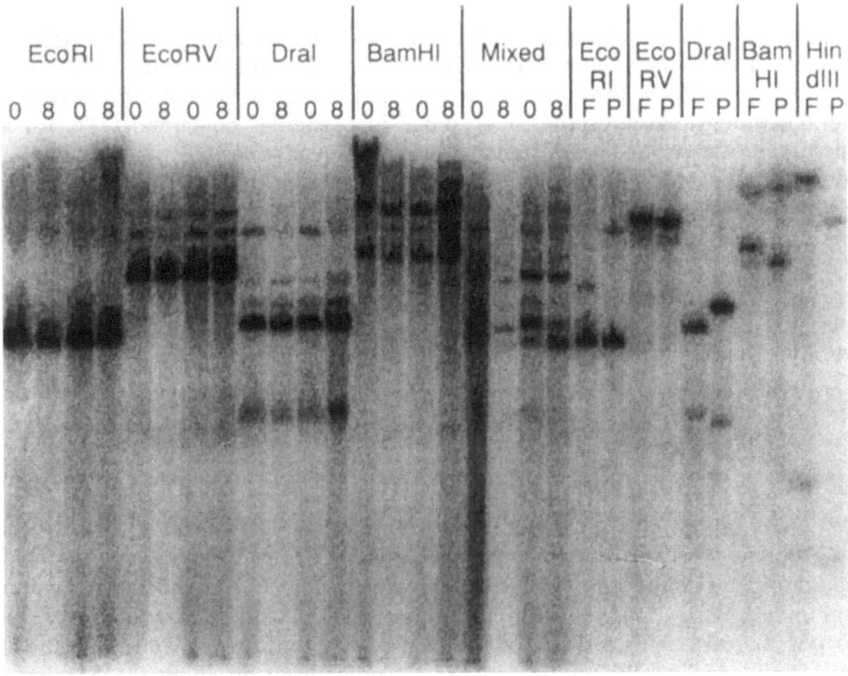

Fig. 3. Hybridization of the probe *umc58* to a filter containing alternating tracks of DNA of C_0 and C_8, Polj17 and F-2 restricted with the enzymes EcoRI, EcoRV, DraI and BamHI. Polj17 and F-2 were also restricted with HindIII

However, 36 probes gave visible differences between the hybridization patterns for the two populations. These differences were subsequently quantified by scanning duplicate tracks of the C_0 and C_8 restricted DNA and measuring changes in peak areas between the tracks. Both quantitative and qualitative changes were found for particular alleles hybridizing with the 36 probes, as shown in Fig. 4. Probes showing consistent changes in allele frequency of more than 10% mapped to each of the ten chromosomes (Fig. 5), with chromosomes 2, 3, 7 and 8 showing a high frequency of significant allelic differences between the C_0 and C_8 populations. Each of the probably stress-related *csu* probes (5, 40 and 133) showed a significant change in allele frequencies in response to selecting for improved yield under drought. However, neither *rab16* nor *rab17* showed a major change (although statistical analysis of the change in allele frequencies was significant for *rab17*, the change amounted to only 9% of the total peak area recorded).

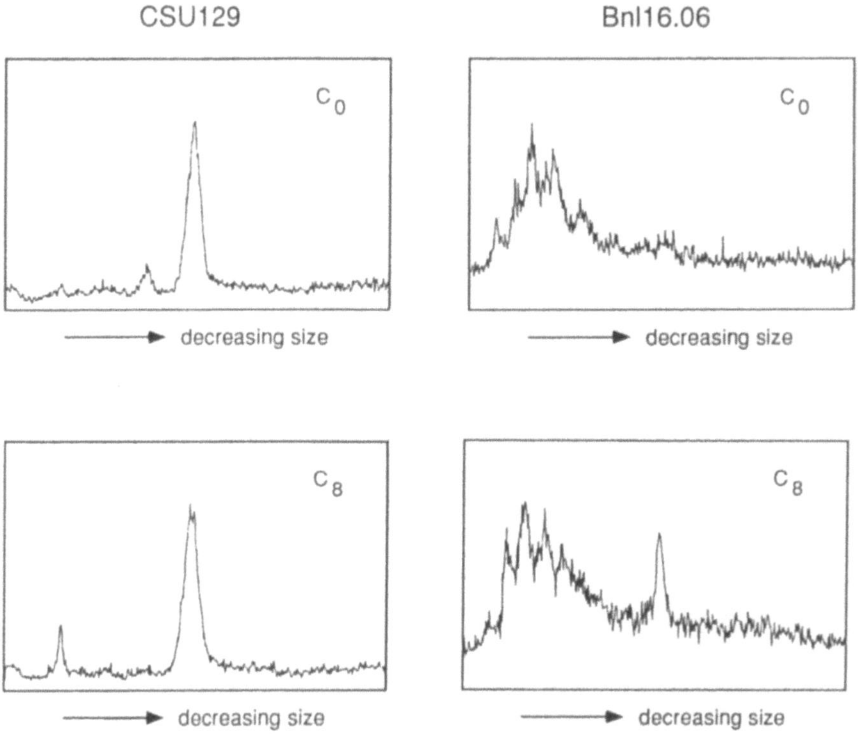

Fig. 4. Examples of line tracings along tracks of C_0 and C_8 restricted DNA showing qualitative (*csu129*) and quantitative (*bnl16.06*) changes in alleles between the two populations

4 Discussion

Both the QTL analysis with a single cross and BSA methods can provide evidence of lack of importance of a particular stress-induced gene in determining stress adaptation. However, because several alleles are usually present at each locus in composite populations, BSA with composite populations is likely to provide stronger evidence than QTL analysis for a particular gene being important in regulating phenotypic adaptation to a stress. The likelihood that a gene showing allelic variation between population bulks is responsible itself for the phenotypic differences will depend on the expected gene function, the number of other markers mapping in the same region of the genome showing similar allelic variation and the degree of linkage disequilibrium achieved during the cycles of recurrent selection to make the bulks.

The BSA technique can be taken further, by carrying out a QTL analysis using individual plants comprising each of the bulks (e.g. Tadić, 1993). Analysis of variance for the traits using each marker allele as a factor will

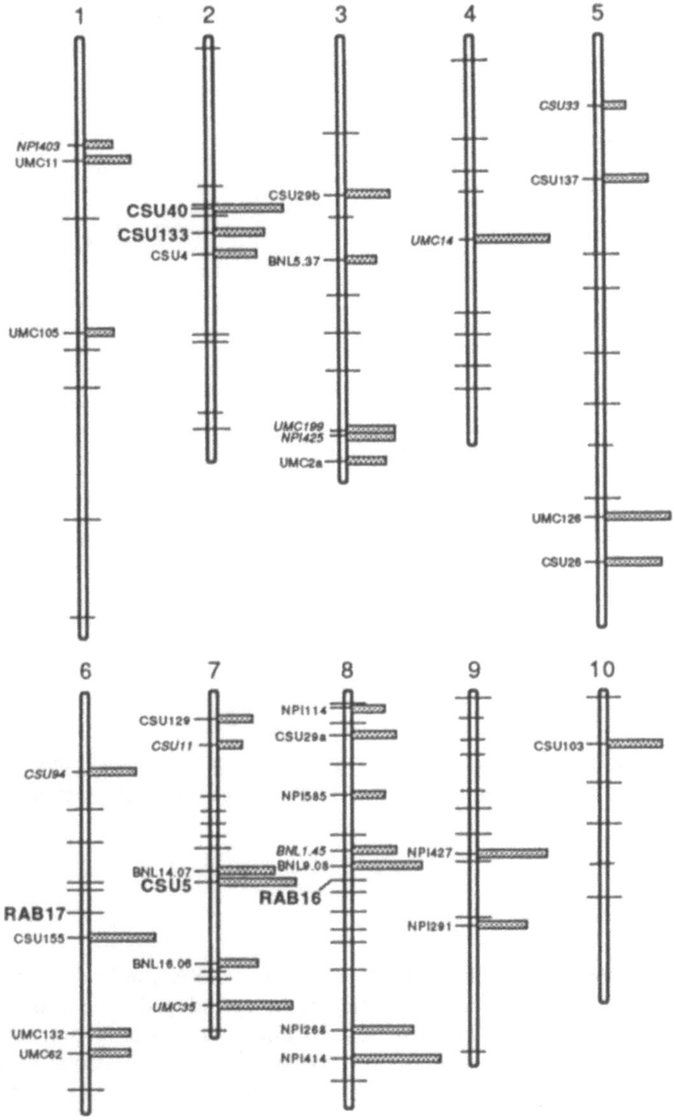

Fig. 5. Maize genetic map showing regions of maize chromosomes with changes in allele frequencies between C_0 and C_8 populations. Bars indicate the likely location of all markers tested. Shaded boxes show significant ($P<0.05$) changes of more than 10% between C_0 and C_8 populations and the length of the box is proportional to the size of the effect. Italicised markers were not mapped in the Polj17 x F-2 population and are located according to other published maize maps

identify those alleles of genes that have beneficial effects on expression of the trait. In this way, the contribution of each allele to determining the phenotype can be assessed.

No evidence was found for any of the *rab* genes having a significant role in the adaptation of maize plants to drought stress, despite the marked increase in transcription of these genes following drought stress (Mundy and Chua, 1988; Pla et al., 1991; Pagès et al., 1993). However, gene induction has usually been tested in seedlings under controlled environment drought conditions.

In contrast, alleles of each of the stress-related genes *csu5*, *csu40* and *csu133* appeared to change in frequency in the Tuxpeño Sequia population by selecting for better drought resistance. The thiol-protease encoded by *csu5* is a member of a class of polypeptides induced by both heat and cold (Schaffer and Fischer 1990) and drought (Guerrero at al.,1990, Williams at al.,1994). These may have a role in degrading proteins denatured by cellular stress. *Csu40* encodes a glutaredoxin, which is functionally similar to thioredoxin (Holmgren, 1989), small proteins that have an active site with a redox-active disulphide group functional in electron transfer. It is possible that these proteins may be important in radical scavenging during drought-induced photo-oxidative stress. The pathogenesis-related protein encoded by *csu133* is structurally similar to osmotin (MaizeDB). The osmotins are antifungal agents and known to be induced also by abscisic acid treatment and abiotic stresses (Bol et al., 1990, Zhu et al., 1995). However, the function of osmotins in maize under drought stress is not yet known. It is also not known whether any of the genes encoding *csu5*, *csu40* or *csu133* is induced by drought stress in maize.

Similar work on the C_0 and C_8 populations from Tuxpeño Sequia is in progress in CIMMYT (Ribau et al., 1995), measuring allele frequencies using individual plants from the two populations. From their results so far, using single-copy probes whose chromosomal locations are unambiguous, it is clear that the BSA and individual plant methods are identifying similar regions of the maize genome. In particular, alleles of *csu133* (*umc371*) and *umc132* (Figs. 1 and 5) were identified by both methods to have been significantly affected by the selection process. In this population, most of the improvement in drought resistance was achieved by reducing the anthesis-silking interval (ASI) in comparison with the unselected population. QTL for ASI in a population derived from Tuxpeño Sequia have been located on chromosomes 1, 2, 5, 6, 8 and 10, with major QTL on chromosomes 1, 2 and 6 (Ribau et al., 1995). Two of these QTL for ASI were in regions of the chromosomes shown by us to differ between the C_0 and C_8 populations (near *csu4*, chromosome 2 and *csu155-umc132*, chromosome 6).

The candidate gene approach has been used successfully by Causse et al. (1995) to dissect the factors responsible for determining genetic variation in growth rates in maize seedlings. Structural genes for key enzymes of carbohydrate metabolism were located within the support intervals for QTL for concentrations of the enzyme product and, in some cases, to plant height and dry matter content.

With the rapid increase in sequence information for the *Arabidopsis* and rice genomes (over 18000 sequences now available in the *Arabidopsis* database and at least 25000 cDNA clones partially sequenced for rice, Delseny et al., 1995;

Sasaki, 1995) it will become increasingly possible to use the candidate gene approach in a wide range of monocot and dicot species to test the likely importance of genes of interest in determining expression of particular traits.

The BSA approach with composite populations is likely to be more powerful than QTL mapping in determining the physiological function of a particular gene. However, this technique relies on out-pollination amongst a large gene pool to ensure that sufficient allelic variation is present to allow effective selection of favourable alleles during the cycles of recurrent selection. Nevertheless, largely autogamous species, such as wheat and soybean, can also be used for BSA with composite populations by introducing male sterile lines into the base population. In this way, phenotypic adaptations of mixed populations to a range of environments can occur, leading to differences in height, flowering time and disease resistances in comparison with the initial populations (see Goldringer and Brabant 1993 for recent review).

These QTL analysis and BSA techniques are useful for testing hypotheses about stress-induced gene functions, and complement methods using transgenic organisms, as proposed by Bohnert et al. (1995).

Acknowledgements. We are grateful to the Royal Society for awarding a Fellowship to V Lazić-Jančić to carry out some of this work.

References

Bohnert HJ, Nelson DE, Jensen RG (1995) Adaptations to environmental stresses. Plant Cell **7**: 1099-1111

Bol JF, Linhorst HJM, Cornelissen BJC (1990) Plant pathogenesis-related proteins induced by virus infection. Annu Rev Phytopathol **28**: 113-138

Bolaños J, Edmeades GO (1993) Eight cycles of selection for drought tolerance in lowland tropical maize. I. Responses in grain yield, biomass, and radiation utilization. Field Crops Res **31**: 233-251

Causse M, Rocher JP, Henry AM, Charcosset A, Prioul JL, De Vienne D (1995) Genetic detection of the relationship between carbon metabolism and early growth in maize, with emphasis on key-enzyme loci. Mol Breeding **1**: 259-272

Close TJ, Fenton RD, Yang A, Asghar R, DeMason DA, Crone DE, Meyer NC, Moonan F (1993) Dehydrin: the protein. *In* Plant Responses to Cellular Dehydration during Environmental Stress. 16th Annual Symposium in Plant Physiology, University of California, Riverside, USA, pp 104-118

Delseny M, Raynal M, Laudie M, Varoquaux F, Comella P, Wu HJ, Cooke R, Grellet F (1995) Sequencing and mapping the *Arabidopsis* genome: a weed model for real crops. SEB Symposium, Unifying Plant Genomes: Comparisons, Conservation and Collinearity, 29-31 August 1995, Abs p 23

Dubcovsky J, Luo M-C, Dvorak J (1995) Linkage relationships among 25 clones from stress-induced genes in wheat. Theor Appl Genet **91**: 795-801

Fischer KS, Edmeades GO, Johnson EC (1989) Selection for the improvement of maize yield under moisture deficits. Field Crops Res **22**: 227-243

Goldringer I, Brabant P (1993) Recurrent selection for breeding self-pollinated species as pure line varieties. 2. A review. Agronomie **13**: 561-577

Gosti F, Bertauche N, Vartanian N, Giraudat J (1995) Abscisic acid-dependent and -independent regulation of gene expression by progressive drought in *Arabidopsis thaliana*. Mol Gen Genet **246**: 10-18

Guerrero FD, Jones JT, Mullet JE (1990) Turgor-responsive gene transcription and RNA levels increase rapidly when pea shoots are wilted. Sequence and expression of three inducible genes. Plant Mol Biol **15**: 11-26

Holmgren A (1989) Thioredoxin and glutaredoxin systems. J Biol Chem **264**: 13963-13966

Keith CS, Hoang DO, Barrett BM, Feigelman B, Nelson MC, Thai H, Baysdorfer C (1993) Partial sequence analysis of 130 randomly selected maize cDNA clones. Plant Physiol **101**: 329-332

Lander ES, Botstein D (1989) Mapping Mendelian factors underlying quantitative traits using RFLP linkage maps. Genetics **121**: 185-199

Lander ES, Green P, Abrahamson J, Barlow A, Daly MJ, Lincoln SE, Newburg L (1987) Mapmaker: An interactive computer package for constructing primary genetic linkage maps of experimental and natural populations. Genomics **1**: 174-181

Lebreton C, Lazić-Jančić V, Steed A, Pekić S, Quarrie SA (1995) Identification of QTL for drought responses in maize and their use in testing causal relationships between traits. J Exp Bot **46**: 853-865

Mäntylä E, Lang V, Palva ET (1995) Role of abscisic acid in drought-induced freezing tolerance, cold acclimation, and accumulation of LT178 and RAB18 proteins in *Arabidopsis thaliana*. Plant Physiol **107**: 141-148

Mundy J, Chua NH (1988) Abscisic acid and water stress induce the expression of a novel rice gene. EMBO J **7**: 2279-2286

Nagao RI, Kimpel JA, Key JL (1990) Molecular and cellular biology of the heat shock response. Adv in Genet **28**: 235-274

Pages M, Vilardell J, Jensen AB, Mar Alba M, Torrent M, Goday A (1993) Molecular biological responses to drought in maize. *In* MB Jackson, CR Black, eds, Interacting Stresses on Plants in a Changing Climate. NATO ASI Series, Vol I 16, Springer-Verlag, Berlin, Heidelberg, pp 583-591

Pla M, Gomez J, Goday A, Pages M (1991) Regulation of the abscisic acid - responsive gene *rab28* in maize viviparous mutants. Mol Gen Genet **230**: 394-400

Ribau JM, Gonzales de Leon D, Edmeades GO, Huerta E, Hoisington D (1995) Changes in allele frequencies in a tropical maize population under selection for drought stress. *In*: Proceedings of Interdrought 1995, an International Congress on Integrated Studies on Drought Tolerance of Higher Plants, 31 Aug - 2 Sept 1995, INRA, Montpellier, pp XIA 1-7.

Sasaki T (1995) Rice cDNAs as a model for expressed genes of plants. SEB Symposium, Unifying Plant Genomes: Comparisons, Conservation and Collinearity, 29-31 August 1995, Abs p 37

Schaffer MA, Fischer RL (1990) Transcriptional activation by heat and cold of a thiol protease gene in tomato. Plant Physiol **93**: 1486-1491

Skriver K, Mundy J (1990) Gene expression in response to abscisic acid and osmotic stress. Plant Cell **2**: 503-512

Tadi B (1993) Identification of quantitative trait loci in maize using molecular markers. PhD Thesis, University of Belgrade

Tanksley SD, Ganal MW, Martin GB (1995) Chromosome landing: a paradigm for map-based gene cloning in plants with large genomes. Trends Genet **11**: 63-68

van Zee K, Chen FQ, Hayes PM, Close TJ, Chen THH (1995) Cold-specific induction of a dehydrin gene family member in barley. Plant Physiol **108**: 1233-1239

Williams J, Bulman M, Huttly A, Phillips A, Neill S (1994) Characterization of a cDNA from *Arabidopsis thaliana* encoding a potential thiol protease whose expression is induced independently by wilting and abscisic acid. Plant Mol Biol **25**: 259-270

Zhu B, Chen THH, Li PH (1995) Activation of two osmotin-like protein genes by abiotic stimuli and fungal pathogen in transgenic potato plants. Plant Physiol **108**: 929-937

Water Stress-Induced Genes in *Arabidopsis thaliana*

Tsuyoshi Mizoguchi[1], Yoshu Yoshiba[2,1], Kazuko Yamaguchi-Shinozaki[3,1] and Kazuo Shinozaki[1]

1 Laboratory of Plant Molecular Biology, The Institute of Physical and Chemical Research (RIKEN), Tsukuba 305, Japan
2 Advanced Research Laboratory, Hitachi Ltd., Hatoyama, Saitama
3 Biological Research Division, JIRCAS, Ministry of Agriculture, Forestry and Fishery, Tsukuba 305, Japan

Abstract. Plants respond to conditions of severe environmental changes or stresses, such as dehydration, low temperature, high-salinity with a number of physiological and developmental changes. We have isolated more than thirty genes that respond to dehydration stresses at transcriptional level. In the present study, we will report the expression and function of different *Arabidopsis* genes involved in the kinase cascade and in the proline biosynthesis. MAP kinases play key roles in the signalling pathway activated by growth factors and in stress response in animals. The mRNA levels of three protein kinases, ATMEKK1 (MAP kinase kinase kinase), ATMPK3 (MAP kinase) and ATPK19 (ribosomal S6 kinase), increased markedly and simultaneously in response to touch, cold and salinity stresses. These results suggest that MAP kinase cascades, which are thought to respond to a variety of extracellular signals, are regulated not only at the post-translational level but also at the transcriptional level in plants. It is also suggested that MAP kinase cascades in plants may function in transducing signals under environmental stresses. Proline is thought to be a compatible osmolyte like glycine betaine, which plays a role in counteracting the effects of osmotic stress in plants. We have isolated two cDNAs from *Arabidopsis* (*cAtP5CS* and *cAtP5CR*) encoding P5CS (Δ^1-pyrroline-5-carboxylate synthetase) and P5CR (Δ^1-pyrroline-5-carboxylate reductase) which play key roles in the biosynthesis of proline in plants under water stress. Northern blot analysis revealed that the *AtP5CS* gene was induced by dehydration, high salt and treatment with ABA. Moreover, a simultaneous accumulation of proline was observed as a result of these treatments of *Arabidopsis* plants. By contrast, the *AtP5CR* gene was not induced to a significant extent by dehydration or high-salt stress. These observations suggest that the *AtP5CS* gene plays a principal role in the biosynthesis of proline in *Arabidopsis* under osmotic stress.

Keywords. ABA, *Arabidopsis thaliana*, MAP kinase (MAPK), MAP kinase kinase kinase (MAPKKK), osmotic stress, proline, protein kinase, P5CR, P5CS

1 MAP Kinase Cascades in *Arabidopsis*

1.1 Introduction

Plants are not mobile and adapt themselves to environmental changes. They are thought to develop unique systems to respond to a variety of environmental stimuli on the process of evolution. Therefore one of the purposes of the present study was to understand how these environmental stimuli are perceived and modulated by the signals that control gene expression and enzyme activities in plants. The number of reports on isolation and characterization of genes that are transcriptionally regulated by signals including changes in light, temperature, and osmolarity has been increasing year by year. Many cDNAs encoding putative trans-acting factors have been isolated from plants. However, we still have little knowledge on the molecular process of the regulation of these transcription factors, that may play key roles in the signal transduction pathways under environmental stress conditions.

Recently, the phosphorylation cascades including MAP kinases (MAPKs), MAP kinase kinases (MAPKKs), and MAP kinase kinase kinases (MAPKKKs) have been reported to function in various signal transduction pathways of organisms from yeasts to vertebrates (Nishida and Gotoh, 1993; Davis, 1994). MAP kinases are activated by the phosphorylation of threonine and tyrosine residues by upstream kinases, MAPKKs, that are themselves activated by the phosphorylation of serine and threonine residues by their upstream kinases, MAPKKKs. Several subgroups of MAP kinase have been reported to exist in mammalian and yeast cells (Fig. 1). In mammalian cells, ERK (also known as MAPK), JNK (also known as SAPK) and p38 MAP kinase have been reported. The ERKs are activated by various growth or differentiating factors, and function in the M-phase-specific phosphorylation cascades. The JNK1 and p38 MAP kinases are activated in response to environmental stress in mammalian cells. In *Saccharomyces cerevisiae* at least five different MAP kinase signaling pathways composed of MAPKKK, MAPKK, and MAPK homologs have been described: the mating-pheromone response pathway (Ste11, Ste7, and Fus3/Kss1), the protein kinase C-dependent signal pathway (Bck1, Mkk1/Mkk2, and Mpk1), the osmoregulatory pathway (Pbs2 and Hog1), the pseudohyphal differentiation pathway (Ste11 and Ste7), and the spore wall assembly pathway (Smk1).

Thus, in order to understand signal transduction pathways in plants, it seems to be important to ascertain whether MAP kinase cascades exist and function in the intracellular signal pathways of plants, because these enzymes appear to play key roles in a wide variety of responses to extracellular signals in animals and yeasts (Nishida and Gotoh, 1993; Jonak et al., 1994). We have first demonstrated that ATMPK1 and ATMPK2 are plant MAP kinases (Mizoguchi et al., 1994). It has also been shown that genes for MAP kinases in *Arabidopsis* constitute a family that contains more than nine members (Mizoguchi et al., 1993, 1994). We have also showed the isolation of the putative upstream and downstream factors, ATMEKK1, ATPK6 and ATPK19, which are structurally most related to MAPKKKs and ribosomal S6 kinases, respectively (Mizoguchi et al., 1996). Finally, we have demonstrated that the mRNA levels of three protein kinases, ATMPK3, ATPK19, and ATMEKK1, increased markedly and simultaneously in response to touch, cold, and salinity stresses (Mizoguchi et al., 1996). On the basis of the characterization of the putative components of MAP kinase cascades in plants, the function of

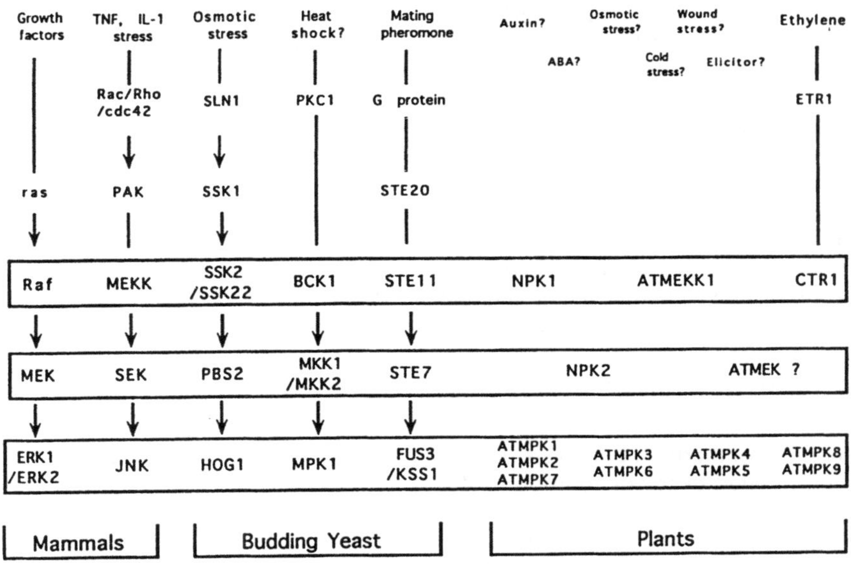

Fig. 1. Structure of MAP kinase signal transduction pathways in mammals, yeasts, and plants. The HOG1 pathway (SSK2/SSK22, PBS2, HOG1) is activated by exposure of yeast to hyperosmolar environments (Davis, 1994). Mammalian MAP kinase, JNK1, is also activated by osmotic stress, and can substitute for HOG1 in *S. cerevisiae* (Davis, 1994). Plant protein kinases that are structurally (and functionally) related to MAPKKKs, as tobacco NPK1 (Banno et al., 1993), *Arabidopsis* ATMEKK1 (Mizoguchi et al., 1996), and *Arabidopsis* CTR1 (Kieber et al., 1993), MAPKKs, as tobacco NPK2 (Shibata et al., 1995) and *Arabidopsis* ATMEK (see text), and MAPKs *Arabidopsis* ATMPK1-9 (Mizoguchi et al., 1993, 1994) are shown. Plant hormones, e.g. auxin (Mizoguchi et al., 1994), ethylene (Kieber et al., 1993) and ABA and stresses such as osmotic stress (Mizoguchi et al., 1996), low temperature (Mizoguchi et al., 1996), wounding (Suzuki and Shinshi, 1995), and elicitor (Suzuki and Shinshi, 1995) are candidates of signals which may activate the MAP kinase cascades in plants

these protein kinases in plants under a variety of environmental stresses, has been discussed

1.2 Characterization of two cDNA Clones that Encode MAP Kinases in *Arabidopsis*

Two cDNA clones, *ATMPK1* and *ATMPK2*, encoding MAP kinases have been cloned from *Arabidopsis* (Mizoguchi et al., 1994). ATMPK1 and ATMPK2 exhibit significant similarity to rat ERK2 (49%) and *Xenopus* MAP kinase (50%). The amino acid residues corresponding to the sites of phosphorylation (Thr-Glu-Tyr) that are involved in the activation of MAP kinases are conserved in ATMPK1 and ATMPK2. Northern blot analysis indicates that the mRNAs that correspond to ATMPK1 and ATMPK2 are significantly present in all the organs except seeds. Purified *Xenopus* MAP kinase kinase (MAPKK) phosphorylates tyrosine as well as threonine/serine residues of ATMPK1 and ATMPK2 proteins that have been expressed in *Escherichia coli*, thus activating these enzymes. Analysis of the activation of MAP kinase using tobacco BY-2 cells has demonstrated that the recombinant ATMPK1 and ATMPK2 proteins become more active as a result of phosphorylation upon incubation with extracts prepared from tobacco BY-2 cells that have been treated with the synthetic auxin, 2,4-dichlorophenoxyacetic acid (2,4-D). These results suggest that auxin may function as an activator of plant MAP kinases, as do various mitogens in animal system (Fig. 1).

1.3 Identification of *Arabidopsis* MAP Kinase Family

Genomic Southern blot analysis has indicated that the *Arabidopsis* genome has several related genes in addition to *ATMPK1* and *ATMPK2* (Mizoguchi et al., 1994). We have screened cDNA libraries under low-stringency conditions with the same probe used to isolate *ATMPK1* and *ATMPK2*, and found that *Arabidopsis* has at least nine different genes for MAP kinases (*ATMPK1-9*) (Mizoguchi et al., 1994; Mizoguchi et al., unpublished). The amino acid residues corresponding to the sites of phosphorylation (Thr-Glu-Tyr) that are involved in the activation of animal MAP kinases are conserved in the seven putative ATMPK proteins (ATMPK1-7). The *Arabidopsis* MAP kinases can be classified into the following four subgroups based on sequence similarities: **subgroup 1**: ATMPK1, ATMPK2 and ATMPK7; **subgroup 2:** ATMPK3 and ATMPK6; **subgroup 3:** ATMPK4 and ATMPK5; **subgroup 4:** ATMPK8 and ATMPK9.

1.4 Identification of the Putative Upstream Factors of MAP Kinases in *Arabidopsis*

We have screened *Arabidopsis* cDNA libraries to isolate cDNAs encoding MAPKK or MAPKKK homologues, which may function as putative upstream factors of ATMPK1-9, by using as probes the PCR fragments or cDNA inserts isolated by the random sequencing project. We have isolated a cDNA (*cATMEKK1*) encoding protein kinase that has a high sequence homology to members of the MAPKKK family (such as *Byr2* from *S. pombe*, *Ste11* from *S. cerevisiae*, *Bck1* from *S. cerevisiae*, *MEKK* from mouse, and *NPK1* from tobacco) (Mizoguchi et al., 1996). Over-expression of *cATMEKK1* complemented a *ste11Δ* mutation, and also weakly restored the function of a *bck1Δ* mutant (Mizoguchi et al., 1996). We have also examined the effect of *cATMEKK1* over-expression in a mutant lacking the *STE7* gene, whose product has been shown to act downstream of *Ste11*. Over-expression of *cATMEKK1* could not suppress the *ste7Δ* mutation. These results indicate that *ATMEKK1* can function as a MAPKKK in *S. cerevisiae*, at least in the Ste11-mediated signaling system. We have isolated an *Arabidopsis* cDNA clone which has a high degree of homology with tobacco NPK1 (Mizoguchi et al., unpublished). Therefore, *Arabidopsis* has at least two types of protein kinases, ATMEKK1-type (Mizoguchi et al., 1996) and NPK1-type (Banno et al., 1993), which are structurally related to MAPKKKs. We have also obtained a PCR-amplified DNA fragment (ATMEK) which has a high degree of homology with MAPKKs using *Arabidopsis* cDNA as a template (Mizoguchi et al., unpublished). Recently, a cDNA clone (*NPK2*) encoding protein kinases related to MAPKKs has been isolated from tobacco (Shibata et al., 1995). Altogether these observations support the probable existence of similar phosphorylation cascades, that include a MAPKKK, a MAPKK and a MAP kinase, in higher plants.

1.5 Identification of the Putative Downstream Factors of MAP Kinases in *Arabidopsis*

We found that ATPK19 and ATPK6 from *Arabidopsis* are closely related to animal $p70^{S6K}$ and $pp90^{rsk}$ (Mizoguchi et al., 1995). Both kinases are activated by serine/threonine phosphorylation, and MAP kinase was shown to phosphorylate and to partially reactivate dephosphorylated $pp90^{rsk}$ (Nishida and Gotoh, 1993; Davis 1994). In animals, $p70^{S6K}$ phosphorylates 40S ribosomal protein S6 and activates protein synthesis. Recently, Zhang et al. (1994) has demonstrated that the Atpk1 protein, which is apparently identical to the ATPK6 protein, can phosphorylate two plant ribosomal proteins.

1.6 Effect of a Variety of Stresses on the Expression of Genes for Putative Components of *Arabidopsis* MAP Kinase Cascades

The expression of three protein kinases ATMEKK1 (structurally related to MAPKKK) (Mizoguchi et al., 1996), ATMPK3 (structurally related to MAPK) (Mizoguchi et al., 1993), and ATPK19 (structurally related to ribosomal S6 kinase) (Mizoguchi et al., 1995) has been examined by northern blot analyses. The mRNA levels of these three protein kinases increase markedly and simultaneously in response to touch, cold, and salinity stress (Mizoguchi et al., 1996). These results suggest that MAP kinase cascades, which are thought to respond to a variety of extracellular signals, are regulated not only at the post-translational level but also at the transcriptional level in plants, and that MAP kinase cascade(s) may function in plants in transducing signals in the presence of environmental stress.

2 Expression of Two Genes for P5CS and P5CR that are Involved in Proline Biosynthesis in *Arabidopsis* under Osmotic Stress

Many plants, including halophytes, accumulate proline to high levels in response to osmotic stress, such as high salinity and water deficit (Szoke et al., 1992). Proline has been shown to be a potent osmoprotectant in bacteria, and proline-overproducing mutants of *E. coli* exhibit clear osmotolerance. It has been suggested that proline, similar to glycine betaine, is a compatible osmolyte that plays a role in counteracting the effects of osmotic stress (Delauney and Verma, 1993). It has also been suggested that proline-overproducing plants constructed by genetic engineering might acquire osmotolerance, namely, the ability to tolerate high-salt conditions and drought.

The pathway for the biosynthesis of proline in *E. coli* has been elucidated. The pathway begins with the ATP-dependent phosphorylation of the γ-carboxy group of L-glutamic acid by γ-glutamyl kinase (γ-GK). The product of γ-GK is reduced to glutamic-γ-semialdehyde (GSA) by GSA dehydrogenase, with which γ-glutamyl kinase forms an obligatory enzyme complex. GSA cyclizes spontaneously to form Δ1-pyrroline-5-carboxylate (P5C), which is finally reduced to proline by P5C reductase (P5CR). In higher plants, it has been suggested that the proline is synthesized from either glutamate or ornithine, and the glutamate pathway is the primary route for proline synthesis during conditions of osmotic stress and nitrogen limitation whereas the ornithine pathway assumes prominence under high nitrogen input (Delauney et al., 1993). The biosynthetic pathway from glutamate is thought to be a reaction that glutamate is converted to proline *via* the intermediates γ-glutamyl phosphate, GSA and P5C, as occurs in *E. coli*, because a cDNA clone for P5C synthetase (P5CS) was isolated from mothbean and the recombinant P5CS protein expressed in *E. coli* had both γ-GK and GSA dehydrogenase activities, as determined by complementation experiments (Hu et al., 1992). Plant P5CR cDNAs were isolated from soybean, pea and *Arabidopsis* (see Delauney and Verma 1993, for review). However, the relationship between gene expression and the accumulation of proline under osmotic stress has not been well characterized in plants. Another question that also remain to be addressed is whether or not plants which accumulated proline become tolerant to osmotic stress, such as high salt and drought.

We have isolated a cDNA for P5CS (*cAtP5CS*), an enzyme involved in the biosynthesis of proline, from a cDNA library prepared from a dehydrated rosette plant of *Arabidopsis* (Yoshiba et al., 1994). Southern blot analysis suggests that only one copy of the corresponding gene (AtP5CS) is present in *Arabidopsis*. The deduced amino acid sequence of the P5CS protein (AtP5CS) from *Arabidopsis* exhibits 75% homology to that of the P5CS from *Vigna aconitifolia.* Northern blot analysis has revealed that the gene for P5CS is induced by dehydration, high salt and treatment with ABA, while it is not induced by heat or cold treatment. Moreover, a simultaneous accumulation of proline is observed as a result of the former treatments in *Arabidopsis*. A cDNA for P5CR (*cAtP5CR*) has also isolated from *Arabidopsis* and used in northern blot analysis. The *AtP5CR* gene is not induced to a significant extent by dehydration or high-salt stress. These observations suggest that the AtP5CS gene plays a principal role in the biosynthesis of proline in *Arabidopsis* under osmotic stress. Recently, it has been demonstrated that tobacco plants over-expressing the mothbean P5CS cDNA in tobacco are able to accumulate proline under stress conditions and that these plants are salt tolerant (Kishor et al., 1995).

3 Conclusions

In summary, it has been demonstrated that MAP kinase cascades are also present in plants (Fig. 1). *CTR1*, which was identified as a gene in which a mutation leads to constitutive activation of the ethylene triple response, encodes a protein kinase that belongs to the Raf family (Kieber et al., 1993), which functions as a MAPKKK in animal cells (Nishida and Gotoh, 1993; Davis, 1994). Recenly, a 47 kD MBP kinase has been reported to be activated in response to fungal elicitor in suspension-cultured tobacco cells (Suzuki and Shinshi, 1995). It has also been shown that the activity of the MBP kinase is regulated by post-translational phosphorylation of tyrosine residues. These results suggest that this protein kinase may be a member of the MAP kinase family in plants. It is suggested that plant hormones, e.g. auxin (Mizoguchi et al., 1994) and ethylene (Kieber et al., 1993) and osmotic stress, low temperature (Mizoguchi et al., 1996), elicitor (Suzuki and Shinshi, 1995) and wounding (Suzuki and Shinshi, 1995) may function as activators of plant MAP kinase cascades in plants (Fig. 1). The yeast HOG1 pathway and the mammalian JNK pathway are activated by osmotic stress (Davis, 1994). Yeast cells produce excessive amounts of glycerol under osmotic stress condition, while some plants accumulate other osmoprotectants such as proline, sugars and glycine betaine. It is easy to speculate that plants may use similar regulatory cascades to control the synthesis of osmoprotectants. We expect that analyses of transgenic plants which over-express the genes for *Arabidopsis* MAPKs, MAPKKs and MAPKKKs, will give us some ideas on how plants respond to and acquire tolerance to environmental stresses.

References

Banno H, Hirano K, Nakamura T, Irie K, Nomoto S, Matsumoto K, Machida Y (1993) NPK1, a tobacco gene that encodes a protein with a domain homologous to yeast BCK1, STE11 and Byr2 protein kinases. Mol Cell Biol **13**: 4745-4752

Davis RJ (1994) MAPKs: new JNK expands the group. Trends Biochem Sci **19**: 470-473

Delauney AJ, Hu C-AA, Kishor PBK, Verma DPS (1993) Cloning of ornithine δ-aminotransferase cDNA from *Vigna aconitifolia* by trans-complementation in *Escherichia coli* and regulation of proline biosynthesis. J Biol Chem **268**: 18673-18678

Delauney AJ, Verma DPS (1993) Proline biosynthesis and osmoregulation in plants. Plant J **4**: 215-223

Hu C-AA, Delauney AJ, Verma DPS (1992) A bifunctional enzyme (Δ1-pyrroline-5-carboxylate synthetase) catalyzes the first two steps in proline biosynthesis in plants. Proc Natl Acad Sci USA **89**: 9354-9358

Jonak C, Heberle-Bors E, Hirt H (1994) MAP kinases: universal multi-purpose signaling tools. Plant Mol Biol **24**: 407-416

Kieber JJ, Rothenberg M, Roman G, Feldmann KA, Ecker JR (1993) CTR1: a negative regulator of the ethylene response pathway in *Arabidopsis*, encodes a member of the Raf family of protein kinases. Cell **72**: 427-441

Mizoguchi T, Hayashida N, Yamaguchi-Shinozaki K, Kamada H, Shinozaki K (1993) ATMPKs: a gene family of plant MAP kinases in *Arabidopsis thaliana*. FEBS Lett **336**: 440-444

Mizoguchi T, Gotoh Y, Nishida E, Yamaguchi-Shinozaki K, Hayashida N, Iwasaki T, Kamada H, Shinozaki K (1994) Characterization of two cDNAs that encode MAP kinase homologues in *Arabidopsis thaliana* and analysis of the possible role of auxin in activating such kinase activities in cultured cells. Plant J **5**: 111-122

Mizoguchi T, Hayashida N, Yamaguchi-Shinozaki K, Kamada H, Shinozaki K (1995) Two genes that encode ribosomal protein S6 kinase homologs are induced by cold or salinity stress in *Arabidopsis thaliana*. FEBS Lett **358**: 199-204

Mizoguchi T, Irie K, Hirayama T, Hayashida N, Yamaguchi-Shinozaki K, Matsumoto K, Shinozaki K (1996) A gene encoding a MAP kinase kinase kinase is induced simultaneously with genes for a MAP kinase and an S6 kinase by touch, cold and water stress in *Arabidopsis thaliana*. Proc Natl Acad Sci USA **93**: 765-769

Nishida E, Gotoh Y (1993) The MAP kinase cascade is essential for diverse signal transduction pathways. Trends Genet **18**: 128-131

Shibata W, Banno H, Itoh Y, Hirano K, Irie K, Usami S, Machida C, Machida Y (1995) A tobacco protein kinase, NPK2, has a domain homologous to a domain found in activators of mitogen-activated protein kinases (MAPKKs). Mol Gen Genet **246**: 401-410

Suzuki K, Shinshi H (1995) Transient activation and tyrosine phosphorylation of a protein kinase in tobacco cells treated with a fungal elicitor. Plant Cell **7**: 639-647

Szoke A, Miao G-H, Hong Z, Verma DPS (1992) Subcellular location of Δ1-pyrroline-5-carboxylate reductase in root/nodule and leaf of soybean. Plant Physiol **99**: 1642-1649

Yoshiba Y, Kiyosue T, Katagiri T, Ueda H, Mizoguchi T, Yamaguchi-Shinozaki K, Wada K, Harada Y, Shinozaki K (1994) Correlation between the induction of a gene for Δ1-pyrroline-5-carboxylate synthetase and the accumulation of proline in *Arabidopsis thaliana* under osmotic stress. Plant Journal **7**: 751-760

Zhang S-H, Broome MA, Lawton MA, Hunter T Lamb CJ (1994) atpk1, a novel ribosomal protein kinase gene from *Arabidopsis*: II. Function and biochemical analysis of the encoded protein. J Biol Chem **269**: 17593-17599

Regulation of Gene Expression during Cellular Adaptation to Water Stress

Stefania Grillo[1], Antonello Costa[1], Marina Tucci[1], Maria Rosaria Amatruda[1], Federica Consiglio[1], László Vigh[2] and Antonella Leone[1]

1 Research Centre for Vegetable Breeding, National Research Council, I-8055 Portici , Naples, Italy

2 Institute of Biochemistry, Biological Research Centre, Hungarian Academy of Sciences, H-6701 Szeged, Hungary

Abstract. Gradual adaptation to water stress was accomplished by a set of metabolic adjustments, including proline accumulation, synthesis of *de novo* proteins, and changes in gene expression. The ability of the adapted cells to maintain the cellular and subcellular membrane integrity under conditions of severe water stress was found to be associated with a reduced level of unsaturation of fatty acids of membrane lipids. By cloning two different desaturase genes (Δ^9-stearoyl-ACP-desaturase and a Δ^{12}-oleoyl-desaturase gene, respectively) it was possible to establish that the reduced level of unsaturation observed in the adapted cells was related to the down-regulation of this class of gene.

Keywords. Acclimation, desaturase genes, low water potential, membrane microviscosity, polar lipids, polyethylene glycol, potato cells, *Solanum tuberosum*

1 Introduction

Injury to cellular membranes is a common phenomenon of environmental stresses. Increased leakage of cytoplasmatic solutes after exposure to water stress are generally indicative of membrane disruption. However, the amount of solute leakage can be significantly decreased by gradual exposure to water deficit (Blum, 1988), indicating that acclimation may trigger structural and biochemical adjustments of biological membranes in order to maintain fluidity. Besides the putative function of some stress proteins (Leone et al., 1993) and osmoprotectants, e.g. sugars, proline, glicine betaine (Delauney and Verma, 1993) in stabilizing biological membranes upon water stress, it has been suggested that an additional mechanism for the maintenance of proper membrane functions under stress is based on changes in lipid composition.

Extensive work has been carried out on the effect of changing external temperature on plant membrane lipid composition. As a general rule, the unsaturation degree of FAs of membrane PLs increases in response to lowering of temperatures and decreases when external temperature raises (Quinn, 1988). These findings have been further supported by recent studies on mutants (Hugly et al., 1989; Miquel et al., 1993) and transgenic plants (Murata et al., 1992) with an altered lipid metabolism. By contrast, controversial evidence is reported for water stress, due to differences in the plant species, the plant growth stage or tissue and the extent and the way of imposing water stress (Liljenberg, 1992 for review).

Here we report on variations in membrane fatty acid composition and microviscosity observed in a population of potato cells gradually habituated to growing under conditions of low potential conditions, created by supplementing the nutrient medium with PEG 8000. By cloning two different desaturase genes, it was possible to establish that the acquisition of the tolerance was associated with changes in the expression of this class of gene.

2 Physiological, Biochemical and Structural Changes of the Adapted Cell Line

In our adapted potato cell line, tolerance to low water potential was acquired by a set of metabolic adjustments, described elsewhere (Leone et al., 1994 a, b) and briefly summarized in Table 1. The changes included proline accumulation, synthesis of *de novo* proteins and changes in gene expression, which ensured the maintenance of active growth under conditions of low water potential which completely inhibited the growth of non-acclimated potato cells. (Leone et al., 1994 a, b).

3 Effects of Gradual Acclimation on Membrane Fatty Acid Composition and Fluidity

As evident from the EM observations, the integrity of cellular and subcellular membranes of adapted cells was completely preserved at values of low water potential which determined an extended membrane disgregation in shocked cells (Leone et al., 1994a). Accumulation of free-proline might have contributed to the maintenance of the membrane structure. We looked at the variation of membrane fatty acid composition. The mol percentage composition of the major extra-plastidial phospholipids (PLs) was basically unchanged, but both phosphatidylcoline (PC) and phosphatidylethanolamine (PE) were more saturated in acclimated cells (Fig. 1). The decrease in unsaturation occurred primarily at the expense of linoleic acid (18:2) in the PE fraction. A significant increase in the percentage content of palmitic acid (16:0) of the PE fraction was also detected in acclimated cells. As is well known, saturated fatty acids have higher melting points than their corresponding unsaturated analogs and a decrease in the degree of unsaturation would be expected to increase membrane microviscosity.

Table 1. Main characteristics of a potato cell population adapted to grow in a PEG-mediated low water potential (see Leone et al., 1994 a and b)

	Adapted cells	**Shocked cells**
•*Cell growth*	Fully recovered	Strong Inhibition
•*Cell viability*	Unaffected	Unaffected
•*ABA content*	No variation	Five-fold increase
•*Proline content*	Twelve-fold increase	Six-fold increase
•*Ultrastructural modifications*	None visible	Strong cellular and subcellular membrane disruption
•*Protein synthesis*	Fully recovered, synthesis of new proteins	Strong inhibition, synthesis of *de novo* proteins, different from those induced during adaptation.

Membrane microviscosity of PEG-acclimated cells, as determined by diphenyl-hexatriene (DPH) labeling, was higher than in control cells (Leone et al., 1995), and therefore, plasma membranes of adapted cells were more rigid. The reduced level of unsaturation observed in PEG-acclimated cells especially in the typical non-bilayer forming phospholipid PE (Gagnè et al., 1985) could be directly linked to membrane stabilization, because it may avoid phase transitions from bilayer to reverse hexagonal II phase. Moreover, the resulting reduction in water permeability of plasma membranes may lead to preservation of cell turgidity, a beneficial trait under water stress (Liljenberg, 1992).

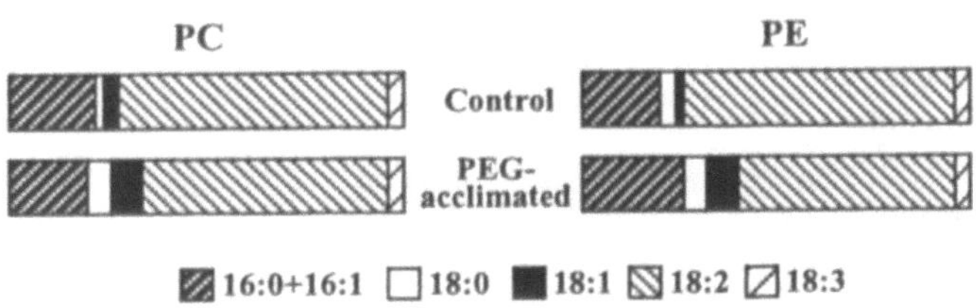

Fig. 1. Fatty acid composition of PC and PEG from control and PEG-acclimated potato cells

4 Cloning of Desaturase Genes

It has been widely proved that membrane fluidity is regulated by the extent of unsaturation of fatty acids of membrane lipids (Quinn, 1988). In plants the FA unsaturation is controlled by a set of desaturases responsible for the introduction of a double bond between two carbons of the acyl chain (Ohlrogge and Browse, 1995). Membrane lipids of higher plants contain mainly C_{16} and C_{18} FAs, which are synthesized in the plastid in the form of palmitoyl (16:0)-ACP and stearoyl (18:0)-ACP. The first step of C_{18} FA unsaturation takes place in the chloroplast by a soluble stearoyl-ACP-desaturase, which introduces a double bond at Δ^9 position of stearoyl-ACP to produce oleoyl-ACP. The subsequent step of unsaturation is catalyzed by another desaturase, Δ^{12}-oleoyl-desaturase, which is bound to the endoplasmic reticulum (ER), only after the transfer of oleic acid to a glycerol moiety. Then, both these enzymes play a pivotal role in controlling respectively the extent of unsaturation and polyunsaturation of membrane FAs. Two full length cDNAs encoding both genes have been recently cloned from a library of a wild potato species, by heterologous hybridization.

Table 2. Sequence homology of a stearoyl-ACP-desaturase (accession no. X78935) and an oleoyl-desaturase (accession no. X92847) of a wild potato species with those of different species. Homology is presented as a percentage of identical amino acid residues

	Identity %	Sequence Accession no.
Δ^9-stearoyl-ACP-desaturase		
Brassica napus	80.5	X63364
Brassica rapa	82.1	X60978
Carthamus tinctorius	84.6	M61109
Cucumis sativus	82.4	M58858
Glycine max	83.0	L34346
Linum usitatissimum	76.6	X70962
Oryza sativa	78.8	D38753
Ricinus communis	85.1	M59857, X56508
Sesamum indicum	84.2	D42086
Simmondsia chinensis	81.6	M83199
Solanum tuberosum	85.7	M91238
Spinacia oleracea	81.9	X62898
Thumbergia alata	81.9	UO7605
Thumbergia alata	83.7	UO7597
Δ^{12}-oleoyl-desaturase		
Arabidopsis thaliana	74.5	L26296
Glycine max	74.3	L43920
Glycine max	76.2	L43921
*Ricinus communis**	67.1	U22378

* Sequence # U22378 corresponds to an oleate hydroxylase gene (van de Loo et al., 1995)

The ORF of clone p*ScDes4.1* (Tucci et al., 1994), is 393 amino acid long, comparable to other stearoyl-ACP-desaturases, including 30 amino acid residues, corresponding to a putative chloroplastic transit peptide. Homology of this clone to other cloned desaturase genes is high at both the nucleotide and amino acid level (Table 2).

The ORF of p*ScD11*, encoding a Δ^{12}-oleoyl-desaturase (Grillo et al., unpublished) is 383 amino acid residue long. The deduced amino acid sequence from the cDNA showed homology to other desaturases from *Arabidopsis*, soybean as well as to an oleate hydroxylase from *Ricinus communis* (Table 2).

The hydropathy plot revealed six-candidate membrane spanning sequences, and sequence analysis identified histidine-rich motifs, highly conserved in other desaturases, which putatively may contribute to an iron binding site in the cytoplasmic domain of the enzyme (Okuley et al., 1994).

5 Regulation of Desaturase Genes during Cellular Adaptation to Water Stress

Since the mol percentage of stearic acid was found to be higher in both PE and PC fractions of PEG-acclimated cells, it was possible to speculate an inhibition of the stearoyl-desaturase.

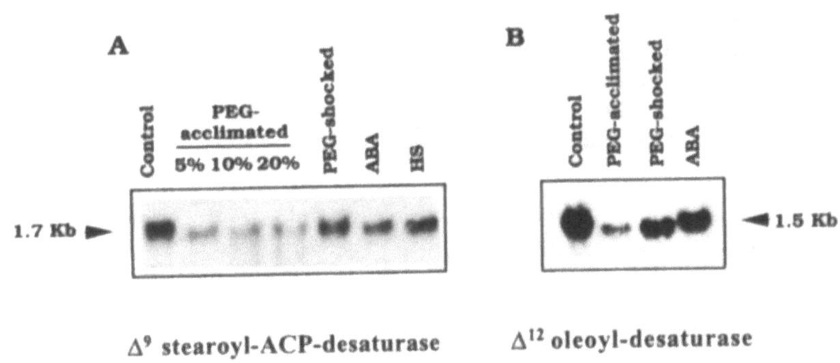

Fig. 2. Expression of desaturase genes in potato cells (*S. tuberosum*), during gradual acclimation to PEG-mediated low water potential, PEG-shock (unadapted cells abruptly transferred into a medium containing 20% PEG), ABA (0.1 M for 24 h) or HS at 37°C for 3 h. Two full length cDNA encoding respectively a stearoyl-ACP-desaturase **(A)** and an oleoyl-desaturase **(B)** as probes

Although an inhibition of the enzymatic activity of the stearoyl-desaturase can not be excluded, the steady state transcript level of a Δ^9-stearoyl-ACP-desaturase gene decreased in PEG-acclimated cells, showing that the increase in stearic acid was at least partially associated with the down-regulation of the transcription of the gene encoding this enzyme and/or decrease in the stabilization of the transcript (Fig. 2A).

However, the decrease in unsaturation level was primarily at the expense of linoleic acid, which determined a subsequent increase in oleic acid. As for the Δ^9-stearoyl-ACP-desaturase, the level of expression of the gene encoding an enzyme responsible of the subsequent step of unsaturation, the Δ^{12}-oleoyl-desaturase, was also found to be inhibited (Fig. 2B). Changes in the expression of these two genes were neither triggered by abrupt water stress nor by heat shock, indicating that biophysical modifications of membrane are time-dependent and only occur after prolonged and gradual exposure to stress.

6 Future Perspectives

The relationship between changes in membrane saturation/unsaturation extent and fluidity and tolerance to environmental stresses will be assessed by manipulating the expression of the two desaturase genes. These genes will be reintroduced in transgenic plants under the control of strong constitutive promoters and the effects of such a manipulation will be examined with regards to changes in the cellular and subcellular membrane physical state and properties and finally on the acquisition of tolerance to physical stress.

Acknowledgements. This work was supported by the Italian Ministry of Agriculture and Forestry in the framework of the project "Resistenze Genetiche delle Piante Agrarie agli stress biotici e abiotici" and by a NATO Collaborative Research Grant (no. CRG 92054) to AL and LV Contribution no. 128 of Research Centre for Vegetable Breeding, CNR, Portici, Italy.

References

Blum A (1988) Plant breeding for stress environments. CRC Press, Inc, Boca Raton, FL

Delauney A, Verma DPS (1993) Proline biosynthesis and osmoregulation in plants. Plant J **4**: 215-233

Gagné J, Stamatatos L, Diacovo T, Hui SW, Yeagle PL, Silvius JR (1985) Physical properties and surface interactions of bilayer membranes containing N-methylated posphotidylethanolamines. Biochemistry **24**: 4000-4008

Hugly S, Kunst L, Browse J, Somerville C (1989) Enhanced Thermal Tolerance of Photosynthesis and Altered Chloroplast Ultrastructure in a Mutant of *Arabidopsis* Deficient in Lipid Unsaturation. Plant Physiol **90**: 1134-1142

Leone A, Nagao RT, Key JL (1993) Use of molecular genetics for stress resistance breeding in cool-season legumes. *In* KB Singh, MC Saxena, eds, Breeding for Stress Tolerance in Cool-Season Food Legumes. Wiley & Sons, Chichester, UK, pp 323-342

Leone A, Costa A, Tucci M, Grillo S (1994a). Adaptation *versus* shock response to PEG-induced low water potential in cultured potato cells. Physiol Plant **92:** 21-30

Leone A, Costa A Tucci M, Grillo S (1994b) Comparative analysis of short- and long-term changes in gene expression caused by low water potential in potato (*Solanum tuberosum*) cell-suspension cultures. Plant Physiol **106**: 703-712

Leone A, Tucci M, Grillo S, Costa A, Vigh L (1995) Adaptation of potato cells to low water potential and changes in membrane fatty acid composition and fluidity. *In* Kader IC and Mazliak P, eds, Plant Lipid Metabolism. Kluwer Academic, Dordrecht, The Netherlands, pp 432-434

Liljenberg CS (1992) The effects of water deficit stress on plant membrane lipids. Progr Lipid Res **31**: 335-343

Miquel M, James JrD, Dooner H, Browse J (1993) *Arabidopsis* requires polyunsaturated lipids for low-temperature survival. Proc Natl Acad Sci USA **90**: 6208-6212

Murata N, Ishizaki-Nishizawa O, Higashi S, Hayashi H, Tasaka Y, Nishida I. (1992) Genetically engineered alteration in the chilling sensitivity of plants. Nature **356**: 710-713

Ohlrogge J, Browse J (1995) Lipid Biosynthesis. Plant Cell **7**: 957-970

Okuley J, Lightnen J, Feldmann K, Yadav N, Lark E, Browse J (1994) *Arabidopsis* FAD2 Gene Encodes the enzyme that is essential for polyunsaturated lipid synthesis. Plant Cell **6**: 147-158

Quinn P (1988) Regulation of membrane fluidity in unicellular organisms. *In* Aloia RC, Curtain CC and Gordon LM, eds, Physiological regulation of membrane fluidity. Alan R. Liss, New York, pp. 293-324

Tucci M, Grillo S, Costa A, Vigh L, Horvarth I., Leone A. (1994) Cloning, Characterization and Expression of a cDNA encoding a Δ^9-stearoyl-Acyl-Carrier-Protein desaturase from cold tolerant *Solanum commersonii*. Proceedings of the 4th International Congress of Plant Molecular Biology 1994 June 19-24, Amsterdam, Abs # 379

van de Loo F, Broun P, Turner S, Somerville C (1995) An oleate 12-hydroxylase from *Ricinus communis* L. is a fatty acyl desaturase homolog. Proc Natl Acad Sci USA **92**: 6743-6747

Specific and General Gene Induction in Limiting Environmental Conditions

Nelson Marmiroli[1], Mariolina Gulli[1], Elena Maestri[1], Cristina Calestani[1], Aliosha Malcevschi[1], Carla Perrotta[3], Steve A. Quarrie[4], Katrien M. Devos[4], Giovanna Raho[2], Hans Hartings[2] and Elisabetta Lupotto[2]

[1] Laboratory of Environmental Biotechnology, Department of Evolutionary Biology, University of Parma, I-43100 Parma, Italy
[2] Experimental Institute for Cereal Research, I-24100 Bergamo, Italy
[3] Department of Biology, University of Lecce, I-73100 Lecce, Italy
[4] John Innes Centre, Norwich NR4 7UH, UK

Abstract. The genetic system that controls the response to environmental stress is characterized by the synthesis of generic and specific stress proteins. The gene *Hvhsp17*, isolated from *Hordeum vulgare* subjected to heat shock at 40°C for 2 h, encodes a cytoplasmic low molecular weight heat shock protein (HSP) (class I). The expression of this gene was monitored in barley, maize and in transgenic tobacco and it was found to be specifically induced by heat but only barely by drought. Other genes, cloned from barley seedlings treated with ABA, most of them encoding DHN- or LEA-type proteins, showed a more specific expression in condition of drought stress. These genes have been mapped in several grasses and the map position utilized to establish the chromosomal localization of a drought related QTL. These stress inducible genes along with others isolated by different molecular genetic techniques (RDD, RDA, subtractive hybridization) could be useful "candidate genes" in the dissection of these Quantitative Traits Loci (QTLs). Stress-induced genes have been proved to be extremely powerful in RFLP analysis, because they can reveal genetic distances between cultivars adapted to different ecogeographic conditions, characterized by different water availability, and could be considered interesting descriptors of what could be termed "useful variation".

Keywords. Abscisic acid, barley, heat shock, *Hordeum vulgare, maize, Nicotiana tabacum*, QTL, stress-induced genes, transformation, *Triticum aestivum*, wheat, *Zea mays*

1 Introduction

Plants have evolved peculiar ways of reacting to environmental stresses, such as temperature extremes, wounding and water deficit. Water stress is a particular severe environmental constraint limiting both survival and reproduction of the plants and affecting yield. In fact, water deficit brings about a whole series of alterations at physiological, morphological, cellular, biochemical and molecular levels, which all together are referred to as the "drought stress response". Some of them may have an adaptive value, conferring protection towards the stress, whereas others might primarily be due to the damage caused by the stress itself (Skriver and Mundy, 1990; Bianchi et al., 1991; Bray, 1993; Bohnert et al., 1994).

A gradual decrease in water availability activates a cascade of events, such as reduction in cell and tissue growth, inhibition of protein synthesis and repression and/or activation of enzymatic activities. Higher intensities of stress are ultimately associated with stomatal closure, which in turn affects photosynthesis and transpiration, and with the increase in abscisic acid (ABA) levels, which is determined by changes in both its biosynthesis and translocation (Harris and Outlaw, 1991; Mac Robbie, 1991). Conjugation, compartmentalization, transport and degradation of different sets of molecules are responsible for interactions among environmental and developmental cues. A type of response present at all levels of the cascade is the activation of the expression of particular genes, whose protein products are essential for modulating or determining the succession of events which may lead to adaptation or tolerance (Assmann, 1994; Nelson et al., 1994).

2 Regulation of the Activity of Generic and Specific Genes during Drought Stress

2.1 Regulation of Protein Synthesis

The accumulation of HSPs during ABA treatment and drought stress has been reported for HSP70 (Heikkila et al., 1984) and HSP16 (Almoguera and Jordano, 1992). Identification and accumulation of drought-related proteins has been reported by several authors (Bray, 1991; Grossi et al., 1995; Mundy and Chua, 1988; Chandler and Robertson, 1994; Gulli et al., 1995).

2.2 Regulation of Transcription of *hsp* genes *in vivo*

Genes for HSPs are not only transcribed in heat shock conditions. There is evidence that HSP70 is accumulated in drought stress conditions and in the presence of high level of ABA (Heikkila et al., 1984). It has also been reported that genes for small HSPs are transcribed during desiccation of seed embryos (Almoguera and Jordano, 1992). We analyzed the regulation of transcription of small HSPs in different barley cell and tissue types, by using a cloned gene for HSP17 (Class I HSP) (Marmiroli et al., 1993). The expression of the cloned *Hvhsp17* was monitored in maize and barley seedlings upon various stress conditions and at different developmental stages. In 20-day-old maize plantlets the *hsp17* homologous genes are induced after heat shock treatment at 40°C for 3 h, in the basal segment and in the apical segment of leaf 1, but are not induced in response to other temperature and osmotic stresses (Fig. 1). We also monitored also the expression in response to heavy metal treatment. *Hvhsp17* homologous transcripts were detected after treatment with 20 mM $CdSO_4$ both in roots and coleoptiles of 5-day-old seedlings (data not shown).

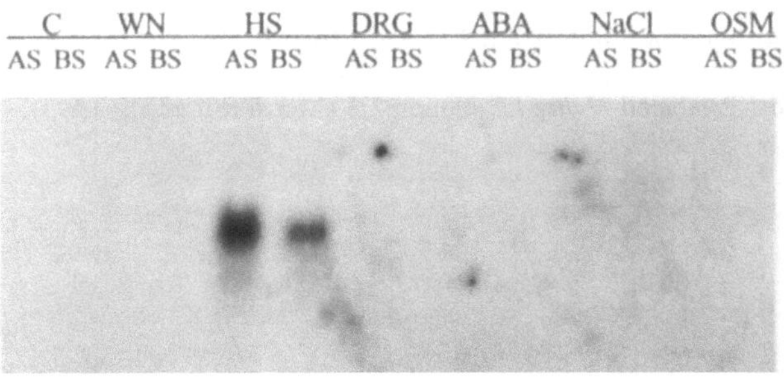

Fig. 1. Northern analysis of total RNA extracted from the apical segment (AS) and the basal segment (BS) of 20-day-old maize plants, and subjected to various stresses. C, control condition; WN, incision 1cm long in the basal segment; HS, 40°C, 3h; DRG, desiccation under air stream, 4.5 h; ABA, 100μM ABA, 48 h; NaCl, 0.4 M NaCl, 3 days; OSM, 0.3M mannitol, 3 days. Hybridization was performed under standard condition with ^{32}P labeled *Hvhsp17*, clone p2.3 (Marmiroli et al., 1993)

Similarly, 10-day-old plantlets of *Hordeum vulgare* (cv. Onice) were subjected to different stresses. Northern analysis of total RNA isolated from barley shoots evidenced that the expression of *Hvhsp17* homologous transcripts was triggered only by heat stress treatment at 40°C (Fig. 2). The transcripts corresponding to *hsp17* were detectable after 30 min of heat shock and further increased after 1 and 2 h. Furthermore, they were still present at high level after 3 h of recovery from the heat stress. However, both in maize and barley plantlets neither drought nor different osmotic stress treatments induced significant *hsp17* transcription.

We also studied the interaction between heat shock and drought stress responses, because it has been reported that drought stress potentiates the heat shock response in soybean (Kimpel and Key, 1985) and these two kinds of stresses usually occur simultaneously in field conditions. Drought stress followed by heat stress induced *Hvhsp17* homologous transcripts in about 30 min and the transcripts were present after 4 h of recovery at 22°C. However, desiccation alone did not have any significant effect on transcription of *Hvhsp17* homologues (Fig. 3).

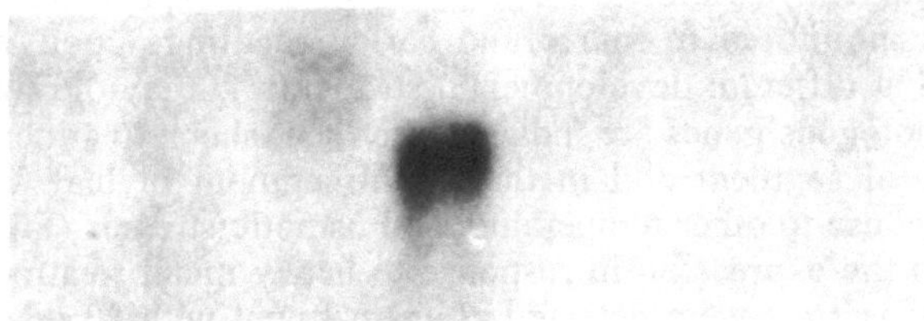

Fig. 2. Northern analysis of total RNA extracted from 10-day-old plantlets of *Hordeum vulgare* (cv. Onice) subjected to the following stress treatments: C, control condition; ABA, 100 μM ABA for 3 h; DRG, desiccation under flowhood for 2 h; OSM, 0.3 M Sorbitol for 3 h; NaCl, 0.4 M NaCl for 3 h; CS, 4°C for 4 h; HS, 40°C for 2 h; PRO, 100 mM Proline for 3 h. Hybridization was performed under standard condition with ^{32}P labeled *Hvhsp17*, clone p2.3 (Marmiroli et al., 1993)

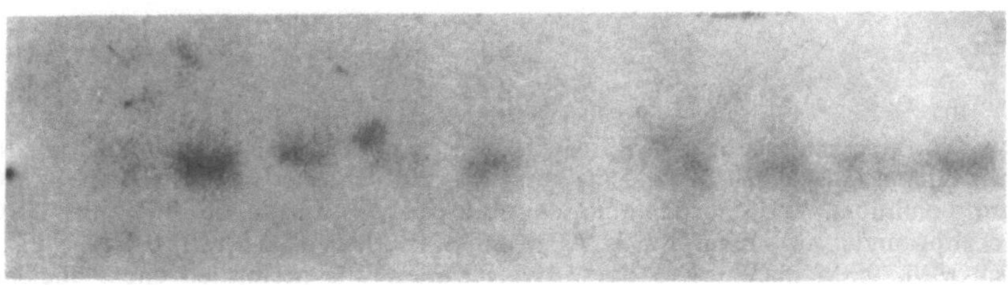

Fig. 3. Northern analysis of total RNA extracted from 10-day-old plantlets of *Hordeum vulgare* (cv. Onice), subjected to the following stress treatments: C, control condition; DRG, desiccation under flowhood for 1 h followed by heat shock at 40°C for the time indicated; DRG, desiccation under flowhood for 2 h; DRG, desiccation under flowhood for 1h followed by heat shock at 40°C for 1 h and by a recovery at 22°C for the time indicated. Hybridization was performed under standard condition with ^{32}P labeled *Hvhsp17*, clone p2.3 (Marmiroli et al., 1993)

2.3 Regulation of Transcription of *hsp* Genes in Transgenic Plants

We also analyzed the regulation of transcription of the gene for a small HSP in transgenic plants transformed with a construct containing the *Hvhsp17* promoter fused with the *gus* reporter gene.

The construct was introduced by PEG-mediated transformation into protoplasts of different monocot and dicot species (Raho et al., 1995). The same construct was also utilized to transform tobacco cells ,from which some stably transformed plants were regenerated. The 5' upstream region (-1700 to +1) of the barley *Hvhsp17* gene was inducible by heat shock in barley protoplasts and in protoplasts of other grasses (Raho et al., 1995). Under heat shock condition this promoter region was responsible for the transcriptional activation of the *Hvhsp17* gene both *in vivo* and *in vitro* (Marmiroli et al., 1993). A genetic construct was developed containing a 1.7 kb fragment spanning the 5' untranslated region of *Hvhsp17* transcriptionally fused with *gus* reporter gene (Fig. 4) and transferred into tobacco by PEG-mediated transformation of leaf mesophyll cell derived protoplasts.

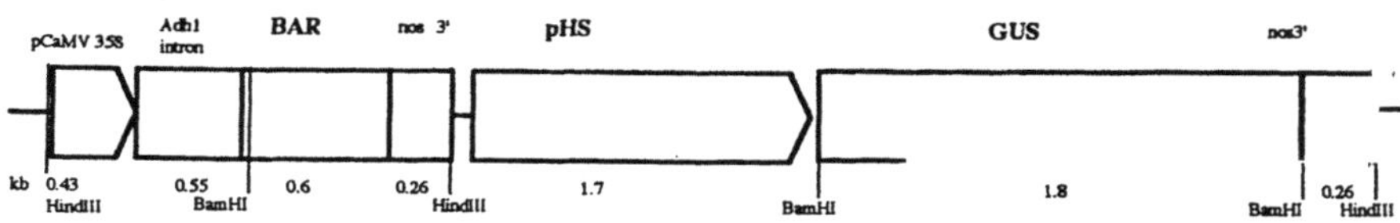

Fig. 4. Physical map of the expression vector used for tobacco transformation

Four transgenic tobacco plants, which expressed the *gus* gene under heat shock conditions, were chosen for further analysis. These plants contained 1 to 5 integrated copies of the chimeric *gus* gene and the histochemical analysis carried out on several tissues showed that the chimeric *gus* gene was particularly heat-inducible in the xylematic vessels of stem and petioles (Lupotto et al., 1993).

The expression of the chimeric gene was also analyzed in transgenic plants subjected to other environmental stresses conditions. In particular we compared the effect of salt stress, ABA treatment, drought stress and heavy metal treatment which can be considered associated or cooperative with heat stress (Table 1).

The results obtained showed that none of the four transgenic tobacco plants analyzed responded directly to drought stress or to salt stress by activating the transcription/translation of the pHSGUS. This result was in agreement with the previous observation obtained *in vivo* by northern analysis of barley plants by using the cloned *Hvhsp17* gene as probe. However, some of the transgenic plants showed a GUS induction when treated with exogenous ABA or with heavy metals (particularly with cadmium).

2.4 Regulation of Transcription of Drought Specific Genes

Several drought inducible genes were isolated from a cDNA library constructed from poly(A)$^+$ of ABA-treated barley plantlets. Seventeen sequences were cloned that were specifically induced by ABA treatment (Gulli et al., 1995). Few clones were selected for sequence and expression analysis. The clone *ABA3* contained an insert of 1016 bp with an open reading frame (ORF) of 615 bp, coding for a protein of 205 amino acids The deduced polypeptide (20.7 kD) shares a strong homology with the protein encoded by the barley dehydrin genes *dhn3* (83.9% identity) and *dhn4* (98% identity), described by Close and Chandler (1990).

Table 1. Comparison of the pHSGUS specific transcription/translation activity in transgenic plants subjected to different environmental stresses. Quantification of the *hsp17* promoter specific activity in the different condition was made by measuring fluorimetrically the GUS activity in control and transgenic plants. + and - indicate approximatively the presence or absence of GUS activity at increasing or decreasing levels upon various conditions. Positive control p35S CaMV-GUS; negative control *N. tabacum* SR1

	Plant				
Treatment	NT2-6-3	NT2-6-7	NT3-1-1	NT3-6-3	positive/ negative control
•Control (26°C)	-	-	-	-	++/-
•Heat shock (2h 40°C)	+	+	+	+	++/-
•Cold shock (4h 4°C)	-	-	-	-	++/-
•Salt treatment (0.4M NaCl)	-	-	-	-	++/-
•Osmotic treatment (0.3M mannitol)	-	-	-	-	++/-
•ABA 100μM	+	-	-	+	++/-
•Metal ions (20mM)					
$CdSO_4$	+	+	+	+	++/-
$ZnSO_4$	+	-	-	+	++/-
$CuSO_4$	-	-	-	-	++/-
$AlCl_3$	+	-	-	+	++/-
$NiCl_2$	-	-	-	-	++/-

The clone *ABA2* was partially sequenced and it was found to be homologous to other cereals genes coding for dehydrins (Fig. 6).

The clone *ABA8* encodes a polypeptide which shows 41% identity to the rat aldose reductase (Boheren et al., 1989), 99.4% identity to the clone pG22-69 isolated from desiccated barley embryos (Bartels at al., 1991) and 90.9% identity to the clone pBGA61 of *Bromus inermis* (Lee and Chen, 1993) also encoding the enzyme aldose reductase.

The nucleotide sequence of clone *ABA7* contains an ORF of 495 bp encoding a protein of 18.9 kD. The molecular weight of the *ABA7* protein product was also confirmed by *in vitro* transcription and translation experiments, which rescued a protein of 19 kD (data not shown). No significant homology with any sequence in data bases was found for ABA7.

The expression of these cDNA clones was determined under various stress conditions, i.e. ABA treatment, cold stress, heat shock, and desiccation. The results of this analysis are summarized in Table 2.

3 Mapping of ABA- and Drought-Responsive Genes in Wheat, Barley and Rice Genomes

Mapping of some of the *ABA* genes was made by hybridizing each clone to digested wheat genomic DNA from 21 nullisomic-tetrasomic lines, ditelosomic lines and wheat/barley addition lines, as well as to rye and rice genomic DNA.

In wheat clone *ABA7* revealed five hybridizing fragments located on chromosome arms 6AS, 2BS and 6DS. The map positions of the loci detected by *ABA7* have been determined more carefully at the distal ends of the respective chromosomes (Devos et al., 1993). In barley, *ABA7* detected one hybridizing fragment located on chromosome 6H, as shown by analysis of the wheat/barley (cv. Betzes) addition lines. In rye genome, *ABA7* was localized on chromosome 4L in a region that is syntenic with chromosome 6AS of wheat.

Table 2. Comparison of the expression of different ABA cDNA clones upon various environmental conditions. Stress treatments were performed as follows: Cont, control condition; +ABA, 10mM ABA for 12 h; Drought, deprivation of water for 24 h; Cold, 4°C for 48 h; Heat, 40°C for 1h. (+/-) indicate level of expression measured as densitometric amount of transcripts. Ribosomal RNA was used as standard control for both hybridization and quantification

Probe	Function	Stress treatment				
		Cont	+ABA	Drought	Cold	Heat
ABA2	Dehydrin	-	++	+++	++	n.t.
ABA3	Dehydrin	-	++	+++	++	-
ABA7	Unknown	-	++	+++	++	+
ABA8	Aldose reductase	-	++	+++	++	-
ABA10	n.s.	-	++	+++	n.t.	n.t.
ABA12	Unknown	-	++	+++	++	n.t.

(n.t.) not tested; (n.s.) not sequenced.

The clone *ABA3* revealed that multiple copies of this sequence are present on the wheat genome, with at least one copy present on homoeologous group 6 and 7 chromosomes.

The clone *ABA8* revealed four fragments, three of which were located on homoeologous group 2 chromosomes in wheat. Due to the synteny of wheat and barley genomes (Moore, 1995) the localization could effectively reflect the position of these genes also in the barley genome.

Clone *ABA2* hybridizes with several fragments of wheat genomic DNA, with at least one copy present on chromosome 5AL.

In barley, genetic analysis of dehydrin gene family evidenced that there are at least 10 different genes encoding for dehydrins and some of them were mapped as follows: *dhn3* and *dhn4* map on the same site of the RFLP map on barley chromosome 6 (6I) (Heun et al., 1991), *dhn6* was precisely localized on chromosome 4 (4I) and *dhn1* and *dhn2* on chromosome 7 (5I) (Close and Chandler, 1989). Interestingly, the clone *ABA2*, which from partial sequencing proved homologous to *dhn1*, maps on a region of wheat chromosome 5A that is homoeologous to barley chromosome 7 (Quarrie et al., 1994).

In rice, the clone *ABA2* has been localized on chromosome 11 that is also homologous to part of chromosome 5 of wheat. Preliminary results gave evidence that *ABA2* hybridizes also with rice chromosome 9 YACs, but this result needs to be confirmed (G Moore, personal communication).

Table 3. Chromosomal location of some ABA-induced clones on cereal genomes

Gene	Copy Number	WHEAT	BARLEY	RYE	RICE
ABA2	multicopy	5AL	7	n.t.	9-11
ABA3	multicopy	6-7	6	n.t.	n.t.
ABA7	one	6AS-2BS-6DS	6	4	n.t.
ABA8	one	2	2	n.t.	n.t.

n.t. not tested

4 The Genetic Variability of Stress Responsive Genes is Very High Among Common Barley Cultivars

Stress-induced genes frequently have the following characteristics: i) they belong to multigenic families, as regards the low molecular weight HSPs, for which twelve different genes for cytoplasmic class I have been identified in wheat (Weng et al., 1993); ii) their amino acid sequences are evolutionary conserved across many different species (Vierling, 1991; Dure et al., 1989). These two aspects have suggested a possible use of these sequences for the evaluation of genetic diversity and in particular as a tool for monitoring useful variations. Genomic DNA was extracted from leaves of 27 barley cultivars, including both winter six-rowed and spring two-rowed genotypes. Five stress responsive genes were used as probes in RFLP analysis: *ABA3*, *ABA7*, *ABA8*, *Hvhsp17* (cloned from barley as previously described) and *Tahsp17.9* (Mcelwain and Spiker, 1989), cloned from wheat. A total of fifteen probe/enzyme combinations yielded 138 different RFLP bands, only three of which were not polymorphic among the 27 varieties. The average number of RFLP bands per probe per individual is 3.4, higher than the number which could be found in RFLP analysis with "classic" monogenic probes.

Some of the probes, such as *ABA7* and *ABA8*, correspond to single copy genes, whereas the genes for small HSPs or genes for dehydrins reveal, on average , up to 10 bands per individual.

The dendrogram obtained (Fig. 5) shows a clear separation of the cultivars into two groups, respectively winter (six-rowed) types, and two-rowed types, mainly spring genotypes. Hence, it seems that stress-related genes have the potential for measuring correctly and more rapidly the differences among gene pools. It must be added, however, that they can be especially useful in measuring differences correlated with the response to environmental temperature and water availability, such as those present in landraces, and which can affect eco-geographical adaptation.

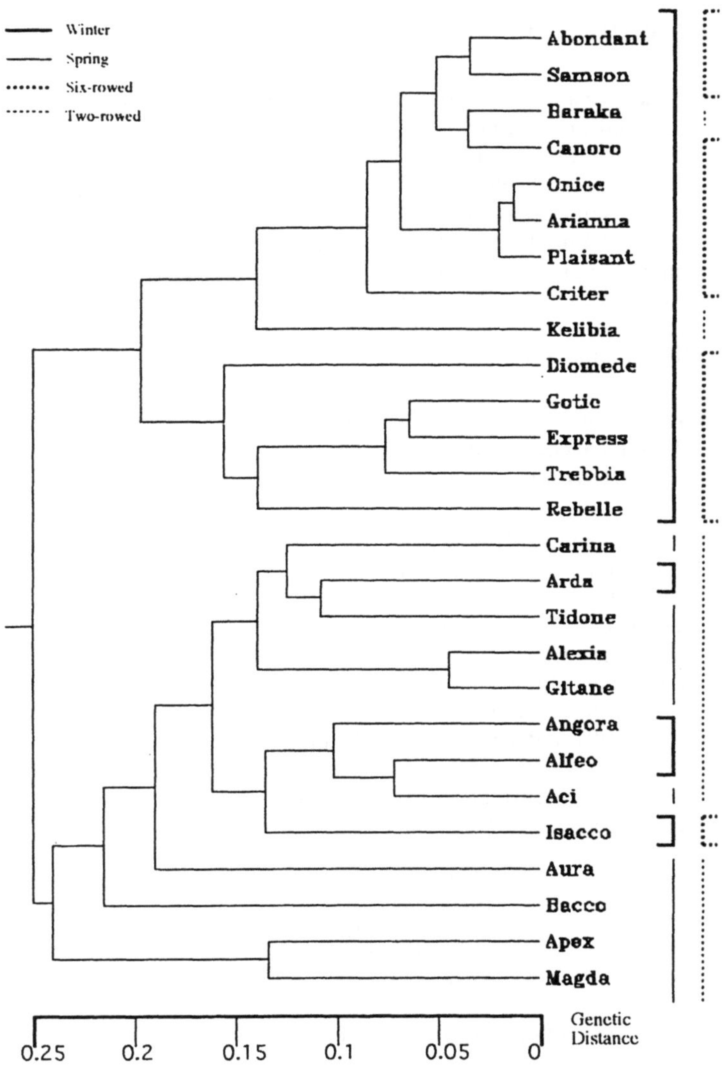

Fig. 5. UPGMA analysis of genetic distances between 27 barley cultivars
Genetic distances between pairs of cultivars were computed starting from restriction fragments polymorphism data, scored as presence/absence, according to Nei and Li (1979). Grouping has carried out by means of UPGMA analysis with the PHYLIP package, version 3.5c (J Felsentein, Univ. of Washington)

5 QTLs are Involved in Drought Responses

In bread wheat a major QTL for ABA accumulation in response to drought stress was localized on chromosome 5AL by RFLP analysis of a F_2 population obtained from the cross Chinese Spring X SQ1 (Quarrie et al., 1994). The affected chromosomal region spans from markers Xpsr575b to Xpsr426a for about 8 cM. The same analysis was performed on 48 dihaploid (DH) lines obtained as described in Quarrie et al. (1994), using ABA content of leaf 4. The results confirmed the presence of the ABA QTL on chromosome 5AL. The same DH lines were used to test ABA accumulation in condition of salt stress, and a significant QTL was located within 2 cM of the drought-induced ABA QTL detected in the F_2 population (Quarrie et al., 1995).

5.1 A cDNA Clone Corresponding to a Class II *lea* Gene has been Localized in the Confidence Interval of a Major QTL for ABA Accumulation in Drought Conditions

A significant LOD score (4.6) was obtained for the locus *XABA2*; this locus corresponds to a gene, *ABA2*, that belongs to the dehydrin family (see Tab. 2-3 for its characteristics). *ABA2* was partially sequenced in the 5' and in the 3' region; the sequenced 5' region is of 124 amino acids and contains part of the coding region. A multiple alignment of the putative first 21 amino acids of ABA2 with the corresponding region of barley dehydrins DHN7 (EMBL accession no. X7136, Robertson et al., unpublished), DHN8 and DHN9 (Close and Chandler., 1990) and rice RAB21 (Mundy and Chua, 1988) is shown in Fig. 6. The sequenced 3' region of 209 bp contained only the untranslated region and it proved identical to the 3' region of *dhn8* of barley and 99% identical to the 3' region of *dhn7* (data not shown).

```
RAB21     MEHQGQHGHVT-SRVDEYGNPV     RICE
DHN7      MEYQGQHGHAT-DKVEEYGQPV     BARLEY
DHN8      MEYQGQHGHAT-DKVEEYGQPV     BARLEY
ABA2      MEYQGQHGHCT-DKVRSTGQPV     BARLEY
          ** ****** *   *   * **
```

Fig. 6. Multiple alignment of dehydrin aminoacidic sequences at the 5' region

6 The Candidate Gene Hypothesis

It has been hypothesized that QTLs could be particular alleles of qualitative genes or alternatively that qualitative genes (candidate genes) could belong to a more complex locus subjected to a quantitative control. The candidate gene hypothesis could be useful for the dissection of quantitative traits (QTs) through the isolation and analysis of qualitative loci which are involved in the response to these environmental factor(s) which are correlated together by that specific QT.

The work we have started on QTL identification and analysis was limited by the availability of polymorphic markers mapping in the region of the QTL. In order to localize exactly the QTL it is necessary to saturate the map in the QTL region. We are currently searching for new molecular markers that could have also a biological function, through the RNA Differential Display (RDD) analysis, a technique that may allow the identification and isolation of genes with a low level of transcription (Liang and Pardee, 1992). In this way, it is possible to obtain genes corresponding to regulatory factors, that are very likely to be involved in QTL structure. We have compared RNA extracted from plants subjected to various stress conditions, namely ABA treatment, drought and cold stress, in order to identify new stress-induced sequences (Fig. 7a).

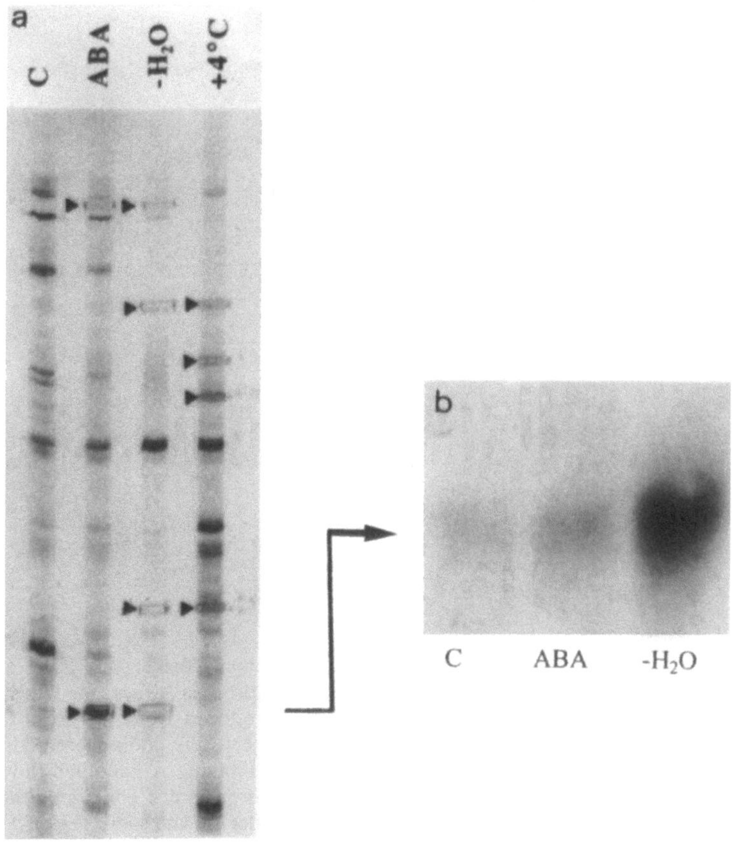

Fig. 7. a. Analysis on 6% polyacrilamide gel of RNA differential display products obtained from total RNA of barley plantlets subjected to different stress treatments: (C) control, ($-H_2O$) deprivation of water for 24 h, (4°C) 4°C for 48 h, (ABA) 10mM ABA for 12 hours. Arrows indicate bands that were isolated from the gel. **b.** Northern hybridization analysis of total RNA obtained from barley plantlets subjected to different stresses using as a probe one of the cDNA fragment isolated from the RDD gel indicated by the long arrow

So far we have been able to isolate several bands differentially expressed and we have analyzed the expression of some of them through northern analysis (Fig. 7a,b). In one case, an increase in the expression of the gene homologous to the new sequence identified by RDD was found upon ABA treatment (2-fold) or drought stress (6-fold) (Fig. 7b).

The search for candidate genes to specific QTL is also currently carried out by comparing gene induction in bulked genotypes differing in ABA accumulation in drought conditions through Bulked Segregant Analysis (BSA)on driver and tester DNAs obtained from DH progenies of two parental types divergent in ABA accumulation in drought condition by Representational Difference Analysis (RDA) (Lisitsyn at al., 1993) (Fig. 8).

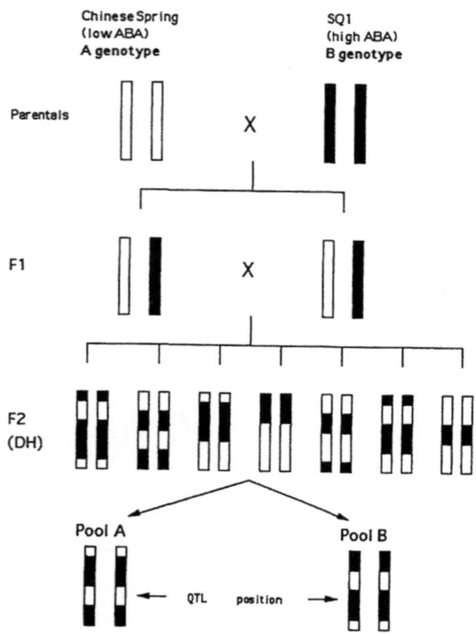

Fig. 8. Construction of the pools of individuals (tester and driver) for RDA analysis for the quantitative trait "ABA accumulation under drought stress" in wheat. Each pool consists of 14 wheat doubled-haploid plants produced from the cross Chinese Spring (low ABA producer) x SQ1 (high ABA producer). The pools were constructed on the basis of an RFLP map base: pool A, individuals having A genotype (low ABA, driver) for the RFLP markers in 8cM region of 5AL chromosome containing the QTL; pool B, individuals having B genotype (high ABA, tester) in the same region

7 Conclusions

Drought stress is probably an environmental condition which is more difficult to define that other environmental stresses, such as heat stress or chemical stress.

In fact, drought is frequently associated with other stresses such as heat, salinity and the increase in internal concentration of toxic ions or molecules specifically accumulated in order to respond to the stress. In contrast with other types of stress which regulate a specific set of genes, exposure to water deficit is characterized by the activation of a number of genes having both generic and specific functions.

This molecular picture fits quite well with the physiological plant response to conditions of water deficit. Many cellular functions change in response to water stress: hormone levels, cell membrane composition, osmotic potential, respiration and photosynthesis rates, accumulation of compatible solutes, and also remodeling of the growth pattern through flowering, senescence and abscission of leaves. Particular interest has been raised by the discovery that heat shock genes are induced in conditions of desiccation (Almoguera and Jordano, 1992) or by the synthesis of antioxidant compounds such as glutathione and ascorbate during heavy metal stress or oxidative stress in the resurrection plant *Boea hygroscopica* (Navari-Izzo et al., 1995). We analyzed the induction of an *hsp* gene in monocot and dicot plants, in different cell types and tissues *in vivo* and *in vitro* as well as in transgenic plants. The results obtained were conflicting because *in vivo* the gene was induced only by heat and some heavy metals, whereas in transgenic tobacco the expression was elicited also by ABA and salt stress.

Some heat shock proteins (high molecular weight? low molecular weight? all of them?) could play a generic role in drought stress response, in chaperoning protein synthesis or protein assembling and transport in anhydrous conditions or by favoring the accumulation of antioxidant molecules which promote the formation of SH bonds between protein molecules. The function of stress-specific proteins and genes so far identified is more peculiar than the role of HSPs. For instance, some *lea* genes have been studied in transgenic plants or expressed in antisense orientation but little evidence of their role has been obtained so far by an *in vivo* assay. A significant contribution to understanding drought response mechanisms comes from genetic analysis and, in particular, from quantitative analysis. Several QTs have been identified and mapped by tagging the relative genetic loci (QTLs) with molecular markers. Traits of enormous practical and theoretical importance such as osmotic adjustment, ASI (anthesis-silking interval), stay green, ABA accumulation and yield have been mapped in several plants. In some cases it has been possible to tag these genes with probes corresponding to stress-inducible genes. The hypothesis that QTLs may be particular alleles of complex Mendelian loci is becoming experimentally attractive.

Acknowledgements. Financial support for this work was provided by National Research Council (Italy) Progetto Finalizzato R.A.I.S.A. assigned to Nelsòn Marmiroli, Paper no. 2469 and by the Italian Ministry of Agricultural, Food and Forest Resources, special programme "Genetic resistance of crop plants to biotic and abiotic stress" project n°10, and by the EU framework III Biotechnology project "Development of rapid novel molecular and cellular tools for the assessment and evaluation of genetic diversity in plants".

References

Almoguera C, Jordano J (1992) Developmental and environmental concurrent expression of sunflower dry-seed-stored low-molecular-weight heat-shock protein and *Lea* mRNAs. Plant Mol Biol **19**: 781-792

Assman SM (1994) Ins and outs of guard cell ABA receptors. Plant Cell **6**: 1187-1190

Bartels D, Engelhardt K, Roncarati R, Schneider K, Rotter M, Salamini F (1991) An ABA and GA modulated gene expressed in the barley embryo encodes an aldose reductase related protein. EMBO J **10**: 1037-1043

Bianchi G, Murelli C, Salamini F, Bartels D (1991) Novel carbohydrate-metabolism in the resurrection plant *Craterostigma plantagineum.* Plant J **3**: 355-359

Boheren KM, Bullock B, Wermuth B, Gabbay KH (1989) The aldo-cheto reductase superfamily: cDNAs and deduced amino acid sequences of human aldehyde and aldose reductases. J Biol Chem **264**: 9547-9551

Bonhert HJ, Thomas JC, De Rocher EJ, Michalowski CB, Breteineder H, Vernon DM, Deng W, Yamada S, Jensen RG (1994) Responses to salt stress in the halophyte *Mesembryanthemum crystallinum. In* JH Cherry, ed, Biochemical and Cellular Mechanisms of Stress Tolerance in Plants. Springer-Verlag, Berlin, pp 415-428

Bray EA (1991) Regulation of gene expression by endogenous ABA during drought stress. *In* WJ Davies, HG Jones, eds, Abscisic acid physiology and biochemistry. Oxford βios Scientific Publishers, pp 81-89

Bray EA (1993) Molecular responses to water deficit. Plant Physiol **103**: 1035-1040

Chandler PM, Robertson M (1994) Gene expression regulated by abscisic acid and its relation to stress tolerance. Annu Rev Plant Physiol Plant Mol Biol **45**: 113-141

Close TJ, Chandler PM (1990) Cereal dehydrins: serology, gene mapping and potential functional roles. Aust J Plant Physiol **17**: 333-344

Devos KM, Millan T, Gale MD (1993) Comparative RFLP maps of the homoeologous group 2 chromosomes of wheat, rye and barley. Theor Appl Genet **85**: 784-792

Dure LIII, Crouch M, Harada J, Ho T-HD, Mundy J, Quatrano R, Thomas T, Sung ZR (1989) Common amino acid sequence domains among the LEA proteins of higher plants. Plant Mol Biol **12**: 475-486

Grossi M, Gulli M, Stanca AM, Cattivelli L (1995) Characterization of two barley genes that respond rapidly to dehydration stress. Plant Sci **105**: 71-80

Gulli M, Maestri E, Hartings H, Raho G, Perrotta C, Devos KM, Marmiroli N (1995) Isolation and characterization of abscisic acid inducible genes in barley seedlings and their responsiveness to environmental stress. Life Sci Adv - Plant Physiol **14**: 89-96

Harris MJ, Outlaw Jr WH (1991) Rapid adjustment of guard-cell abscisic acid levels to current leaf-water status. Plant Physiol **95**: 171-173

Heikkila JJ, Papp JET, Schultz GA, Bewley JD (1984) Induction of heat shock protein messenger RNA in maize mesocotyls by water stress, abscisic acid, and wounding. Plant Physiol **76**: 270-274

Heun M, Kennedy AE, Anderson JA, Lapitan NLV, Sorrells ME, Tanskley SD (1991) Construction of a restriction fragment length polymorphism map for barley (*Hordeum vulgare*). Genome **34**: 437-447

Kimpel JA, Key JL (1985) Presence of heat shock mRNA in field grown soybean. Plant Physiol **79**: 672-678

Lee SP, Chen THH (1993) Molecular cloning of abscisic acid-responsive element mRNAs expressed during the induction of freezing tolerance in bromegrass (*Bromus inermis* Leys) suspension culture. Plant Physiol **101**: 1089-1096

Liang P, Pardee AB (1992) Differential display of eukaryotic messenger RNA by means of the polymerase chain reaction. Science **257**: 967-970

Lisitsyn N, Lysitsyn N, Wigler M (1993) Cloning the differences between two complex genomes. Science **259**: 946- 951

Lupotto E, Raho G, Hartings H, Della Torre PA, Perrotta C, Gulli M, Marmiroli N (1993) Expression of genes related to environmental stress in maize. *In* Breeding and Molecular Biology: accomplishments and future promises Proceedings XVI Conference EUCARPIA Maize and Sorghum, Bergamo, pp 224-230

Mac Robbie EAC (1991) Effect of ABA on ion transport and stomatal regulation. *In* WJ Davies, HG Jones, eds, Abscisic acid physiology and biochemistry. Oxford ßios Scientific Publishers, pp 39-52

Marmiroli N, Pavesi A, Di Cola G, Hartings H, Raho G, Conte MR, Perrotta C (1993) Identification, characterization, and analysis of cDNA and genomic sequences encoding two different small heat shock proteins in *Hordeum vulgare*. Genome **36**: 1111-1119

Mcelwain EF, Spiker S (1989) A wheat cDNA clone which is homologous to the 17kD heat shock protein gene family of soybean. Nucl Acid Res **17**: 1764

Moore G (1995) Grasses line up and form a circle. Curr Biol **5**: 737-739

Mundy J, Chua NH (1988) Abscisic acid and water-stress induce the expression of a novel rice gene. EMBO J **7**: 2279-2286

Navarri-Izzo F, Loggini B, Bochicchio A, Sgherri CLM (1995) The response of the glutathione system to drought in *Boea hygroscopica*: protection against oxidation of -SH groups of proteins. Proceedings of International Congress of Integrated Studies on Drought Tolerance of Higher Plants, Montpellier 30 August-2 September, V17

Nelson D, Salamini F, Bartels D (1994) Abscisic acid promotes novel DNA-binding to a desiccation related promoter of *Craterostigma plantagineum*. Plant J **5**: 451-458

Nei M, Li WH (1979) Mathematical model for studying genetical variation in terms of restriction endonucleases. Proc Natl Acad Sci USA **74**: 5267-5273

Quarrie SA, Gulli M, Calestani C, Steed A, Marmiroli N (1994) Location of a gene regulating drought-induced abscisic acid production on the long arm of chromosome 5A of wheat. Theor Appl Genet **89**: 794-800

Quarrie SA, Tuberosa R, Stefanelli S, Melchiorre R, Calestani C, Steed A, Semikhodski A (1995) QTL for leaf abscisic acid content in stressed wheat plants and association with other traits. Proceedings of International Congress of Integrated Studies on Drought Tolerance of Higher Plants, Montpellier 30 August-2 September, XI 3

Raho G, Lupotto E, Della Torre P, Hartings H, Perrotta C, Marmiroli N (1995) Functional analysis of the temperature dependent expression of the barley Hvhsp17 gene promoter in monocot and dicot cell system. Plant Sci **106**: 63-69

Skriver K, Mundy J (1990) Gene expression in response to abscisic acid and osmotic stress. Plant Cell **2**: 503-512

Vierling E (1991) The role of heat shock proteins in plants. Annu Rev Plant Physiol Plant Mol Biol **42**: 579-620

Weng J, Wang Z-T, Nguyen HT (1993) Molecular cloning and sequence analysis of cDNAs encoding cytoplasmic low molecular weight heat shock proteins in hexaploid wheat. Plant Sci **92**: 35-46

Enhanced Desiccation Survival by Engineering Osmolyte Biosynthesis in Plants

E. Tapio Palva[1], Kjell-Ove Holmström[1], Einar Mäntylä[1], Björn Welin[1], Abul Mandal[1], Outi E. Tunnela[2,3] and John Londesborough[2]

1 Department of Molecular Genetics, Uppsala Genetic Center, Swedish University of Agricultural Sciences, S-750 07 Uppsala, Sweden
2 Primalco Ltd, Biotec, FIN-05200 Rajamäki, Finland
3 Present address: National Public Health Institute, FIN-00300 Helsinki, Finland

Abstract. Plant growth, productivity and distribution are severely limited by environmental stress factors such as drought, salinity and freezing temperatures, all of which disturb the water balance of the cell. Plants as well as other organisms have evolved different strategies to alleviate the adverse effects of these stresses. A common adaptive response to stresses that results in water deficit is the accumulation of osmolytes or osmoprotectants that help cells to maintain their water balance and, in addition, protect macromolecules in stressed cells. The simplicity of the metabolic pathways leading to osmolyte biosynthesis makes them amenable to genetic engineering. We will discuss engineering the biosynthetic pathways of two different types of osmoprotectants, the quaternary ammonium compound glycine betaine, and the nonreducing disaccharide trehalose. Biosynthesis of both compounds is a two-step process in prokaryotes as well as in eukaryotes. For biosynthesis of glycine betaine we employed the bacterial genes *betA* and *betB* encoding choline dehydrogenase and betaine aldehyde dehydrogenase, respectively. These genes have been expressed in transgenic tobacco and were shown to result in biosynthesis of glycine betaine. For biosynthesis of trehalose we used the *TPS1* gene from yeast encoding the first enzyme of the pathway, trehalose-6-phosphate synthase. Transgenic tobacco plants expressing *TPS1* were shown to produce active TPS1 and accumulate trehalose in the leaves. Production of trehalose was associated with enhanced desiccation survival both in primary transformants and TPS1-positive progeny. The results suggest that engineering osmolyte biosynthesis may provide an efficient strategy to generate crop plants with enhanced tolerance to water deficit as well as improved post harvest storage properties.

Keywords. *Arabidopsis thaliana*, desiccation survival, drought tolerance, glycine betaine, osmolyte, osmoprotectant, salt tolerance, trehalose, water deficit

1 Adaptation to Water Deficit

Most plants are at least temporarily exposed to conditions of water deficit during their life cycle. Drought, high salinity and freezing temperatures are environmental stress factors that all perturb the water balance of the cell and the stress caused by these factors constitutes the most serious limitation to plant growth, productivity and distribution (Boyer, 1982). Hence, breeding for increased tolerance to stresses resulting in water deficit is of prime importance in areas regularly affected by these stresses. Such breeding programs could be greatly facilitated by incorporation of genetic engineering approaches, provided that the molecular basis for stress tolerance has been characterized.

Plants have evolved different types of adaptive strategies to alleviate the detrimental effects of dehydration stress. As discussed by McCue and Hanson (1990) developmental, structural or physiological adaptations to stress are often based on complex mechanisms involving a number of yet uncharacterized genes. Such traits are not amenable to genetic engineering. However, some of the described responses to dehydration stress appear to be based on relatively simple metabolic traits governed by a limited number of genes. Many of the genes induced upon water deficit or anticipated water deficit (e.g. during cold acclimation) encode polypeptides with putative protective roles in the stressed cells, such as ion sequestration, membrane stabilization, and chaperone functions (Bray, 1993; Palva, 1994). Such protective proteins, including LEA proteins and dehydrins, have been shown to accumulate in plants during seed maturation as well as in vegetative tissues in response to drought, salt and low temperature stress (Baker et al., 1988; Skriver and Mundy, 1990; Welin et al., 1994; Mäntylä et al., 1995). Another category of protective substrates that accumulate during response to stresses that result in water deficit is composed of small organic molecules called compatible solutes, osmolytes or osmoprotectants (Yancey et al., 1982). These types of molecules can accumulate to high levels in response to stress in most living organisms from bacteria to man without disrupting metabolic processes or cellular structures and functions. Only a limited number of different types of organic compounds are commonly employed as osmoprotectants including quaternary ammonium compounds such as glycine betaine, some amino acids like proline, polyols like inositol and mannitol, and sugars including sucrose and trehalose (Yancey et al., 1982; McCue and Hanson, 1990; Strøm and Kaasen, 1993; Potts, 1994; Bohnert et al., 1995).

Protection by osmolytes seems to be by two separate but overlapping mechanisms, osmotic adjustment and osmoprotection. In osmotic adjustment or osmoregulation, osmolyte accumulation helps the cells to maintain their water balance by facilitating water retention. As osmoprotectants these molecules appear to stabilize macromolecules and to maintain the integrity of cellular structures in cell suffering from stress caused by water loss (Yancey et al., 1982; Crowe et al., 1992; Potts, 1994). This stress protective role of osmolytes has been clearly established both by *in vivo* studies correlating osmolyte accumulation with stress tolerance (Csonka, 1989; Wiemken, 1990; Delauney and Verma, 1993; Strøm and Kaasen, 1993) and by *in vitro* studies demonstrating protection of isolated enzymes from stress inactivation. Osmolyte biosynthesis has been well characterized in many cases and shown to involve only a limited number of steps. The genes required for biosynthesis of several of the osmolytes have been cloned from both prokaryotic and eukaryotic organisms (McCue and Hanson, 1990; Lamark et al., 1991; Delauney and Verma, 1993; Strøm and Kaasen, 1993). Consequently, genetic engineering of osmolyte biosynthesis is becoming feasible and there are already examples of successful application of this approach including mannitol and proline overproduction in transgenic tobacco (Tarczynski et al., 1993; Kisher et al., 1995). Introduction of the bacterial *mtlD* gene into tobacco resulted in mannitol production and improved salinity tolerance (Tarczynski et al., 1992, 1993). Similarly, over-expression in tobacco of the mothbean gene for 1-pyrroline-5-carboxylate synthase, catalyzing a limiting step in proline biosynthesis, resulted in increased proline biosynthesis and conferred enhanced osmotolerance to the transgenic plants (Kisher et al., 1995). Our own efforts have been directed to engineer the biosynthetic pathways of two different types of osmolytes, glycine betaine and trehalose in plants (Holmström et al., 1994; Holmström, unpublished results). The current status of this research will be reviewed below.

2 Production of Glycine Betaine in Transgenic Plants

The quaternary ammonium compound glycine betaine is one of the most commonly employed osmolytes in bacteria, cyanobacteria, algae and higher plants as well as animals. Glycine betaine accumulation has been primarily associated with high salinity stress but is also induced by drought and low temperatures (Yancey et al., 1982; McCue and Hanson, 1990; Ishitani et al., 1993). However, many important crop plants including potato, tomato and rice do not accumulate this osmolyte and hence provide the targets for crop improvement (Weretilnyk et al., 1989; McCue and Hanson, 1990).

Glycine betaine biosynthesis is a two-step oxidation of choline in both prokaryotes and eukaryotes. In the first step choline is converted to betaine aldehyde by a chloroplastic choline mono-oxygenase (CMO) in plants while bacteria like *Escherichia coli* employ a membrane associated choline dehydrogenase (CDH) for the same task (Landfald and Strøm, 1986; Brouquisse et al., 1989; Weretilnyk and Hanson, 1990; Lamark et al., 1991). The bacterial CDH is also able to catalyze the second step, the conversion of betaine aldehyde to glycine betaine. This reaction is facilitated, in both pro- and eukaryotes, by the second enzyme in the pathway, betaine aldehyde dehydrogenase (BADH), a chloroplast protein in plants (Weigel et al., 1986) and soluble protein in bacteria (Landfald and Strøm, 1986; Boyd et al., 1991; Lamark et al., 1991). The *E. coli* genes, *betA* and *betB* encoding CDH and BADH, respectively, have been cloned and characterized (Lamark et al., 1991). To express these genes in plants they were placed under the control of the strong light-responsive promoter of the *Arabidopsis thaliana ats1A* gene, which codes for ribulose 1,5-bisphosphate carboxylase (Rubisco) small subunit (Krebbers et al., 1988). In addition to the promoter we employed in some of the constructs the chloroplast targeting signal of *ats1A* (Fig. 1 A).

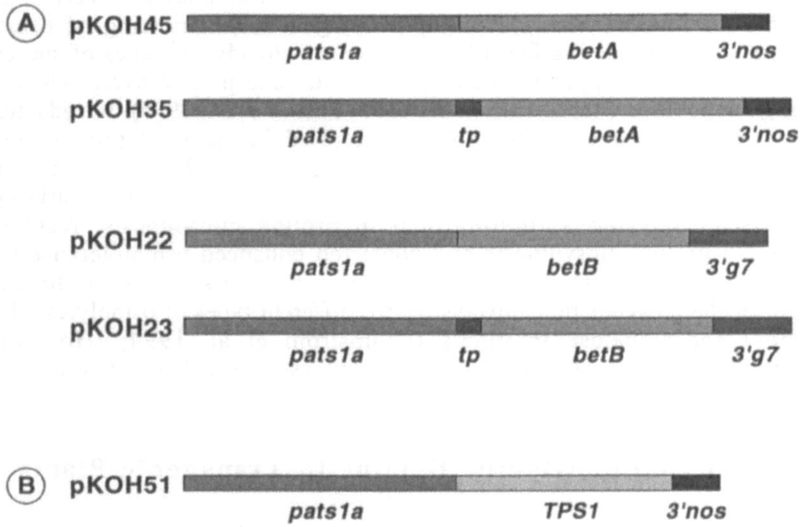

Fig. 1 A. The chimeric genes carrying *E. coli betA* and *betB* coding sequences. **B**. The chimeric gene carrying yeast *TPS1* coding sequence. The promoter of *ats1A* is indicated by *pats1A*, the 122 bp sequence from *ats1A* encoding the transit peptide for chloroplast targeting is indicated by *tp*, and the 3' untranslated region of the T-DNA genes *g7* and *nos* by 3'*g7* and 3'*nos*, respectively.

We have previously demonstrated that the introduction of the chimeric *betB* genes to tobacco resulted in accumulation of the corresponding transcripts and synthesis of the BetB polypeptide which was localized in the cytoplasm, or was successfully targeted to the chloroplast if provided with the transit peptide (Holmström et al., 1994). Similarly, expression of the chimeric *betA* genes in transgenic tobacco resulted in the accumulation of the BetA polypeptide, either as cytoplasmic or chloroplast targeted forms (Fig. 2). The *betB* encoded BADH was shown to be active in the transgenic tobacco lines as demonstrated by activity assays as well as by the ability of BetB-positive transgenic lines to grow on inhibitory concentrations (10 mM) of betaine aldehyde (Holmström et al., 1994) indicating that these lines indeed were able to accumulate glycine betaine. Similar experiments with *betA* transgenic plants indicated that also BetA was active, as the transgenic lines were able to grow on inhibitory concentrations of choline (30 mM) which was clearly toxic to non-transformed wild type tobacco. Preliminary experiments with BetA-positive transgenic lines indicate that these plants appear to have increased tolerance to salt stress. Our efforts are now directed in generating double transgenic *betA*$^{+}$ *betB*$^{+}$ lines by genetic crosses and to use these for evaluation of the contribution of increased glycine betaine levels to stress tolerance in a plant species (tobacco) that does not normally synthesize this osmolyte.

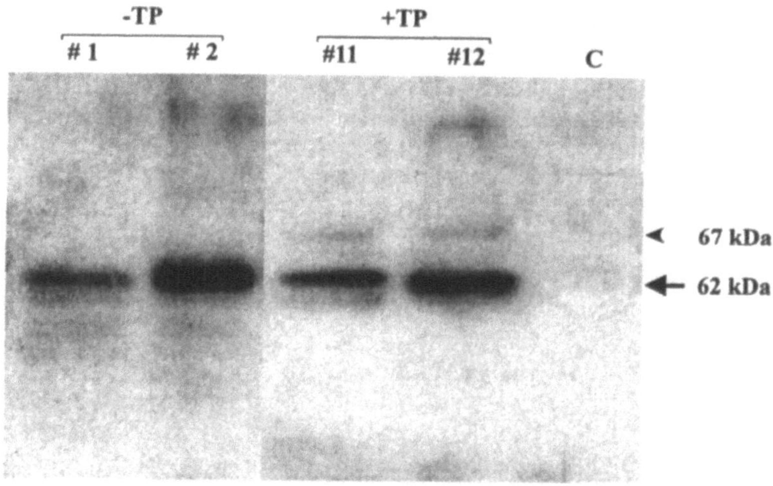

Fig. 2. Accumulation of BetA polypeptide in transgenic tobacco. Western analysis of protein extracts from transgenic tobacco lines harbouring chimeric *betA* genes. The 62 kD BetA polypeptide is indicated by an arrow. The presence or absence of the chloroplast targeting signal (transit peptide) is indicated by +TP and -TP, respectively. The weak 67 kD band indicated by an arrowhead is the non-processed form of BetA including the transit peptide. Non-transformed control plants are analyzed in C

3 Engineering Drought Tolerance by Trehalose Biosynthesis

Trehalose, the non-reducing disaccharide of glucose, is employed as an osmoprotectant by many anhydrobiotic organisms capable of surviving complete dehydration. Accumulation of trehalose has been associated with desiccation survival in yeast, fungi, cyanobacteria and some resurrection plants (Crowe et al., 1984; Clegg, 1986; Wiemken, 1990; Crowe et al., 1992; Bianchi et al., 1993; Drennan et al., 1993; Eleutherio et al., 1993). Trehalose is, however, rare in vascular plants and it has been suggested that sucrose may replace trehalose in desiccation tolerant seed and pollen (Hoekstra et al., 1989). Thus, most crop plants would provide useful targets for engineering stress tolerance by introduction of trehalose biosynthesis. Trehalose is a very efficient protectant of enzymes, liposomes and membranes *in vitro* (Colaço et al., 1992) and its ability to stabilize enzymes and cellular structures may explain its role in desiccation survival (Crowe et al., 1992; Crowe and Crowe, 1993). In addition to its role as an osmoprotectant, trehalose accumulation in yeast appears to be associated also with enhanced heat stress tolerance (De Virgilio et al., 1994) and is also induced by other stress factors including heavy metals and hydrogen peroxide (Wiemken, 1990). This more general stress protective function of trehalose could also be derived from its ability to stabilize macromolecules in stressed cells.

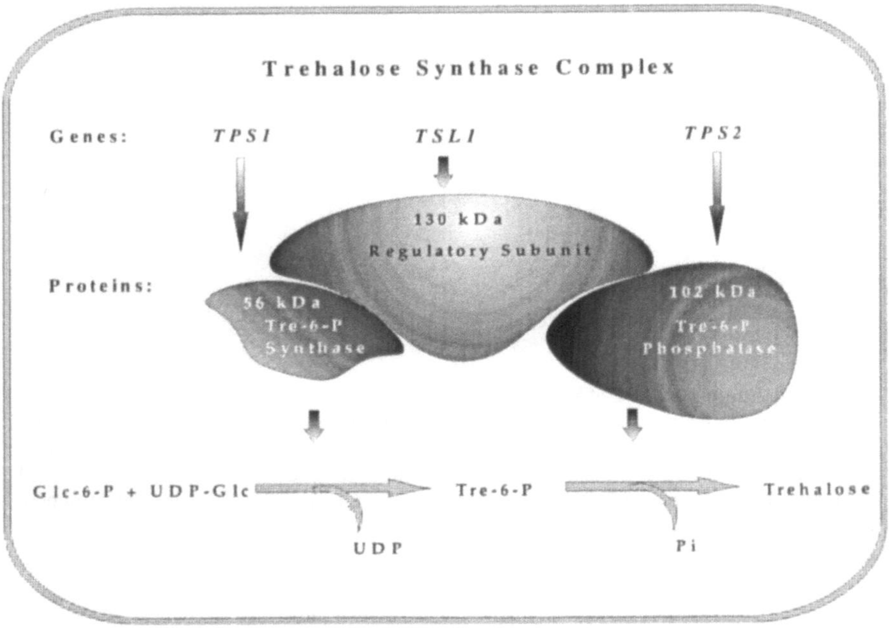

Fig. 3. The trehalose synthase complex in yeast. The *TPS1* gene encodes the trehalose-6-phosphate synthase subunit of the complex which is required for condensation of UDP-glucose and glucose-6-phosphate to trehalose-6-phosphate (Tre6P). *TPS2* encodes a specific Tre6P phosphatase that removes the phosphate from Tre6P yielding trehalose. TSL1 encodes a regulatory subunit of the trehalose synthase complex

Biosynthesis of trehalose is a two-step process both in yeast and in *E. coli*, starting with a condensation of UDP-glucose and glucose-6-phosphate into trehalose-6-phosphate (Tre6P) catalyzed by trehalose-6-phosphate synthase. A specific trehalose-6-phosphate phosphatase dephosphorylates Tre6P to yield trehalose (Cabib and Leloir, 1958). The genes coding for the enzymes involved in trehalose biosynthesis have been characterized from *E. coli* (Kaasen et al., 1994) and bakers yeast (De Virgilio et al., 1993; Vuorio et al., 1993). In yeast, the trehalose biosynthetic enzymes form a large complex, trehalose synthase, is probably located in the cytoplasm (Londesborough and Vuorio, 1991). This complex appears to be composed of the Tre6P synthase (TPS1), Tre6P phosphatase (TPS2) and a regulatory subunit of 130 kD (TSL1) (Fig. 3).

The regulatory subunit encoded by *TSL1* does not appear to be necessary for trehalose biosynthesis (Vuorio et al., 1993), hence engineering of trehalose biosynthesis into target plants would only require introduction of the *TPS1* and *TPS2* genes. Furthermore, the required substrates UDP-glucose and glucose-6-phosphate are products of basal carbohydrate metabolism and available in the plant cell cytoplasm.

To investigate the potential of trehalose in improving desiccation tolerance in plants in tobacco the *TPS1* gene from yeast, encoding the first enzyme in the trehalose biosynthetic pathway was expressed. To accomplish this, *TPS1* was placed under the control of the *ats1A* promoter of *A. thaliana* (Fig. 1B) and the chimeric gene transferred to tobacco (Holmström, unpublished results). Transgenic TPS1 positive tobacco lines were isolated and shown to exhibit TPS1 activity as well as to produce detectable amounts of trehalose (Fig. 4).

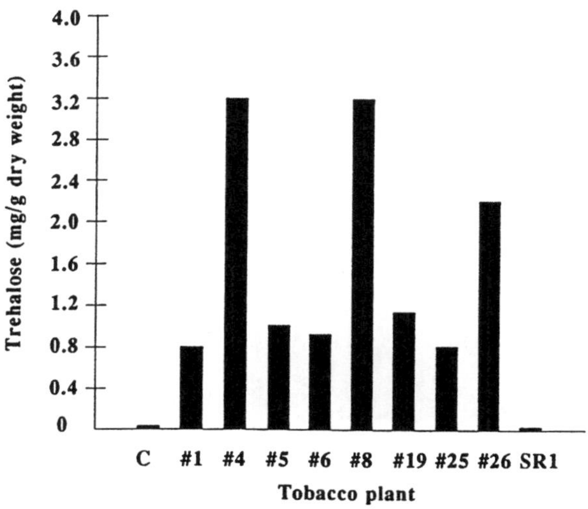

Fig. 4. Accumulation of trehalose in transgenic tobacco. The amount of trehalose was determined chromatographically in different transgenic TPS1-positive lines (#1-26) of tobacco (Holmström, unpublished results). The non-transformed and vector transformed control tobacco are indicated by SR1 and C, respectively

Fig. 5. Enhanced desiccation survival of trehalose-producing tobacco. A vector transformed control line (C) and a trehalose-producing transgenic line (#4) of tobacco propagated *in vitro* were exposed to air drying (30% relative humidity). The plants were exposed to this dehydration stress (DH) for 16 h and then rehydrated (RH) for times indicated

Apparently, introduction of the first enzyme of the pathway, TPS1, was sufficient for trehalose production suggesting that plants may possess unspecific phosphatases capable of converting the Tre6P formed by TPS1 to trehalose.

The production of trehalose in tobacco could be correlated to enhanced desiccation survival (Holmström, unpublished results). Exposure of trehalose-producing transgenic tobacco plants to air drying, demonstrated that the transgenic plants did have improved ability for water retention when compared with non transformed or vector-transformed controls. Only the trehalose-producing plants were able to survive a prolonged drought exposure, resuming growth when the dried plants were rehydrated (Fig. 5).

These results demonstrate that trehalose can, even at relatively low concentrations, enhance desiccation survival of transgenic plants producing this osmolyte. The mechanism of this enhanced desiccation tolerance is not clear but does not seem to be due to osmotic adjustment of the cell cytoplasm, for which the observed concentrations of trehalose are too low. Rather, the improved stress tolerance could be due to the osmoprotective properties of trehalose. Such protection comparable to those observed in the transgenic plants has previously been observed in yeast at relatively modest trehalose concentrations (De Virgilio et al., 1990). Thus the protective effects of trehalose on cellular structures such as membranes and macromolecules may underlie the observed improvement in water retention and desiccation survival. In conclusion, production of trehalose clearly results in enhanced drought tolerance and may also provide protection against the adverse effects of other types of stress that result in water deficit, namely freezing and high salinity.

The results presented here demonstrate that engineering plant tolerance to water deficit is feasible and can be achieved by introducing relatively simple metabolic pathways for osmoprotectant biosynthesis. Consequently, such novel approaches should be incorporated into traditional breeding programs and would provide new avenues for breeding plants for improved water stress tolerance.

Acknowledgements. This research was supported, in part, by grants from the Swedish Natural Sciences Research Council, the Swedish Research Council for Engineering Sciences and the Technology Development Centre of Finland. A grant from the Nilsson-Ehle foundation is acknowledged.

References

Baker J, Steele C, Dure III L (1988) Sequence and characterization of 6 Lea proteins and their genes from cotton. Plant Mol Biol **11**: 277-291

Bianchi G, Gamba A, Limiroli R, Pozzi N, Elster R, Salamini F, Bartels D (1993) The unusual sugar composition in leaves of the resurrection plant *Myrothamnus flabellifolia*. Physiol Plant **87**: 223-226

Bohnert HJ, Nelson DE, Jensen RG (1995) Adaptations to environmental stresses. Plant Cell **7**: 1099-1111

Boyd LA, Adam L, Pelcher LE, McHughen A, Hirji R, Selvaraj G (1991) Characterization of an *Escherichia coli* gene encoding betaine aldehyde dehydrogenase (BADH): structural similarity to mammalians ALDHs and a plant BADH. Gene **103**: 45-52

Boyer JS (1982) Plant productivity and environment. Science **218**: 443-448

Bray EA (1993) Molecular responses to water deficit. Plant Physi **103**: 1035-1040

Brouquisse R, Weigel P, Rhodes D, Yocum CF, Hanson AD (1989) Evidence for ferredoxin-dependent choline monooxygenase from spinach chloroplast stroma. Plant Physiol **90**: 322-329

Cabib E, Leloir FL (1958) The biosynthesis of trehalose-6-phosphate. J Biol Chem **231**: 259-275

Clegg JS (1986) The physical properties and metabolic status of *Artemia* cysts at lower water contents: the "water replacement hypothesis". *In* AC Leopold, ed, Membranes, Metabolism and Dry Organisms. Cornell University Press, Ithaca, pp 169-187

Colaço C, Sen S, Thangavelu M, Pinder S, Roser B (1992) Extraordinary stability of enzymes dried in trehalose: simplified molecular biology. Biotechnology **10**: 1007-1011

Crowe JH, Crowe LM, Chapman D (1984) Preservation of membranes in anhydrobiotic organisms: the role of trehalose. Science **223**: 701-703

Crowe JH, Hoekstra FA, Crowe LM (1992) Anhydrobiosis. Annu Rev Physiol **54**: 579-599

Crowe JH, Crowe LM (1993) Evidence for direct interaction between disaccharides and dry phospholipids. Cryobiology **30**: 226-227

Csonka LN (1989) Physiological and genetic responses of bacteria to osmotic stress. Microb Rev March: 121-147

De Virgilio C, Simmen U, Hottiger T, Boller T, Wiemken A (1990) Heat shock induces enzymes of trehalose metabolism, trehalose accumulation, and thermotolerance in *Schizosaccharomyces pombe*, even in the presence of cycloheximide. FEBS Letters **273**: 107-110

De Virgilio C, Bürckert N, Bell W, Jenö P, Boller T, Wiemken A (1993) Disruption of *TPS2*, the gene encoding the 100-kDa subunit of the trehalose-6-phosphate synthase/phosphatase complex in *Saccharomyces cerevisiae*, causes accumulation of trehalose-6-phosphate and loss of trehalose-6-phosphate phosphatase activity. Eur J Biochem **212**: 315-323

De Virgilio C, Hottiger T, Dominguez J, Boller T, Wiemken A (1994) The role of trehalose synthesis for the acquisition of thermotolerance in yeast: genetic evidence that trehalose is a thermoprotectant. Eur J Biochem **219** (1-2): 187-193

Delauney AJ, Verma DPS (1993) Proline biosynthesis and osmoregulation in plants. Plant J **4**: 215-223

Drennan PM, Smith MT, Goldsworthy D, Van Staden J (1993) The occurrence of trehalose in the leaves of the desiccation-tolerant angiosperm *Myrothamnus flabellifolius* Welw. J Plant Physiol **142** (4): 493-496

Eleutherio ECA, de Araujo PS, Panek AD (1993) Role of the trehalose carrier in dehydration resistance of *Saccharomyces cerevisiae*. Biochim Biophys Acta **1156**: 263-266

Hoekstra FA, Crowe LM, Crowe JH (1989) Differential desiccation sensitivity of corn and *Pennisetum* pollen linked to their sucrose contents. Plant, Cell and Envir **12**: 83-92

Holmström K-O, Welin B, Mandal A, Kristiansdottir I, Teeri T, Lamark T, Ström AR, Palva ET (1994) Production of the *Escherichia coli* betaine-dehydrogenase, an enzyme required for the synthesis of the osmoprotectant glycine-betaine, in transgenic plants. Plant J **6**: 749-758

Ishitani M, Arakawa K, Mizuno K, Kishitani S, Takabe T (1993) Betaine aldehyde dehydrogenase in the *Gramineae*: levels in leaves of both betaine-accumulating and non-accumulating cereal plants. Plant Cell Physiol **34**: 493-495

Kaasen I, McDougall J, Strøm AR (1994) Analysis of the *ots BA* operon for osmoregulatory trehalose synthesis in *Escherichia coli* and homology of the OtsA and OtsB proteins to the yeast trehalose-6-phosphate synthase/phosphatase complex. Gene **145**: 9-15

Kisher PBK, Hong Z, Miao G-H, Hu C-AA, Verma DPS (1995) Over-expression of 1-pyrroline-5-carboxylate synthetase increases proline production and confers osmotolerance in transgenic plants. Plant Physiol **108**: 1387-1394

Krebbers E, Seurinck J, Herdies L, Cashmore AR, Timko MP (1988) Four genes in two diverged subfamilies encode the ribulose-1,5-biphosphate carboxylase small subunit polypeptides of *Arabidopsis thaliana*. Plant Mol Biol **11**: 745-759

Lamark T, Kaasen I, Eshoo MW, Falkenberg P, Mcdougall J, Strøm AR (1991) DNA sequence and analysis of the *bet* genes encoding the osmoregulatory choline-glycine betaine pathway of *Escherichia coli*. Mol Microbiol **5**: 1049-1064

Landfald B, Strøm AR (1986) Choline-glycine betaine pathway confers a high level of osmotic tolerance in *Escherichia coli*. J Bacteriol **165**: 849-855

Londesborough J, Vuorio O (1991) Trehalose-6-phosphate synthase/phosphatase complex from baker's yeast: purification of a proteolytically activated form. J Gen Microbiol **137**: 323-330

Mäntylä E, Lång V, Palva ET (1995) Role of abscisic acid in drought-induced freezing tolerance, cold acclimation, and accumulation of LTI78 and RAB18 proteins in *Arabidopsis thaliana*. Plant Physiol **107**: 141-148

McCue KF, Hanson AD (1990) Drought and salt tolerance: towards understanding and application. Trends in Biotech **8**: 358-362

Palva ET (1994) Gene expression under low temperature stress. *In* AS Basra, ed, Stress Induced Gene Expression in Plants. Harwood Academic Publishers, New York, pp 103-130

Potts M (1994) Desiccation tolerance of prokaryotes. Microbiol Rev **58**(4): 755-805

Skriver K, Mundy J (1990) Gene expression in response to abscisic acid and osmotic stress. Plant Cell **2**: 503-512

Strøm AR, Kaasen I (1993) Trehalose metabolism in *Escherichia coli*: stress protection and stress regulation of gene expression. Mol Microbiol **8**: 205-210

Tarczynski MC, Jensen RG, Bohnert HJ (1992) Expression of a bacterial *mtlD* gene in transgenic tobacco leads to production and accumulation of mannitol. Proc Natl Acad Sci USA **89**: 2600-2604

Tarczynski MC, Jensen RG, Bohnert HJ (1993) Stress protection of transgenic tobacco by production of the osmolyte mannitol. Science **259**: 508-510

Vuorio E, Kalkkinen N, Londesborough J (1993) Cloning of two related genes encoding the 56-kDa and 123-kDa subunits of trehalose synthase from the yeast *Saccharomyces cerevisiae*. Eur J Biochem **216**: 849-861

Weigel P, Weretylnik EA, Hanson AD (1986) Betaine aldehyde oxidation by spinach chloroplasts. Plant Physiol **82**: 753-759

Welin BV, Olson Å, Nylander M, Palva ET (1994) Characterization and differential expression of *dhn/lea/rab*-like genes during cold acclimation and drought stress in *Arabidopsis thaliana*. Plant Mol Biol **26**: 131-144

Weretilnyk EA, Bednarek S, McCue KF, Rhodes D, Hanson AD (1989) Comparative biochemical and immunological studies of the glycine betaine synthesis pathway in diverse families of dicotyledons. Planta **178**: 342-352

Weretilnyk EA, Hanson AD (1990) Molecular cloning of a plant betaine-aldehyde dehydrogenase: an enzyme implicated in adaptation to salinity and drought. Proc Natl Acad Sci USA **87**: 2745-2749

Wiemken A (1990) Trehalose in yeast, stress protectant rather than reserve carbohydrate. Antonie Leeuwenhoek Intl J Gen Microbiol **58**: 209-217

Yancey PH, Clark ME, Hand SC, Bowlus RD, Somero GN (1982) Living with water stress: evolution of osmolyte systems. Science **217**: 1214-1222

Part 5

Light Stress

Transduction of Light Signals in Plants

Chris Bowler[1], Gunther Neuhaus[2] and Nam-Hai Chua[3]

[1] Stazione Zoologica, I-80121 Napoli, Italy
[2] Institut für Pflanzenwissenschaften, ETH-Zentrum, CH-8092 Zürich, Switzerland
[3] Laboratory of Plant Molecular Biology, The Rockefeller University, NY10021-6399, USA

Abstract. The phytochromes are the best studied plant photoreceptors, controlling a wide variety of responses at both whole plant and single cell levels. Three signal transduction pathways, dependent on cGMP and/or calcium, are used by phytochrome to control the expression of genes required for chloroplast development and anthocyanin biosynthesis. Control of the relative inputs into these different transduction pathways may be a means by which plant cells can adjust their biochemical status, in particular for controlling the levels of photoprotectants (such as anthocyanins) and the relative amounts of the different photosynthetic complexes. If this were the case one would predict the existence of cross-talk mechanisms between the different phytochrome pathways, which would allow the activities of each pathway to be known by the others and which would allow regulatory signals to pass between them. Experiments will be presented that have elucidated some of these negative regulatory mechanisms. In addition to these studies of light-activated processes, we have also performed some experiments to understand how phytochrome can down-regulate other events. For example, the expression of several genes is known to be negatively-regulated by light. We have determined how such genes are regulated by phytochrome in the context of calcium and cGMP and have identified an 11 bp cis-element present within the promoters of such genes that is both necessary and sufficient to mediate light down-regulation.

Keywords. Light signal, *Lycopersicon esculentum*, photoperception, phytochrome, signal transduction, tomato

1 Introduction

Plants detect changes in ambient light to optimize photosynthetic reactions and to regulate their growth and development. Light is perceived through four distinct light receptors that respond to particular wavelengths and intensities of light - phytochrome, which absorbs red and far red light, the blue, UV-A, and UV-B photoreceptors (Quail et al., 1995). Of these, the best characterized biochemically and physiologically is phytochrome. Phytochrome consists of a polypeptide and a covalently bound chromophore, and light perception is mediated by the photoisomerization of the chromophore. Etiolated and green tissue contain different pools of phytochrome, respectively referred to as type I and type II. Most plants contain small multigene families which encode the proteins that make up these two pools; *PHYA* encodes the type I phytochrome (also known as PHYA) whilst other genes (e.g., *PHYB*) encode type II phytochrome (Quail et al., 1995). Two spectrophotometrically different, but photo-interconvertible, forms of these phytochromes exist in the plant: the red light-absorbing P_r, and the far red-absorbing P_{fr}. P_{fr} is generally thought to be the active form.

There has recently been much progress in the elucidation of *cis*- and *trans*-acting factors that mediate phytochrome-regulated gene expression (Gilmartin et al., 1990). In contrast, there has been little real progress in our understanding of the molecular transduction events that couple photoperception by phytochrome in the cytosol to changes in gene expression in the nucleus. In the current work, our aim was to develop a new approach for elucidating phytochrome signal transduction intermediates. Specifically, we wanted to address the following questions: 1) What are the components of the phytochrome phototransduction machinery, and how are they ordered? 2) Because phytochrome regulates such a large number of molecular targets, are the transduction chains separate or branched, and are any of the intermediates shared? 3) Is phytochrome signalling cell autonomous or is there cell to cell signalling?

In order to address these questions we developed single cell assays employing microinjection to introduce large or small molecules into cells in an attempt to identify components that could modulate reactions normally carried out by phytochrome. As recipient material, we chose to use hypocotyl cells of a PHYA-deficient mutant of tomato known as *aurea* (Koornneef et al., 1981; Sharma et al., 1993). We used this mutant because, whilst wild type plant cells develop chloroplasts and anthocyanins in response to light, *aurea* hypocotyl cells do not.

This report summarizes data presented more thoroughly in other articles (Neuhaus et al., 1993; Bowler and Chua, 1994; Bowler et al., 1994a, b), and the interested reader is advised to consult these previous reports for further information. Essentially, we show that microinjection of an exogenous PHYA can restore chloroplast development and anthocyanin biosynthesis within an injected cell. Furthermore, we show that these responses require the participation of one or more heterotrimeric G proteins, together with calcium, calmodulin, and cGMP acting further downstream. Although G protein activation can mediate a full cellular response, equivalent to that initiated by PHYA, the efficacy of calcium and calmodulin is more restricted: no anthocyanin biosynthesis and chloroplast development is incomplete, due to the absence of two of the five major photosynthetic complexes, photosystem I (PSI) and cytochrome b_6f (Cyt. b_6f). cGMP, on the other hand, can stimulate anthocyanin biosynthesis and acts together with calcium and/or calmodulin to complete chloroplast development.

2 Results

2.1 Identification of Phytochrome Signalling Intermediates

When dark-grown *aurea* seedlings are placed in the light for 48 h, there is neither significant development of chloroplasts nor anthocyanin pigment biosynthesis in hypocotyl cells, in striking contrast to wild type cells treated in the same way. Since *aurea* lacks PHYA, we therefore reasoned that it should be possible to biochemically rescue wild type traits by microinjecting PHYA protein into *aurea* cells. Indeed, when oat PHYA was injected into such cells we observed within 48 h the development of anthocyanin pigments within the vacuole and the appearance of mature chloroplasts (Table 1). Interestingly, the injection of PHYA into a cell resulted only in the expression of PHYA-regulated processes in that cell and not in neighbouring cells (data not shown). This demonstrates that phytochrome signalling is cell autonomous.

To use this system to biochemically dissect the signal transduction pathways downstream of phytochrome, we used oat PHYA to initiate the molecular responses and attempted to identify pharmacological reagents that could antagonize these processes. A very common mechanism of signal transduction in animal cells involves the activation of a heterotrimeric G protein. We speculated that an initial event in phytochrome signalling may be the activation of such a component, and this was tested by coinjecting inhibitors of G proteins (either GDP-ß-S or pertussis toxin) with PHYA. This resulted in the loss of all detectable phytochrome responses, i.e., no anthocyanin or chloroplast development (Table 1). This result suggested that phytochrome signal transduction requires the participation of a G protein very early in the signalling process, because its inhibition resulted in the loss of all phytochrome-mediated events observed in a single cell.

The importance of calcium as a second messenger for phytochrome was assessed by first analysing the effect of PHYA injection in cells that were subsequently bathed in nifedipine, which blocks L-type calcium channels in animal cells. Interestingly, nifedipine was found to block chloroplast development but had no effect on anthocyanin biosynthesis, indicating specificity (Table 1). The same results were obtained with verapamil (another L-type calcium channel blocker, with an unrelated chemical structure) and also with the calmodulin antagonists trifluoroperazine and W7 (which are also structurally distinct) (Table 1). These results suggested that calcium, probably acting through calmodulin, is essential for regulating events during chloroplast development but not anthocyanin biosynthesis. Hence, it appears that phytochrome signalling may involve at least two pathways, one of which requires calcium whereas the other does not. Since GDP-ß-S and pertussis toxin are able to block all PHYA-regulated events, heterotrimeric G protein activation most likely precedes the branch point identified by its calcium requirements.

Table 1. Summary of microinjection experiments in *aurea* hypocotyl cells

Injected material	Conditions	Conc.	Number of injections	Number of activations of anthocyanin (A) and Chloroplasts (C)	Efficiency (%)
PHYA		> 10,000	628	48 (A and C)	7.6
Injection Buffer			1210	0	-
BSA		10,000	709	0	-
PHYA	+ G protein inhibitors	> 10,000	891	0	-
PHYA	+ Ca^{2+} or CaM inhibitors	> 10,000	823	37 (A)	4.5
GTP-γ-S		30 - 100μM	463	32 (A and C)	6.9
GTP-γ-S	+ Ca^{2+} or CaM inhibitors	30-100μM	225	18 (A)	8.0
CTX	+ GTP-γ-S	>1,000	280	15 (A and C)	5.3
Ca^{2+}		0.5-5μM	415	32 (C)	7.8
Ca^{2+}	+ CaM inhibitors	0.5-5μM	784	0	-
CaM	Ca^{2+} activated	> 5000	620	37 (C)	5.9
CaM	Not activated	> 5000	420	0	-
cGMP		30-80μM	800	58 (A)	7.3
PHYA	Rp-cGMPS	> 10,000	820	71 (C)	8.7
GTP-γ-S	Rp-cGMPS	30-100μM	630	72 (C)	11.4
Ca^{2+} + cGMP		as above	560	48 (A and C)	8.6
CaM + cGMP		as above	610	46 (A and C)	7.5

Notes: All injections and subsequent incubations (for 48 hours) were done in white light. All concentrations are expressed as final estimated intracellular concentrations. Concentrations are in number of molecules, unless indicated otherwise. G protein inhibitors were GDP-β-S (>1mM) and pertussis toxin (>5,000 molecules); Ca^{2+} inhibitors were nifedipine (125mM) and verapamil (130mM); CaM inhibitors were trifluoroperazine (200mM) and W7 (210mM). Rp-cGMPS was used at 50mM. When a reaction was observed inside an injected cell, it was always the same, eg., all positive cells injected with >10,000 molecules PHYA in the light showed anthocyanin (A) and chloroplasts (C), and never just one; all positive cells injected with >5,000 molecules Ca^{2+} activated calmodulin showed only chloroplasts (C) and never anthocyanin.

Although this antagonist data provided some hints as to the nature of the signal transduction chain, it was necessary to perform positive "gain-of-function" experiments to show that injection of proposed transduction intermediates could stimulate a response in the absence of the photoreceptor. This is especially important for plant cells, because, although it is assumed that the effects of these inhibitory drugs parallel those in animal cells, their action has not been well characterized in plants.

If a heterotrimeric G protein was involved early in the pathway we would predict that injection of G protein activators (such as GTP-γ-S and cholera toxin (CTX), would potentiate expression of all the markers for phytochrome signalling. Indeed, this was found to be the case (Table 1). In addition, we found that CTX and GTP-γ-S effects on chloroplast development were antagonized by nifedipine and trifluoroperazine, whilst anthocyanin production was unaffected by these treatments (Table 1). This again suggests that the calcium-defined branch point is downstream of the putative G protein.

A suggestion following these antagonist results (using nifedipine, verapamil, W7 and trifluoroperazine) was that calcium and activated calmodulin might be active participants in a transduction pathway that leads from PHYA to chloroplast development, but play no signalling role in regulating PHYA-mediated anthocyanin biosynthesis. As such, injection of calcium or calcium-activated calmodulin ought to stimulate processes required for chloroplast development but not those required for anthocyanin biosynthesis. The results in Table 1 clearly show this to be true.

The observed responses of cells injected with calcium could be antagonized in the presence of trifluoroperazine or W7 (Table 1), but not by nifedipine or verapamil, suggesting that calcium most likely acts by activating calmodulin. Indeed, we found that cells injected with calcium and incubated in trifluoroperazine or W7 could be rescued (i.e., to show chloroplast development) by removing the antagonist and injecting a second time with activated calmodulin (data not shown), albeit at low efficiency. The low efficiency is due to the low survival rate of cells injected two times.

As previously shown, PHYA or GTP-γ-S injection could induce only anthocyanin biosynthesis in cells bathed in nifedipine, verapamil, trifluoroperazine or W7 (Table 1). If the antagonists were removed and the cells injected a second time with either calcium (in the case of nifedipine- or verapamil-treated seedlings) or activated calmodulin (in the case of trifluoroperazine- or W7-treated seedlings), chloroplast development could be restored, again at low efficiency (data not shown). This suggests that in the normal physiological response, PHYA, acting via a heterotrimeric G protein, can somehow stimulate entry of calcium into the plant cells, which then activates calmodulin. Presumably, calmodulin then mediates other signal transduction events, possibly involving phosphorylation.

To identify the signalling molecule(s) involved in anthocyanin biosynthesis, we screened a variety of low mol wt compounds. Interestingly, we found that cGMP could effectively replace PHYA as an agonist of anthocyanin biosynthesis (Table 1), although by itself it had no effect on chloroplast development.

Table 2. Immunofluorescence analysis of polypeptides present in plastids induced by microinjection of PHYA, calcium, and calcium plus cGMP

Antibody	Chloroplast complex	Genome	Reaction in cells injected with PHYA	Reaction in cells injected with Ca^{2+} or CaM	Reaction in cells injected with Ca^{2+} and cGMP
RbcS	RUBISCO	N	+	+	+
LHCII	PS II	N	+	+	+
OEE1	PS II	N	+	+	+
D1	PS II	C	+	+/-	+
D2	PS II	C	+	+/-	+
Ferredoxin	PS I	N	+	-	+
Plastocyanin	PS I	N	+	-	+
PsaD	PS I	N	+	-	+
PsaF	PS I	N	+	-	+
LHCI	PS I	N	+	-/+	+
Rieske FeS	$Cytb_6f$	N	+	-	+
Cyt. b_6	$Cytb_6f$	C	+	-	+
Cyt.f	$Cytb_6f$	C	+	-	+
γ Subunit	ATP Synthase	N	+	+/-	+
α Subunit	ATP Synthase	C	+	+/-	+

Abbreviations: N, nuclear; C, chloroplastic. + denotes a strong positive reaction; +/- denotes a clearly positive reaction; -/+ denotes weak immunoreactivity; - denotes no observable staining. Antibody staining was examined in 10 independent cells injected with each compound and the staining pattern for each was reproducible in all cases.

The role of cGMP in PHYA signalling was confirmed by showing that coinjection of PHYA with Rp-cGMPS, an inactive cGMP analog, resulted in chloroplast development but not anthocyanin biosynthesis (Table 1). Furthermore, coinjection of Rp-cGMPS with GTP-γ-S also blocked anthocyanin biosynthesis but not chloroplast development, indicating that cGMP, like calcium, acts downstream of the G protein in the PHYA signal transduction pathway (Table 1). Indeed, injection of both calcium (or calmodulin) and cGMP produced a response identical (i.e., both chloroplasts and anthocyanin) to that generated by PHYA and GTP-γ-S (Table 1). It therefore appeared that calcium and cGMP are the minimal signals required to mediate PHYA effects in this experimental system. Neither cAMP nor GMP had any effect (data not shown), demonstrating specificity of the response for cGMP.

2.2 Role of Calcium, Calmodulin and cGMP in Chloroplast Development

In the experiments described above we used chloroplast development as a gross phenotype of phytochrome signalling. In experiments where calcium or activated calmodulin were injected into cells, however, the chloroplasts appeared to be slightly malformed (data not shown), and this prompted us to examine them in more detail. To do this we used immunofluorescence to study the appearance of individual chloroplast proteins. The proteins examined represented all five of the major photosynthetic complexes: RUBISCO, photosystem II (PSII), photosystem I (PSI), cytochrome b_6f and ATP synthase. Table 2 shows that all 15 proteins examined were synthesized in *aurea* cells injected with PHYA (and GTP-γ-S; data not shown). Although not all of them appear following calcium or calmodulin injection, we were only able to clearly detect RbcS (from RUBISCO), LHCII, OEE1, D1, D2 (from PSII), and the α and γ subunits from ATP synthase. Whilst most of the proteins examined are nuclear-encoded, the D1, D2, and ATP synthase α subunit are chloroplast-encoded. Hence, in addition to their role in controlling nuclear gene expression, calcium and calmodulin are also able to modulate the expression of chloroplast genes.

Interestingly, none of the proteins from PSI or the cytochrome b_6f complex were present in calcium- or calmodulin-injected cells (Table 2). However, coinjection of cGMP with calcium or calmodulin resulted in the generation of chloroplasts that, by our immunological criteria, were fully developed (Table 2). Taken collectively, these results demonstrate that cGMP and calcium are all that are required to initiate the formation of a fully mature chloroplast.

3 Discussion

In this work we have been able to identify several components that are necessary for the transduction and amplification of the signal initially generated by phytochrome following photoperception. Likely an early event in the signal transduction pathway is the activation of a heterotrimeric G protein, identified on the basis of antagonist data from GDP-ß-S and pertussis toxin and agonist data from GTP-γ-S and CTX. G protein activation is able to reproduce fully the effects of PHYA, perhaps suggesting that this step is very close to the initial event of phytochrome activation.

Downstream of G protein activation there appear to be three different pathways. One of these requires the participation of cytosolic calcium and activated calmodulin and results in (partial) chloroplast development, another requires cGMP and controls anthocyanin biosynthesis, and the third requires both calcium/calmodulin and cGMP and is responsible for completing chloroplast development. The hypothesis that calcium and cGMP are the only signalling molecules used by PHYA is strongly supported by the facts that the result of PHYA and nifedipine coinjection is identical to cGMP injection, and that PHYA and Rp-cGMPS coinjection is identical to injection of calcium alone.

The chloroplasts induced by calcium or calmodulin injection were not as well developed as those that were formed by PHYA or GTP-γ-S stimulation. The reason for this was deduced from immunofluorescence analysis of 15 different chloroplast proteins, members of the five photosynthetic complexes: RUBISCO, PSII, PSI, cytochrome b_6f, and ATP synthase. Whilst GTP-γ-S could fully substitute for PHYA to trigger the synthesis of all these proteins, calcium and activated calmodulin could only promote the synthesis of RbcS (from RUBISCO) , LHCII, OEE1, D1, D2 (from PSII), and the ATP synthase α and γ subunits. Other proteins examined, from the PSI and cytochrome b_6f complexes, were undetectable in calcium- or activated calmodulin-injected cells. These results suggest that the calcium-dependent pathway is only able to direct the synthesis of the RUBISCO, PSII and ATP synthase complexes, and that the whole PSI and cytochrome b_6f complexes are absent in calcium- or activated calmodulin-injected cells.

Results from coinjection experiments with calcium (or calmodulin) and cGMP reveal, however, that cGMP provides the inductive signal for PSI and cytochrome b_6f formation, in addition to mediating anthocyanin biosynthesis.

These results therefore reveal that the calcium- and cGMP-dependent pathways are able to stimulate the synthesis of chloroplast-encoded proteins (see Table 2), in addition to nuclear-encoded proteins. The fact that the synthesis of chloroplast-encoded proteins is largely dependent on post-transcriptional and translational regulatory mechanisms, whereas known nuclear genes for photosynthetic proteins appear to be regulated primarily at the level of transcription (Mullet, 1988; Gruissem, 1989; Rochaix, 1992), demonstrates the astonishing effectiveness of calcium and cGMP in being able to modulate cellular processes in plants at several fundamental levels. Equally impressive is the effectiveness of PHYA and GTP-γ-S in mediating chloroplast development. The biosynthesis of the photosynthetic apparatus alone requires the participation of several hundred genes, and is dependent upon the concerted action of the nuclear and chloroplast genomes. Functional complexes also require the binding of numerous cofactors such as chlorophyll, carotenoids, and plastoquinones. A long-standing question has been how light can generate signals that can initiate and control all these processes. Through the studies reported here we have shown that light acting through PHYA activation can be sufficient to mediate these events. Furthermore, since GTP-γ-S and a combination of calcium and cGMP appear to be equally effective in stimulating this myriad of cellular responses, it would appear that PHYA mediates these processes in just one way - by first activating one (or more) heterotrimeric G proteins, and then generating increases in cellular calcium and cGMP.

Taken collectively these data begin to show the complexity of the phototransduction events mediated by phytochrome, and reveal that single signal intermediates are not always sufficient to promote a full cellular response. Notwithstanding, we believe that the continued use of this system will yield further information about phytochrome signal transduction pathways. This will be all the more valuable when mutants with defective phytochrome signalling can be characterized at a similar level.

Acknowledgements. This research was supported, in part, by Human Frontier Science Program grant # RG-362/95 M to C.B., G.N., and N.-H.C., National Institutes of Health grant # 44640 to N.-H.C., and by Huber Kudlich Stiftung to G.N.

References

Bowler C, Chua N-H (1994) Emerging themes of plant signal transduction. Plant Cell **6**: 1529-1541

Bowler C, Neuhaus G, Yamagata H, Chua N-H (1994a) Cyclic GMP and calcium mediate phytochrome phototransduction. Cell **77**: 73-81

Bowler C, Yamagata H, Neuhaus G, Chua N-H (1994b) Phytochrome signal transduction pathways are regulated by reciprocal control mechanisms. Genes & Dev **8**: 2188-2202

Gilmartin PM, Sarokin L, Memelink J, Chua N-H (1990) Molecular light switches for plant genes. Plant Cell **2**: 369-378

Gruissem W (1989) Chloroplast gene expression: how plants turn their plastids on. Cell **56**: 161-170

Koornneef M, Van Der Veen JH, Spruit CJP, Karssen CM (1981) The isolation and use of mutants with an altered germination behaviour in *Arabidopsis thaliana* and tomato. *In* Induced Mutations, a Tool for Crop Plant Improvement. Vienna, International Atomic Energy Agency, pp 227-232

Mullet JE (1988) Chloroplast development and gene expression. Annu Rev Plant Physiol Plant Mol Biol **39**: 475-502

Neuhaus G, Bowler C, Kern R, Chua N-H (1993) Calcium/calmodulin-dependent and -independent phytochrome signal transduction pathways. Cell **73**: 937-952

Quail PH, Boylan MT, Parks BM, Short TW, Xu Y, Wagner D (1995) Phytochromes: photosensory perception and signal transduction. Science **268**: 675-680

Rochaix J-D (1992) Post-transcriptional steps in the expression of chloroplast genes. Annu Rev Cell Biol **8**: 1-28

Sharma R, López-Juez E, Nagatani A, Furuya M (1993) Identification of photo-inactive phytochrome A in etiolated seedlings and photo-active phytochrome B in green leaves of the *aurea* mutant of tomato. Plant J **8**: 1035-1042

Post-Translational Stabilization of ELIPs and Regulation of other Light Stress Genes under Prolonged Light and Cold Stress in Barley

Marie-Hélène Montané[1], Silke Dreyer[2], and Klaus Kloppstech[2]

[1] Department of Ecophysiology and Microbiology, CEA Cadarache, F-13108 St Paul lez Durance, Cedex, France

[2] Institute of Botany, Hannover University, D-30419 Hannover, Germany

Abstract. In nature light and temperature stress occur very often in combination. When high light and temperature were applied together this treatment led to an increase in the expression of Early Light Inducible Proteins (ELIPs) together with other light stress proteins. To study in more detail the expression of light stress proteins at low temperatures barley plants were grown for several days under various light intensities (100, 500, 1000 and 1500 µE m^{-2} s^{-1}) and temperatures (5-25°C). It was observed that in plants grown at high light intensities and low temperatures the amount of ELIP protein increased considerably above the levels that could be obtained by a short term (2-8 h) high light experiment at 25°C. The highest protein levels were reached at the lowest temperature coupled to high light intensities (500 to 1500 µE m^{-2} s^{-1}). In the latter case, the protein levels were almost at the maximal level already in the morning, while at the lowest light intensity the ELIP protein accumulates during the light phase. It appeared that the ELIP accumulation was primarily the result of a restored capacity of protein synthesis at low temperatures but also of a reduced degradation of ELIPs during the night phase. At 25°C the ELIPs were almost completely decomposed during the dark period.

Keywords. Barley, cold, ELIPs, gene expression, *Hordeum vulgare*, light stress, low temperature, photoinhibition

1 Introduction

Light intensities which exceed the saturation point of photosynthesis can be considered as stressful to plants. There are basically two situations possible: permanent high light stress, such as in dry and hot open areas, and short duration stresses. Upon a long duration stress plants may adapt and develop morphological changes, whereas a light stress of short duration cause the plant, in the first instance, to respond by transient changes in gene expression. During the last few years genes have been cloned and sequenced, whose transcription has been found to be enhanced under the influence of strong light. It is also known that low temperatures can enhance the negative effects of high light fluxes. In this latter case, gene expression was also found to be affected, as reported for some light stress-regulated genes, such as chalcone synthase (CHS) (Leyva et al., 1995), phenylalanine ammonium lyase (PAL) (Kreuzahler et al., 1983), and genes of the carotenoid biosynthetic pathway (Pecker et al., 1992). However, the most studied light stress genes are those coding for ELIPs. ELIPs were originally discovered as products of genes transiently up-regulated during greening of etiolated seedlings in several species (Grimm et al., 1989). They belong to nuclear gene families coding for proteins of 13.5 and 17 kD in barley and of 17 kD in pea, which are localized in the stroma thylakoids but also in the immediate surrounding of D1 protein. Taken together, these two properties indicate a function of ELIPs during assembly or repair of photosystem II (PS II) reaction centers. The idea gets further support from the fact that ELIPs have always been found to be induced under conditions favourable for the decay of D1 and photoinhibition of PS II, which are frequently reported as a consequence of light stress. In accordance with what has been stated above, ELIP transcripts accumulate to higher levels in temperature downshift experiments when they are carried out in the presence of high light fluxes. In pea plants, this effect can be detected down to a temperature of about 10°C, but below this temperature the levels of high-light-inducible ELIP mRNAs decrease rapidly (Adamska and Kloppstech, 1994). Interestingly, in short-term stress experiments, synthesis of ELIPs does not increases despite the fact that ELIP mRNA levels are enhanced several times. Only after return to the ambient growth temperature of 25°C is ELIP synthesis resumed even under conditions of low light («memory» effect). However, under these conditions the ELIP mRNAs are degraded with a $\tau/2$ of about 1 h but the resumption of translation of ELIPs is only transient.

The apparent discrepancy between high levels of ELIP mRNA accumulation and low rates of protein synthesis can be explained as the result of the sudden decrease in temperature. By run-off transcription experiments, it could be shown that at 10°C transcription continued, but that the level of protein synthesis was very low. Therefore, we assumed that these conditions were not favourable for the adaptation of the protein synthesis machinery during the time of experimentation. The question was addressed as to whether there might exist a gradual long-term mechanism of adaptation leading to the accumulation of ELIPs under light and cold stress. To prove this assumption, it was decided to carry out long-term experiments with plants acclimated to low temperatures and high light fluxes. Barley was chosen as the experimental system because it quite often has to face low temperatures during germination in the spring, especially in Mediterranean climates. Additionally, genes for ELIPs have been cloned and characterized in barley and other clones, encoding barley genes known to be up-regulated under light stress, were available in the laboratory, allowing an investigation of changes in gene expression on a broader basis. Barley is a rather cold resistant species and it is able to resume a slow growth rate even at low temperatures. It could be expected that, based on its cold tolerance, either the rate of protein synthesis might be restored and/or that the stability of proteins would increase proportionally in cold-stressed barley plants. The experimental conditions chosen were 5, 10 and 25°C at light fluxes between 100 and 1500 µE m^{-2} s^{-1}. As shown in preliminary experiments, these conditions are close to those occurring frequently during cold and sunny days in spring, in the Northern hemisphere but more likely in Mediterranean regions.

2 Methods

2.1 Growth of Plants

Barley seeds (*Hordeum vulgare*, cv Apex) were grown in pots for 5 d at 25°C and 100 µE m^{-2} s^{-1} in vermiculite. One pot was used for every time point. Subsequently, the plants were exposed to temperatures of 5°C for 7 d or 10°C for 5 d under light conditions of 100, 500, 1000, or 1500 µE m^{-2} s^{-1}, as indicated in the legends of the figures. For the experiments at 25°C, seeds were grown directly at the indicated light intensities in order to reach a comparative stage of development. In all cases, the photoperiod was 12 h of light. Samples were harvested and immediately frozen in liquid nitrogen and stored at -70°C.

2.2 Isolation and Quantification of Proteins

Total proteins were isolated by homogenization of 0.5 g of frozen plant material in 2.5 ml sample buffer and separated on 15% PAGE gels, according to Pötter and Kloppstech (1993). After transfer to PVDF membranes, the proteins were detected using anti-ELIP and an alkaline phosphatase conjugated secondary antibody. The linearity of this reaction was checked in western blots and, whenever necessary, dilutions of the protein samples were used. All data refer to the ELIP level detected in control etiolated barley plants illuminated for 8 h at 100 µE m^{-2} s^{-1}. The amount of ELIPs on the blot was estimated by using a Biorad scanner (Gel doc 1000) and the corresponding software.

2.3 Isolation and Quantification of Poly(A)$^+$RNAs

Poly(A)$^+$RNA from 2.5 g of frozen material was isolated as described (Pötter and Kloppstech, 1993), but with only one phenol extraction. After chloroform extraction the RNA was precipitated with isopropanol and resuspended in binding buffer. The poly(A)$^+$RNA was separated on denaturing agarose gel, transferred to nylon membranes, and hybridized to homologous cDNAs coding for chalcone synthase (CHS), light-harvesting chlorophyll a/b protein (LHC), small subunit ribulose-1,5-biphosphate carboxylase (SSU), glycine decarboxylase (GDC), HL#2, large and small ELIPs and analyzed according to

Pötter and Kloppstech (1993). Densitometric scanning of the films was performed as described for proteins.

3 Results

3.1 Characterization of Light Stress-Induced mRNAs

The comparison between the mRNA levels observed upon short-term illumination of barley leaves at 100 and 1500 µE m^{-2} s^{-1} showed clearly that in addition to ELIPs also clones HL#2 of unknown function and GDC (Turner et al., 1992) belong to the class of mRNAs which were induced by light stress (Fig. 1). Their levels remained high for as long as light stress was imposed (up to 9 h and presumably even longer). In contrast, the mRNA for LHC II decreased considerably during almost all conditions of light stress. After 9 h at a light intensity of 2000 µE m^{-2} s^{-1} this mRNA was no longer detectable while the levels of other housekeeping mRNAs remained relatively unaffected by light stress. The levels of the above-mentioned genes remained high during the entire experiment (Pötter et al., unpublished).

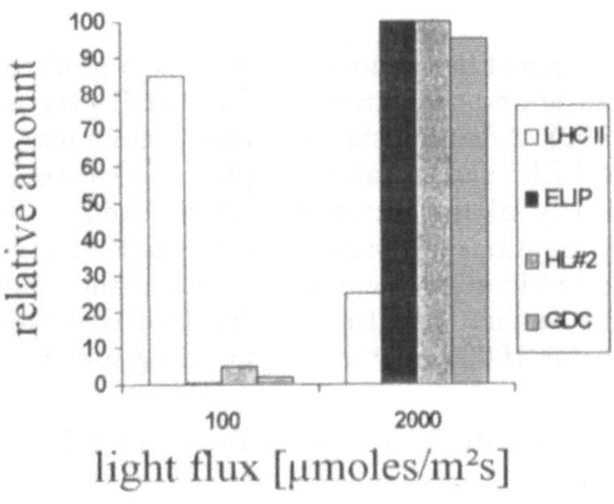

Fig. 1. Induction of mRNAs by light stress. 7-day-old barley plants were illuminated for 4 h at the indicated light intensities and the mRNA quantified by hybridization (see methods)

3.2 Temperature Dependence of ELIP Accumulation

In pea plants, it was previously shown that a reduction in the ambient temperature during light stress leads to a considerable increase in the level of ELIP mRNAs (Adamska and Kloppstech, 1994), while a parallel effect of temperature on protein accumulation could not be observed. In fact, the *de novo* synthesis of ELIPs and protein synthesis in general declined according to a decrease in the temperature. Since ELIPs were accumulated in photoinhibited plants at 25°C, this prompted us to ask how a relatively cold-resistant plant like barley would behave upon acclimation to a low temperature. The plants were allowed to adapt for 5 d at 10°C and 7 d at 5°C, before any measurement was taken. The effect of acclimation on protein accumulation was remarkable (Fig. 2). At 100 µE m^{-2} s^{-1} and 25°C ELIPs accumulated to approximately 20% of the amount observed after 8 h of greening at 100 µE m^{-2} s^{-1}. After 5 d at 10°C the level of accumulated ELIPs was twice as high as in the greening samples, which was 10-fold higher than at 25°C. After 7 d at 5°C and at 1500 µE m^{-2} s^{-1} of light stress the ELIP level exceeded 8 times the amount found in etiolated control, which was 40 times higher than at 25°C. This raised the question of how these high levels might be reached.

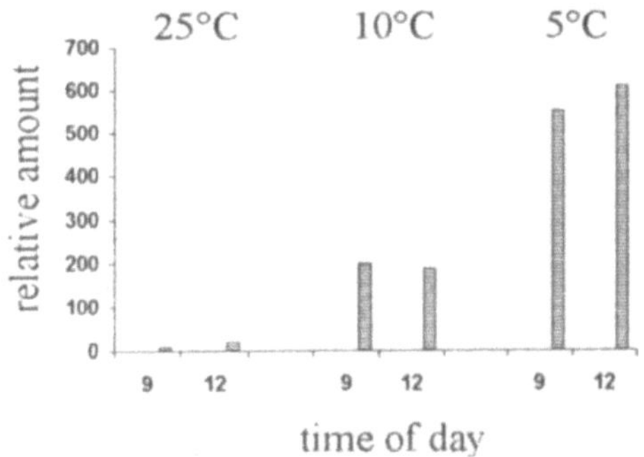

Fig. 2. Accumulation of ELIP under light stress according to temperature. After growth for 5 d at 25°C and 100 µE m^{-2} s^{-1} plants were transferred to high light at the indicated temperatures for 5 or 7 d (5°C) and ELIPs determined

One assumption is that ELIPs might be stabilized by an unknown mechanism, but in addition it is possible to assume that when gradually exposed to low temperatures plants are able to synthesize ELIPs, presumably due to a process of adaptation of protein synthesis to low temperatures. Evidence for the first assumption, i.e. that ELIPs might be stabilized, came from experiments in which the samples were taken at various times during the day (Fig. 3). At 25°C the plants had very low levels of ELIPs in the morning (start of the light). During the day ELIPs accumulated accordingly to the light intensity. The maximum was reached at noon time, i.e. 6 h after the beginning of day (9 am). The amount of ELIPs started to decrease in the afternoon. At the highest light intensity, the level of ELIPs increased to values which were about

20% of those detected in etiolated control plants illuminated for 8 h. As the following morning the ELIP level was close to zero, the proteins might have been degraded during the night. At 10°C the ELIP level was at least 10-fold higher and stayed elevated throughout the day. At this temperature a light intensity of 500 μE m^{-2} s^{-1} was apparently experienced as a light stress, since at a 3-fold higher light intensity (1500 μE m^{-2} s^{-1}), the ELIPs increased by only 50%. At 5°C the ELIPs were accumulated to levels similar to those observed in the etiolated controls, even at a light intensity of 100 μE m^{-2} s^{-1}. This means that cold *per se* favours ELIP accumulation. However, at a light intensity of 1500 μE m^{-2} s^{-1}, the amount of ELIPs was 8-fold above the maximal level observed in the etiolated and illuminated controls at 25°C. In addition, at 5°C there was no apparent degradation during the night and thus at this temperature the ELIP amount did not vary during the day as at 25°C, but instead tended to increase linearly during the light phase.

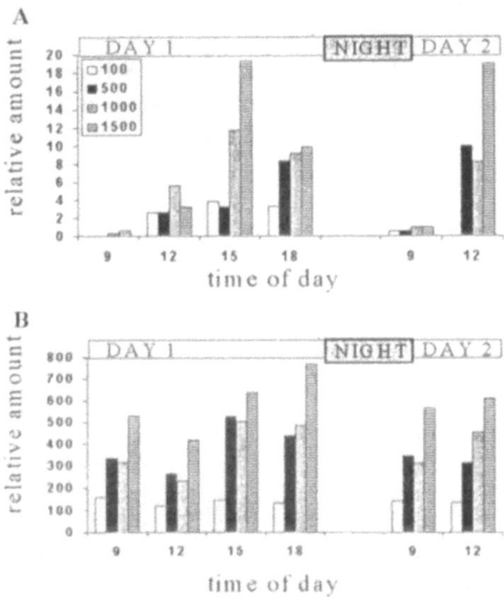

Fig. 3. ELIP accumulation in dependence of light intensity and growth temperature. Plants were grown at 25°C and high light of different intensities for 5 days at 25°C (A) or 7 days at 5°C (B)

3.3 Accumulation of ELIPs within the Barley Leaf Gradient

In barley leaves, gene expression depends on the developmental cell stage, which varies along the leaf. The youngest cells are located in the lower third part of the leaf, where chloroplasts mature and photosystems I and II are assembled. ELIPs have been found to be transiently associated with PS II particles and it thus was expected that ELIPs should accumulate in the lower third of the leaf, where the maximum mRNA expression has been observed. At 25°C and 100 μE m^{-2} s^{-1} accumulation of ELIPs occurred preferentially in the leaf segments 1 and 2, i.e. simultaneously to or even before assembling of PS II, while in mature cells (leaf segments 3 to 6) ELIP levels decreased almost to zero (Fig. 4).

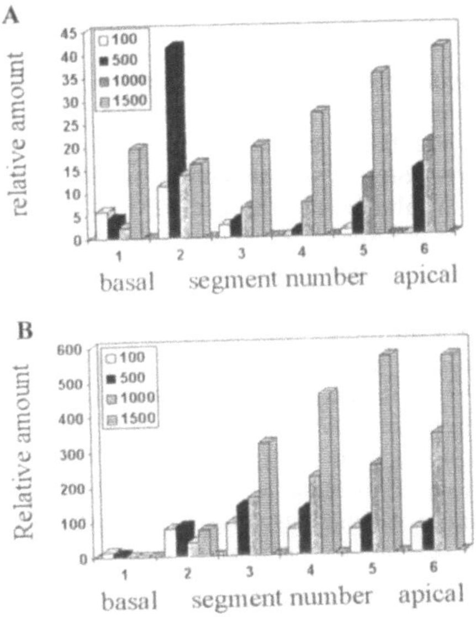

Fig. 4. ELIP accumulation in the leaf gradient according to light intensity and growth temperature. Plants were grown as described for Fig. 3 A (25°C) and B (5°C), respectively. The leaves were cut in 6 segments of equal length

In previous experiments light stress was applied on isolated leaves with an angle of 90°. The present results were accomplished by illuminating the entire plants from the top, as occurs in nature. However, under light stress a different pattern of expression was observed. At a light intensity of 500 µE m^{-2} s^{-1}, which corresponds to an optimal growth irradiance in nature, a considerable increase in the ELIP content was found in the leaf segment 2, but ELIPs started to accumulate also in the most apical segments. At even higher light intensities the maximum ELIP expression was shifted to the top of the leaves indicating that under light stress ELIP accumulation was resumed in mature cells, as already found in pea plants (Adamska and Kloppstech, 1994). At 5°C ELIP accumulation showed a different pattern of expression. In this case, a low light intensity of 100 µE m^{-2} s^{-1} was experienced as a stress by the plant, since ELIP accumulation was found along the entire leaf, in contrast to the gradient of expression observed at 25°C. Only in the leaf basal segment 1 was ELIP accumulation low, indicating that leaf cells at that particular developmental stage did not permit ELIP expression even at the highest light intensities. At 5°C the maximal expression of ELIP was not found in leaf segment 2, as observed at 25°C, but in segments 3 to 6. The accumulation appeared to be strongly dependent on light intensity between 500 and 1500 μE m^{-2} s^{-1}.

3.4 The Effect of Short-Term Light Stress *versus* Long-Term Adaptation on ELIP Expression

The data so far gathered were obtained by exposing barley plants to the light stress for several days or under conditions allowing the plant to adapt to light stress. That the process of adaptation was one component of the process, could be deduced from the observation that the phenotype of the plants had changed. As a consequence of the simultaneous presence of high light and cold, waxy substances were deposited at the tips of the leaves and anthocyanins appeared at the bases of the leaves. These effects were best observed at 5°C. The question was asked whether there would be a difference in the levels of ELIP accumulated under short-term and long-term light stress of plants. For this purpose, 7-day-old plants were grown at 25°C and under different light intensities and then transferred to 1500 μE m^{-2} s^{-1} in the morning. The leaves were harvested 6 h later. At the end of the experiment, the ELIP levels were different between high light-shocked and high light-adapted plants (data not shown). At a given light intensity, the high light imposed as shock raised the ELIP levels 5 times above the level observed in plants exposed to high light for a longer time, thus indicating an overshooting reaction in response to light.

3.5 High Light-Induced and Temperature-Dependent Regulation at the Level of ELIP mRNA

Poly $(A)^+$ RNA was isolated at different times of the day and hybridized to an ELIP probe in dot blot experiments. It was found that the effect of high light on the mRNA levels was not related to the changes in terms of proteins previously observed. While ELIP accumulation was 40-fold higher at 5°C compared to 25°C, the level of mRNAs only differ by a factor of two at the temperatures tested (data not shown). Another observation was that the dependency on light of ELIP mRNA levels was very strong especially at 25°C, but later on was reduced (Fig. 5). This indicated a higher stability of ELIP mRNAs at the very high light intensity of 1500 μE m^{-2} s^{-1}. We observed in these experiments that the lower the temperature, the later during the day the maximum in the mRNA levels occurred.

3.6 Light Stress-Induced mRNAs Coding for other Proteins

In addition to ELIPs, the expression of other genes is known to be enhanced in response to high light intensities, such as PAL and CHS. More recently other light stress proteins have been described. In our laboratory, clones HL#2 and GDC have been cloned from a high light subtractive cDNA library. The clone HL#2 has been found to share homology with two clones encoding proteins from rice (Claes et al., 1990; DePater et al., 1992). Under normal conditions, in

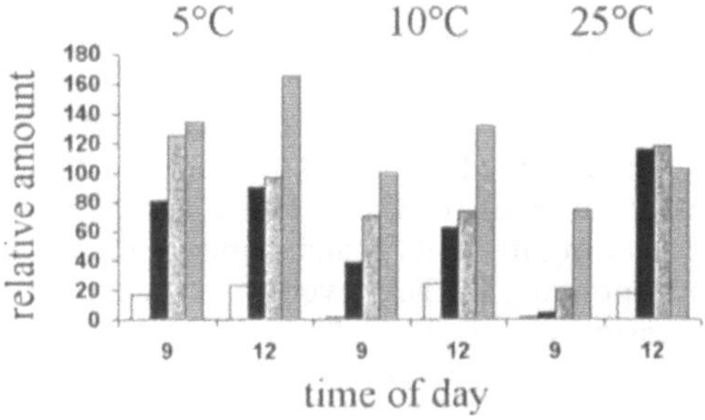

Fig. 5. Light and temperature dependence of mRNA levels for small ELIP. Plants were grown as described in Fig. 3. Poly $(A)^{+}$RNA extracted and determined by dot hybridization. Light intensities as in Fig. 3

Table 1. Differential expression of mRNAs of light-regulated genes relative to small ELIP mRNA

Poly $(A)^{+}$RNA from plants grown as in Fig. 3 (at 12 am) were hybridized with barley cDNA of light regulated genes. All membranes for one probe were hybridized in the same solution and were scanned. The ratios of the signal for each dot and that of small ELIP were calculated

t°C $\mu E\ m^{-2}\ s^{-1}$	5 100	5 1500	10 100	10 1500	25 100	25 1500
L-ELIP	0.4	0.30	0.90	0.30	1.70	0.5
SSU	1.1	0.30	0.90	0.10	16.8	4.3
LHC	0.5	0.06	0.40	0.05	9.80	0.7
CHS	0.5	0.25	0.25	0.20	3.60	0.4
GDC	0.3	0.13	0.38	0.10	16.4	2.0
HL#2	0.2	0.40	0.51	0.20	5.10	0.6

barley leaves, the expression of this gene with an unknown function was found to be very similar to the expression of ELIP genes. The other clone codes for GDC, a protein localized in the mitochondria and mRNAs for GDC accumulated in the mature cells of the apical section of the leaf.

The cDNA clones of large ELIP, GDC, HL#2 and CHS together with two photosynthetic genes, coding for LHC II and Rubisco SSU, were used as probes to quantify their corresponding mRNA levels in response to light stress (Table 1). The level of expression was quantified and correlated with that observed for the small ELIP mRNAs, previously shown to be positively controlled by high light and low temperature (Fig. 5). Except for the gene encoding the large ELIP, the expression of all other genes was higher at 25°C compared to 5°C at a low light intensity of 100 $\mu E\ m^{-2}\ s^{-1}$. At 5°C the expression decreased by a factor of 10 to 20, with housekeeping proteins like LHC II, SSU and GDC being the most down-regulated genes. A second group of light stress-specific genes (HL#2, CHS) showed less variation with temperature, but do not change with respect to light stress at low temperatures. The exception was LHC II which dramatically decreased at 5°C and light stress. These data showed that in plants acclimated to stress conditions for about one week mRNA status was different from that observed under a short term stress (Fig. 1).

4 Discussion

When applied as shock, low temperatures reduce the activity of proteins involved in the repair of enzymes in the chloroplast, while photoreactions continue to take place. As a consequence, the damage due to over-reduction of the electron transport chain increases and causes an increase in the damage. The uncoupling of the LHC II antennae is known to reduce the efficiency of photon entrapment and the uncoupled antennae might even serve as protection shields of the reaction centers. The surplus of light energy will be emitted as light of longer wavelength or transformed into heat by carotenoids, e.g. the xanthophyll cycle (Adamska et al., 1992). Upon long-term exposure, other protection mechanisms are triggered, which are not yet fully explored. In this regard, an observation of interest was that under light stress conditions, inducing ELIP mRNA accumulation, LHC II mRNAs declined rapidly. This may reduce LHC II synthesis and after a long-term light stress may lead to a decline of the LHC II antennae, establishing a new equilibrium. This raises the question as to which strategies plants have developed to optimize LHC II synthesis in response to change in the incidence of photosynthetic light. Our data indicate that the plants program the synthesis of ELIPs and LHC II according to the levels used during the previous day. Another equilibrium may be established only if external conditions change.

Several observations have provided information on the mode of regulation of ELIPs. Firstly, the levels of ELIP mRNAs were enhanced according to decreasing temperature (at least to 5°C). Furthermore, at this temperature the extent of the daily variation in the levels of mRNA declined dramatically. This may allow protein synthesis, normally less efficient at low temperatures, to continue for a longer time during the day, leading to a partial increase in the ELIP level. While it remains to be ascertained whether this type of regulation is sufficient to explain the high ELIP levels observed at low temperature, it has also to be added that degradation of ELIPs was also retarded. The unique temperature dependency of an ELIP degrading protease might support this possibility (Adamska, unpublished).

Another question concerns the function of the increased steady state level of ELIPs and whether this might help in protecting the plants against light stress at low temperature. Although we do not have an unequivocal answer to this question, some observations and considerations might be helpful to open new ways for unravelling the ELIP function. One of these observations was that above a temperature of 25°C ELIPs accumulated only barely as result of light stress. Consequently, it can be stated that above 25°C the need for a functioning ELIP seems to be low. In contrast to previous findings, it was also found that plants entirely illuminated from above did not experience a strong reduction in photosynthetic activity at a light flux of 1500 $\mu E\ m^{-2}\ s^{-1}$ even after several hours, as measured by chlorophyll fluorescence (Pötter and Kloppstech, 1993). The difference with previous data might be explained by the fact that previously leaves were excised and exposed to light stress at an angle of 90°. Another interesting result came from the study of the ELIP distribution along the leaf. The maximum ELIP expression was found to be at two major sites: segments 2 and 6, which correspond to two different stages of chloroplast differentiation, respectively the etiochloroplasts (where PS II is assembled) and the mature chloroplasts (photosynthetically active and then most sensitive to photoinhibition). The data presented in this paper suggest that the involvement for assembly is more predominant at 25°C while at low temperatures the involvement of ELIPs for repairing cold- and light- induced deleterious structural function is predominant, since these conditions are highly photoinhibitory.

At 1500 $\mu E\ m^{-2}\ s^{-1}$ and at 25°C a ratio of one molecule of ELIP to 20 D1 proteins was determined in pea plants (Adamska, unpublished). If we consider that as an effect of long-term exposure at 5°C a 40-fold increase in the level of ELIP can be obtained, this would mean that two ELIPs would be available for each reaction center. Since the amount of photons given to a cell depends on its location inside the leaf this ratio could be higher in cells close to the epidermis. Unfortunately, it is not yet known precisely where ELIPs are located in the thylakoids of plants kept at 5°C and high light. It was previously shown that ELIPs are integrated and located within the stroma lamellae. However, it is not known whether this distribution is also true for cold-stressed plants or also whether after light stress in the cold there is retention of the grana to stroma membrane distribution which we usally associate with the chloroplast structure. Whether this is the case will have to be shown by future experiments.

Acknowledgements. We gratefully acknowledge financial support by the DFG and Procope.

References

Adamska I, Ohad I, Kloppstech K (1992) Synthesis of the early light-inducible protein is controlled by blue light and related to light stress. Proc Natl Acad Sci USA **89**: 2610-2613

Adamska I, Kloppstech K (1994) Low temperature increases the abundance of early light-inducible transcript under light strss conditions. J Biol Chem **269**: 30221-30226

Claes B, DeKeyser R, Villarroel R, Van der Bulcke M, Bauw G, Van Montagu M, Caplan A (1990) Characterization of a rice gene showing organ specific expression in response to salt stress. Plant Cell **2**: 19-27

DePater S, Hensgens LAM, Schilperoort RA (1992) Structure and expression of a light-inducible shoot specific rice gene. Plant Mol Biol **18**: 161-164

Grimm B, Kruse E, Kloppstech K (1989) Transiently expressed early light-inducible thylakoid proteins share transmembrane domains with light-harvesting chlorophyll binding proteins. Plant Mol Biol **13**: 583-593

Kreuzahler F, Ragg H, Fautz E, Kuhn DN, Hahlbrock K (1983) UV-induction of chalcone synthase mRNA in cell suspension cultures of *Petrosilenum hortense*. Proc Natl Acad Sci USA **80**: 2591-2593

Leyva A, Jarillo JA, Salinas J, Martinez-Zapater JM (1995) Low temperature induces the accumulation of phenylalanine ammonia lyase and chalcone synthase mRNA of *Arabidopsis thaliana* in a light dependent manner. Plant Physiol **108**: 39-46

Pecker I, Chamovitz D, Linden H, Sandmann G, Hirschberg J (1992) A single polypeptide catalyzing the conversion of phytoens to zeta-carotene is transcriptionally regulated during tomato fruit ripening. Proc Natl Acad Sci USA **89**: 4962-4966

Pötter E, Kloppstech K (1993) Effects of light stress on the expression of early light-inducible proteins in barley. Eur J Biochem **214**: 779-786

Turner SR, Ireland R, Rawsthorne S (1992) Cloning and characterization of the P-subunit of glycine decarboxylase from pea (*Pisum sativum*). J Biol Chem **267**: 5355-5360

Expression of Catalase and its Relation to Light Stress and Stress Tolerance

Jürgen Feierabend, Peter Streb, Matthias Schmidt, Silvia Dehne and William Shang

Botanisches Institut, JW Goethe-Universität, D-60054 Frankfurt am Main, Germany

Abstract. The peroxisomal enzyme catalase is inactivated by visible light in the presence of O_2 both *in vitro* and *in vivo*. The photoinactivation is mediated by blue light absorbed by the prosthetic heme groups. In addition, at higher photon flux, in leaves catalase is photoinactivated in red light absorbed by chlorophylls, through activated oxygen products transmitted from the chloroplasts. In mature green leaves with constant catalase activity the catalase apoprotein undergoes a permanent degradation and turnover in light which increases with the photon flux, while the heme groups remain largely undamaged. The light-dependent changes in the rate of catalase apoprotein synthesis are not related to changes of its mRNA but they must be determined by translational control mechanisms. Under various stress conditions which enhance oxidative stress or impair translation (e.g. low temperature or heat-shock, salt) a rapid loss of catalase activity is observed as early symptom together with the photoinhibition of photosystem II. The decline of catalase in stress-exposed, e.g. NaCl-treated, leaves contributes to enhanced oxidative damage in light, which is indicated by the bleaching of chlorophyll as well as increases in the proportions of the oxidized forms of ascorbate and glutathione, but no H_2O_2 accumulation was detected. Plants have developed adaptive mechanisms to avoid photoinactivation of catalase *in vivo*. Leaves of cold-hardened rye or several alpine high-altitude plants are tolerant against catalase photoinactivation in spite of high light and low temperature.

Keywords. Adaptation, alpine plants, antioxidants, catalase, heme, photoinactivation, rye, *Secale cereale*, stress

1 Introduction

Light acts as stress factor for plants because it initiates oxidative damage in photosynthetic cells and tissues. The production of reactive activated oxygen species in light is mainly mediated by the triplet excitation state of chlorophyll and subsequent singlet dioxygen formation and by the photoreduction of dioxygen through the photosynthetic electron transport chain (Lidon and Henriques, 1993). Plants have developed a major arsenal of efficient protective mechanisms which help them either to avoid or to repair oxidative damage, since they depend on living in the light. Protective mechanisms include carotenoids as scavengers of triplet chlorophyll and singlet oxygen and, in particular, the xanthophyll zeaxanthin, which is thought to mediate excess light energy dissipation as heat (Demmig-Adams and Adams III, 1993). In addition, plants contain a whole series of antioxidative enzymes and substances which are engaged in the detoxification of $O_2^{\cdot -}$ and H_2O_2 (Asada, 1994). Among these antioxidative enzymes is also the peroxisomal enzyme catalase. Irrespective of the fact that the demand for this enzyme increases with rising photon flux, catalase has quite paradoxal and enigmatic properties in that it is itself light-sensitive.

2 Catalase Photoinactivation as Light Stress Response

2.1 Photoinactivation of Catalase and Control of its Expression in Leaves

Catalase is a tetrameric heme-containing enzyme. It is well known that the purified enzyme is inactivated by visible light *in vitro* (Mitchell and Anderson, 1965; Cheng et al., 1981; Feierabend and Engel, 1986). Photoinactivation of the pure enzyme is mediated by blue light in the presence of oxygen (Cheng et al., 1981) and presumably involves the oxidation of histidine residues (Nakatani, 1961). Photoinactivation of catalase appears to occur also *in vivo* in intact leaves in a wide variety of plants so far investigated. Photoinactivation of catalase is observed even in moderate light under conditions to which plants are adapted. However, under non stress conditions it is usually latent because the loss of enzyme activity is continuously compensated for by a synchronous new synthesis. Therefore, under unstressed conditions, photodegradation of catalase can only be detected in leaves when protein synthesis is blocked, e.g. by a translation inhibitor (Feierabend and Engel, 1986), or in pulse-chase labeling experiments (Hertwig et al., 1992).

In Fig. 1 the *in vivo* photoinactivation of catalase observed in the presence of the translation inhibitor cycloheximide in rye leaves is compared in light of different spectral quality and of different photon flux. In addition, non-green etiolated and mature green leaves are compared. At a low photon flux of 70 $\mu mol\ m^{-2}\ s^{-1}$ a substantial decline of catalase activity was induced in blue light, particularly in etiolated leaves, while red light had little effect.

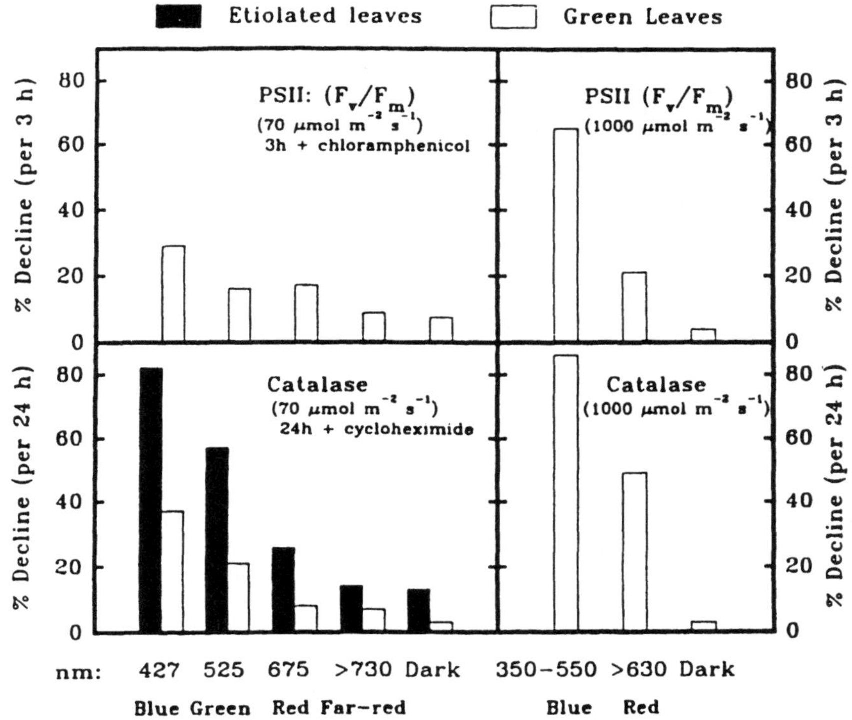

Fig. 1. Decline of catalase activity during a 24 h incubation in 35 µM cycloheximide and decline of the ratio of variable to maximal fluorescence (F_v/F_m) during a 3 h incubation in 6.3 mM chloramphenicol in middle sections (5 cm) excised from etiolated or green rye leaves and exposed in the presence of the inhibitors to light of different spectral quality and different photon flux. Primary leaves of 6-day-old rye seedlings grown in darkness or in light of 90 µmol m^{-2} s^{-1} were used

At a higher light intensity of 1000 µmol m^{-2} s^{-1} photoinactivation of catalase was accelerated and in green leaves substantial inactivation was also observed in red light, which is not absorbed by the catalase-heme. Remarkably, blue light was also very efficient to induce photoinhibition of PSII when the repair of the D1 protein was inhibited by chloramphenicol (Fig. 1). Because of the efficiency of red light, catalase inactivation at high light intensity must also be mediated through light absorbed by chlorophylls and by resulting the activated oxygen products transmitted from the chloroplasts. This was demonstrated by an *in vitro* model system containing purified catalase and isolated chloroplasts (Fig. 2). In a mixture of catalase and isolated chloroplasts the enzyme was readily inactivated by red light which had virtually no effect when only the pure enzyme was irradiated. This procedure may serve as an assay for quantifying the activated oxygen production by chloroplasts from different conditions.

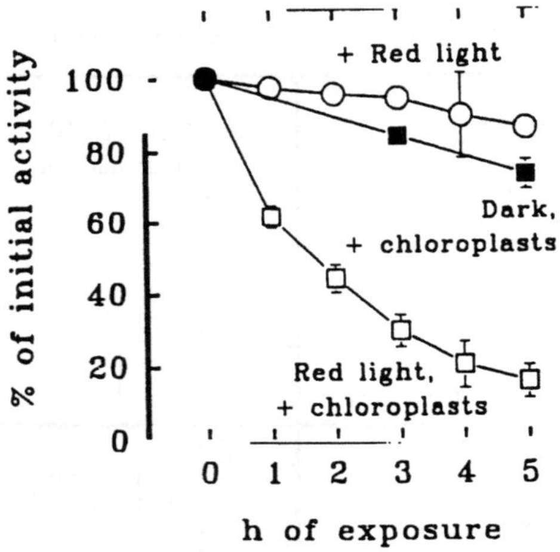

Fig. 2. Effect of red light of 800 µmol m^{-2} s^{-1} on the activity of pure bovine liver catalase (10 µkat ml^{-1}) or of bovine liver catalase in a mixture with Percoll gradient-purified intact chloroplasts (50 µg chlorophyll ml^{-1}) from rye leaves. Change of catalase activity during dark incubation in a mixture with chloroplasts is shown as control

In mature green leaves, which have a constant steady state level of catalase activity, the light-dependent degradation of the enzyme was demonstrated by pulse labeling with [^{35}S]methionine and subsequent chase experiments (Hertwig et al., 1992). During chase experiments both radioactive incorporation into the catalase polypeptide and its turnover greatly increased with light intensity. In this respect the behavior of catalase is similar to that of the D1 reaction center protein of PSII (Aro et al., 1993), however, the turnover of the latter is twice as rapid as compared to catalase. Under conditions (520 µmol m^{-2} s^{-1} PAR) where catalase had a half life of 3-4 h in rye leaves, that of the D1 protein was about 90 min. During *in vitro* photoinactivation the catalase protein is not cleaved (Feierabend and Engel, 1986). Also *in vivo* the enzyme appears to be first inactivated and only subsequently degraded by proteolysis. This was demonstrated by irradiation of leaves at low temperature where the catalase activity was lost but the degradation of its protein was retarded until plants were exposed to a higher temperature (Feierabend et al., 1992).

Several lines of experimental evidence suggest that the heme moiety of catalase is not damaged during its photoinactivation and may be available for the new synthesis of active enzyme. During the *in vitro* photoinactivation of purified catalase from bovine liver and from rye leaves we assayed the amount of functional heme by its ability to restore peroxidase activity to apoperoxidase (Thomas and Weinstein, 1990). According to this assay the heme remained largely undamaged during complete *in vitro* photoinactivation of catalase (Feierabend and Dehne, 1996). Furthermore, in pulse-chase experiments after labeling mature green rye leaves with d-[^{3}H]aminolevulinic acid the half time of the catalase-heme was much longer than that of the apoprotein (Fig. 3). Even the slow turnover of the heme which was observed must not reflect degradation but presumably represents dilution by newly synthesized unlabeled heme, because free hemes seem to mix in a common pool (Thomas and Weinstein, 1990; Feierabend and Dehne, 1996). Finally, new synthesis of catalase in high light did not require new heme synthesis because it was not impaired by the inhibitor gabaculine. By contrast, under identical conditions the repair cycle for PSII was clearly sensitive to gabaculine and new synthesis of the D1 protein appeared to require new synthesis of chlorophyll even in mature green leaves (Feierabend and Dehne, 1996).

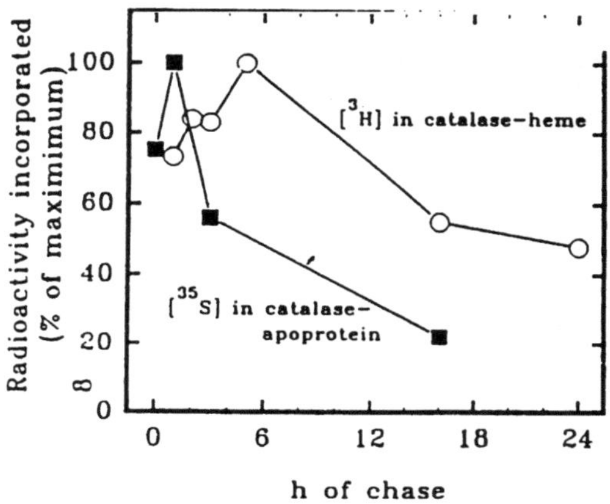

Fig. 3. Change in the radioactivity incorporated into the heme group or the apoprotein of catalase within 4 h of application of d-[^{3}H]aminolevulinic acid or L-[^{35}S]methionine to sections of rye leaves during a subsequent chase period, at 520 µmol m^{-2} s^{-1} PAR in the absence of radioactivity. Data from Hertwig et al. (1992) and Feierabend and Dehne (1996)

In order to investigate the regulation of catalase synthesis during its light-dependent turnover and to assay catalase transcript levels in rye leaves a cDNA for rye catalase was isolated and sequenced (unpublished results). Southern hybridization clearly indicated that only one catalase gene existed in rye. A second catalase polypeptide form with an apparent mol wt of 61 kD, which accumulated in rye leaves during prolonged exposure to darkness (Kar et al., 1993) in addition to the 57 kD polypeptide which usually occurs in light, appeared to represent a post-translational oxidative modification. Its generation from the 57 kD form could be mimicked by incubation of crude leaf extracts with H_2O_2 (unpublished results).

The rye cDNA was used as a probe for northern hybridizations. Transcript levels were compared under identical conditions. Although apparent catalase synthesis was about 6-fold higher in light (520 µmol m^{-2} s^{-1}) than in darkness

darkness, the transcript levels were even higher in darkness than in light and were not related to the rate of catalase synthesis (Fig. 4). Therefore, control of catalase synthesis by light must be exerted at translational level. Ongoing investigations about the distribution of catalase mRNAs in polysome fractions suggest that the rate of elongation is modulated by light intensity.

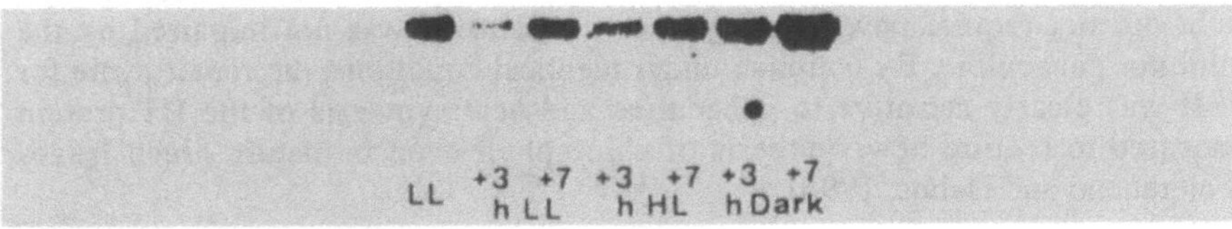

Fig. 4. Northern hybridization with a rye catalase cDNA-probe. Total RNA was extracted from rye leaves grown at 90 µmol m^{-2} s^{-1} PAR (LL) which were, in addition, exposed for 3 or 7 h to the same light intensity (LL), to a higher light intensity of 520 µmol m^{-2} s^{-1} (HL), or to darkness

In Fig. 5 our present knowledge of the sequence of events involved in the inactivation and repair cycle of catalase is summarized. Conceivably the availability of heme may control the rate of translation of the catalase-apoprotein during its turnover in leaves, as demonstrated for catalase synthesis in *Saccharomyces* (Hamilton et al., 1982) or for cytochrome synthesis (Howe and Merchant, 1994).

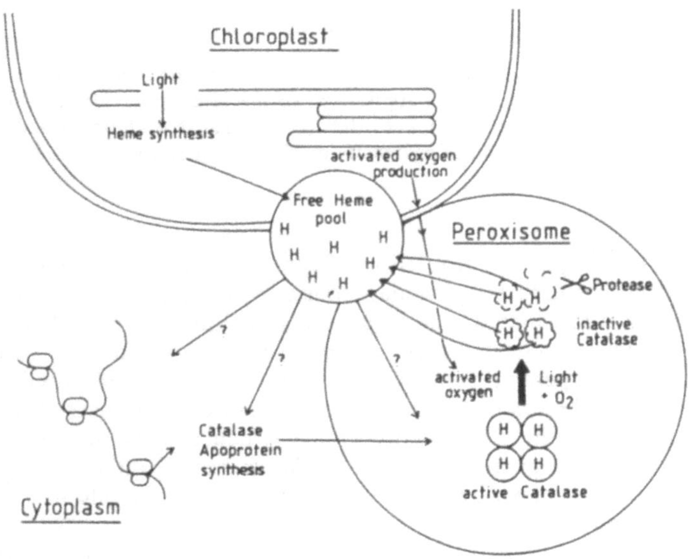

Fig. 5. Scheme to illustrate the life cycle of catalase during its light-dependent turnover in leaves. H, heme

The free heme pool is presumably membrane-bound (Thomas and Weinstein 1990; Feierabend and Dehne, 1996). While it had been believed for some time that the assembly of the tetrameric catalase molecule and heme addition occur in the peroxisome after import of polypeptide monomers, evidence is recently accumulating that catalase is fully assembled in the cytoplasm and transported as oligomeric protein (Middlekoop et al., 1993; McNew and Goodman, 1994). This would allow that the attachment of the heme could directly control the elongation and completion of the catalase polypeptide during its synthesis on cytosolic ribosomes.

2.2 Catalase-Deficiency under Stress Conditions

Since catalase undergoes a continuous turnover in light, its steady state level can be rapidly lowered under all stress conditions which either enhance degradation, such as excessive light, or which inhibit translation, such as low temperature (Volk and Feierabend, 1989), heat-shock temperature (Feierabend et al., 1992), salt (Streb et al., 1993) or aging (Kar et al., 1993). Under such conditions declines of catalase activity and photoinhibition of PSII were observed as early symptoms of stress damage even in moderate light. We addressed the question of whether the early loss of catalase might be responsible for the initiation of more severe photooxidative damage which is usually observed during prolonged stress exposures in light. Exposure of sections of rye leaves to increasing concentrations of NaCl under high light intensity allowed to gradually decrease the catalase activity to lower levels (Streb et al., 1993), because salt severely interferes with protein synthesis. The inhibition of new synthesis of catalase in the presence of NaCl has been documented by radioactive labeling (Hertwig et al., 1992). The oxidative stress accompanying the loss of catalase activity was clearly indicated by increasing photodegradation of chlorophyll (Streb et al., 1993; Streb and Feierabend, 1996) and by marked increases of the proportions of the oxidized forms of the antioxidants ascorbate and glutathione in NaCl-treated leaves (Fig. 6) (Streb and Feierabend, 1996). The suppression of protein synthesis in NaCl-treated leaves not only caused a depletion of catalase but simultaneously prevented increases of alternative antioxidative systems, which were usually observed when leaves were transferred to a higher photon flux or when catalase activity was eliminated by the inhibitor aminotriazole. Thus the pools of antioxidants became not only more oxidized but even declined and thus could not provide sufficient protection (Fig. 6). All symptoms of enhanced oxidative stress seen in salt-treated leaves could be closely mimicked by the application of the translation inhibitor cycloheximide. However, using the procedure of Patterson et al. (1984) which applies the reaction with a Ti(IV)-4-(2-pyridylazo)resorcinol complex, we were unable to detect any accumulation of H_2O_2 in catalase depleted rye leaves treated with NaCl or with aminotriazole. By contrast, in NaCl-treated leaves the H_2O_2-levels even declined (Fig. 7).

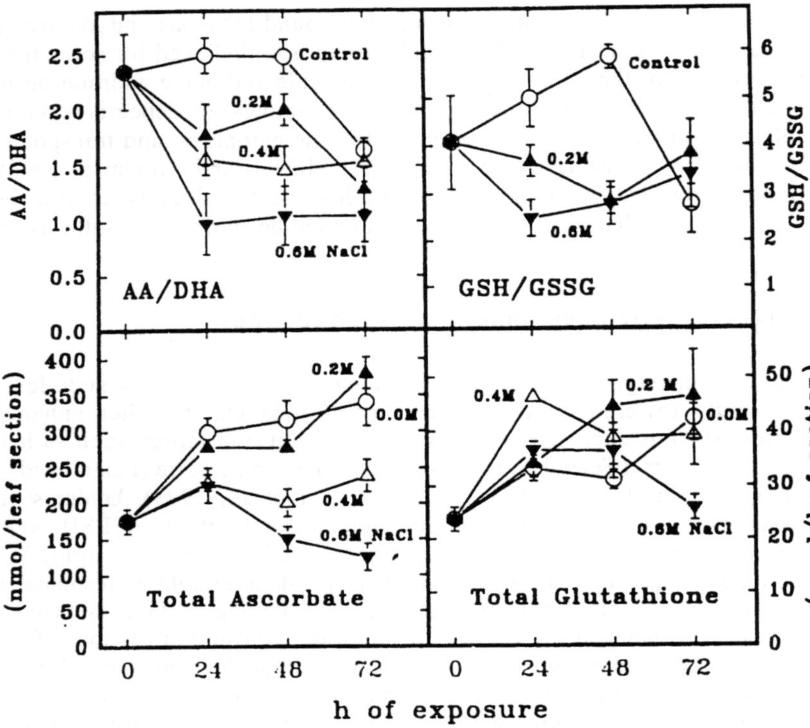

Fig. 6. Changes of the contents of total ascorbate and total glutathione and of the ratios of reduced to oxidized ascorbate (AA/DHA) or glutathione (GSH/GSSG) in middle sections of the primary leaves of 6-day-old rye seedlings (grown at 90 µmol m^{-2} s^{-1} PAR) during 3 d of exposure to light of 500 µmol m^{-2} s^{-1} PAR in the presence of different concentrations of NaCl. Extraction and HPLC-analysis of the antioxidants were performed as described (Kar et al., 1993)

The inhibition of catalase by salicylic acid in tobacco leaves has been implicated as the initial step for the induction of pathogen defense genes in systemic acquired resistance (Chen et al., 1993). Similarly, the declines of catalase seen as early symptoms under various stress conditions may also be expected to induce pathogen defense genes, if this was a general response. This would endow the phenomenon of photoinactivation of catalase with a significant biological role. The rapid early loss of catalase might serve as a signal that the plant begins to be severely weakened by stress damage and needs an enhancement of its defense system. Such response to oxidant stress may, however, represent a more complex defense syndrome which, in addition to pathogen defense genes, presumably includes also the enhancement of alternative antioxidative systems. Thus a most striking accumulation of catalase-mRNAs was observed in salt-stressed plants, particularly at a low concentration of 0.2 M NaCl.

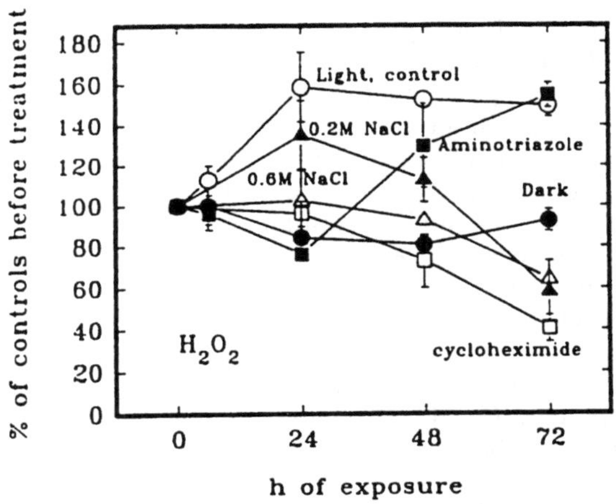

Fig. 7. Changes of the H_2O_2 contents in middle sections of the primary leaves of 6-day-old seedlings (grown at 90 µmol m^{-2} s^{-1} PAR) during 3 d of incubation in darkness (●) or in light of 500 µmol m^{-2} s^{-1} PAR (all other treatments). During the light exposure the leaf sections were incubated with 2 mM aminotriazole (■), 35 µM cycloheximide (□), or 0.2 M (▲) or 0.6 M NaCl (△), ○, control without further addition

At higher concentrations all biosynthetic events are presumably increasingly retarded. While we could not confirm the accumulation of H_2O_2 in catalase-depleted leaves which was described by Chen et al. (1993), the increases in the oxidation states of antioxidants may alternatively have a regulatory role. According to the results of Smith et al. (1985) particularly the proportion of oxidized cytosolic glutathione increased from 3% to 76% after inhibition of catalase by aminotriazole.

2.3 Stress Tolerance

Many plants have adapted to live under severe stress conditions leading to grow permanent photoinhibition of catalase or of PSII. As examples for tolerance to photoinhibition and light stress, we have studied cold acclimation in winter cereals and the adaptation of alpine high-altitude plants to their biotope characterized by extremely high irradiation and prevailing low temperature, in combination also with other physical stresses. In both instances marked tolerance to light stress was observed. When grown at 22°C rye leaves were subsequently exposed to 4°C in light of 500 µmol m^{-2} s^{-1} catalase and PSII were readily inactivated (Volk and Feierabend, 1989). However, when rye seedlings were, cold-hardened by growing them at 4°C catalase and PSII were fairly stable when exposed to the same light intensity at 4°C. Catalase photoinactivation appeared to be avoided in intact rye leaves from cold-hardened plants and also its repair at low temperature appeared to be more efficient than in grown leaves at 22°C (Streb, 1994). Remarkable resistance to photoinactivation of catalase and of PSII was also observed in leaves of several alpine plants, which had no specific morphological protection, such as *Soldanella alpina, Homogyne alpina* and *Ranunculus glacialis.*

Mechanisms contributing to the tolerance against photodamage in cold-hardened rye or in alpine high altitude plants were apparently not uniform among different species. In cold-hardened rye and in *Soldanella* many classical components of antioxidative protection were greatly increased. Particularly high levels were found for total and zeaxanthin cycle carotenoids, superoxide dismutase, glutathione reductase and ascorbate. However, in *Homogyne* and particularly in *Ranunculus glacialis* most components of antioxidative systems were low, even lower than in low-light grown rye leaves. Also in rye leaves experimental changes of the contents of antioxidants or of antioxidative enzymes did not appear to correlate with changes in the sensitivity of catalase to photodamage *in vivo* (Streb, 1994 and unpublished results). While the antioxidative systems probably support protection, they are thus unlikely to provide the main mechanism of tolerance to photodamage in these plants. According to Huner et al. (1993) resistance to photoinhibition of photosystem II in cold-hardened leaves depends on their increased capacity to prevent over-reduction of the photosynthetic electron transport chain under high light conditions and low temperature. When over-reduction of the electron transport chain is avoided also the probability of oxygen activation is reduced and the chloroplasts can be expected to transmit less oxidant stress to the surrounding cells that could contribute to catalase inactivation. Using the assay system shown in Fig. 2, it was indeed observed that catalase was considerably less inactivated in red light *in vitro* in the presence of isolated chloroplasts from the adapted alpine plant *Ranunculus glacialis* than in the presence of chloroplasts from non-acclimated plants, such as non-hardened rye or *Taraxacum*. In conclusion, the *in vivo* resistance of catalase to photoinactivation is expected to largely depend on metabolic or molecular adaptations in the chloroplasts of light stress -tolerant plants. An interesting further observation was, that the *in vitro* photoinactivation of cell-free catalase extracted from leaves of two alpine plants, *Soldanella* and, in particular, *Homogyne*, appeared to be remarkably slower, relative to catalases from non-alpine plants (Streb, 1994). If cDNA sequences for such catalases with increased light-stability would become available, they might provide quite valuable tools to further analyze the role of catalase and the potential significance of its common light-sensitivity in plants.

Acknowledgements. Financial support by the Deutsche Forschungsgemeinschaft and the experimental contribution by Astrid Nellner are greatly appreciated. We are grateful to R. Bligny and R. Douce, Grenoble, for using the research facilities of the alpine laboratory at the Col du Lautaret.

References

Aro E-M, Virgin I, Andersson B (1993) Photoinhibition of photosystem II. Inactivation, protein damage and turnover. Biochim Biophys Acta **1143**: 113-134

Asada K (1994) Production and action of active oxygen species in photosynthetic tissues. *In* CH Foyer, PM Mullineaux, eds, Causes of Photooxidative Stress and Amelioration of Defense Systems in Plants. CRC Press, Boca Ration, pp 77-104

Chen Z, Silva H, Klessig DF (1993) Active oxygen species in the induction of plant systemic acquired resistance by salicylic acid. Science **262**: 1883-1886

Cheng L, Kellog III EW, Packer L (1981) Photoinactivation of catalase. Photochem Photobiol **34**: 125-129

Demmig-Adams B, Adams III WW (1993) The xanthophyll cycle. *In* RG Alscher and JL Hess, eds, Antioxidants in Higher Plants. CRC Press, Boca Raton, pp 91-110

Feierabend J, Dehne S (1996) Fate of porphyrin cofactors during the light-dependent turnover of catalase and of the photosystem II reaction-center protein D1 in mature rye leaves. Planta **198**: 413-422

Feierabend J, Engel S (1986) Photoinactivation of catalase *in vitro* and in leaves. Arch Biochem Biophys **251**: 567-576

Feierabend J, Schaan C, Hertwig B (1992) Photoinactivation of catalase occurs under both high- and low-temperature stress conditions and accompanies photoinhibition of photosystem II. Plant Physiol **100**: 1554-1561

Hamilton B, Hoffaner R, Ruis H (1982) Translational control of catalase synthesis by hemin in the yeast *Saccharomyces cerevisiae.* Proc Natl Acad Sci USA **79**: 7609-7613

Hertwig B, Streb P, Feierabend J (1992) Light dependence of catalase synthesis and degradation in leaves and the influence of interfering stress conditions. Plant Physiol **100**: 1547-1553

Howe G, Merchant S (1994) Role of heme in the biosynthesis of cytochrome c_6. J Biol Chem **269**: 5824-5832

Huner NPA, Öquist G, Hurry VM, Krol M, Falk S, Griffith M (1993) Photosynthesis, photoinhibition and low temperature acclimation in cold tolerant plants. Photosynth Res **37**: 19-39

Kar M, Streb P, Hertwig B, Feierabend J (1993) Sensitivity to photodamage increases during senescence in excised leaves. J Plant Physiol **141**: 538-544

Lidon FC, Henriques FS (1993) Oxygen metabolism in higher plant chloroplasts. Photosynthetica **29**: 249-279

McNew JA, Goodman JM (1994) An oligomeric protein is imported into peroxisomes *in vivo.* J Cell Biol **127**: 1245-1257

Middelkoop E, Wiemer EAC, Schoenmaker DET, Strijland A, Tager JM (1993) Topology of catalase assembly in human skin fibroblasts. Biochim Biophys Acta **1220**: 15-20

Mitchell RL, Anderson IC (1965) Catalase photoinactivation. Science **150**: 74

Nakatani M (1961) Studies on histidine residues in hemeproteins related to their activities. J Biochem **49**: 89-102

Patterson BD, McRae EA, Ferguson IB (1984) Estimation of hydrogen peroxide in plant extracts using titanium (IV). Anal Biochem **139**: 487-492

Smith IK, Kendall AC, Keys AJ, Turner JC, Lea PJ (1985) The regulation of the biosynthesis of glutathione in leaves of barley (*Hordeum vulgare* L.). Plant Sci **41**: 11-17

Streb P, Michael-Knauf A, Feierabend J (1993) Preferential photoinactivation of catalase and photoinhibition of photosystem II are common early symptoms under various osmotic and chemical stress conditions. Physiol Plant **88**: 590-598

Streb P (1994) Lichtschäden und Streßwirkungen in Blättern und antioxidative Schutzmechanismen. Doctoral Thesis, Department of Biology, JW Goethe-Universität, Frankfurt am Main

Streb P, Feierabend J (1996) Oxidative stress responses accompanying photoinactivation of catalase in NaCl-treated rye leaves. Botan Acta (in press)

Thomas J, Weinstein JD (1990) Measurement of heme efflux and heme content in isolated developing chloroplasts. Plant Physiol **94**: 1414-1423

Volk S, Feierabend J (1989) Photoinactivation of catalase at low temperature and its relevance to photosynthetic and peroxide metabolism in leaves. Plant Cell Environ **12**: 701-712

Part 6

Oxidative Stress and Plant Signal Stress

Post-Transcriptional Regulation by Anoxia of a *myb* Gene in Rice (*Oryza sativa*)

Giovanna Giovinazzo[1], Flavio Magaraggia[2], Emilio Menguzzato[2], Giorgio Valle[3] and Immacolata Coraggio[2]

[1] IRBA, C.N.R., I-73100 Lecce, Italy
[2] Istituto Biosintesi Vegetali, C.N.R., I-20133 Milano, Italy
[3] Università degli Studi di Padova, Dipartimento di Biologia, I-35121 Padova, Italy

Abstract. We isolated a *myb* coding cDNA expressed in anaerobic rice coleoptiles. Sequence analysis revealed some peculiar features, suggesting the presence of post-transcriptional regulation events: an upstream ORF (uORF) in the long leader sequence, an unspliced intron and the presence of a short putative leucine zipper in the ORF encoded by the unspliced RNA. Analyses of mRNA in rice roots and coleoptiles by RT-PCR demonstrated that both RNA forms, spliced and unspliced, are present *in vivo* and that anaerobiosis strongly induced the presence of spliced transcript in roots. Northern blot analysis indicated that in rice roots anoxia affects expression of a set of genes belonging to the *myb* family. Moreover, we demonstrated the negative influence of the *myb7* uORF on the expression of the downstream ORF, and that the inhibition was depended, at least in part, on the uORF-AUG.

Keywords. Anoxia, leucine zipper, *myb*, *Oryza sativa*, regulated splicing, rice, upstream ORF

1 Introduction

The expression of transcription factors is likely to play an important role in determining the kind of response to environmental stress sequence. Analysis of transcription factors has revealed the presence of conserved domains (leucine zipper, zinc finger, homeobox, MYB domain) involved in DNA binding or in determining important structural features.

The MYB conserved domain consists of imperfect repeats of 50-53 amino acids, known as "tryptophan cluster", because of high conserved tryptophan residues involved in stabilizing the structure of the DNA binding domain. It has been proposed that each repeat folds in a helix-turn-helix structure, which is present also in other transcriptional factors (Ogata et al., 1992). Three conserved domains are found in animal transcription factors, two in viruses and plants, whereas some MYB proteins present one complete and one incomplete repeat. In one case the presence of just one incomplete repeat has been reported (Ju et al., 1990; England et al., 1992; Baranowskij et al., 1994).

In animals *myb* genes are present in single copy or as a small gene family (1-3 members) and are involved in cellular division (Guarante and Berminghan, 1992). Conversely, in plants *myb* genes are present as large families (6-30 members), involved in the control of a wide range of biochemical pathways (Marocco et al., 1989; Jackson et al., 1991; Avila et al., 1993).

Expression of transcription factors can be regulated at the transcriptional level, by post-translational modifications and by interaction with other factors (Nakabeppu and Nathans, 1991; Sun and Baltimore, 1991; Bengal et al., 1992; Hsu et al., 1992; Li et al., 1992). Alternative splicing has been also reported in the regulation of the expression of several transcriptional factors, with mechanisms such as the alternative use of exons, the alternative use of intron acceptor sites and the inclusion of coding "intronic" sequences (ON/OFF use) (Grotewold et al.; 1991, Kuhn et al., 1994; Springhorn et al., 1994). In addition, many mRNAs encoding critical regulatory proteins do not appear to be designed for efficient translation because of the presence of uORF (Hinnebusch, 1990; Lohmer et al., 1993), or because of the presence of a non-AUG translation-start codon (Kozak, 1991).

Changes in expression deriving from post-transcriptional mechanisms can occur rapidly and in the absence of protein synthesis, and this may provide a pool of transcriptional factors available for a prompt response to environmental stress.

In the present work we report the isolation of a rice cDNA clone, *myb7*, encoding a MYB factor, expressed during anoxia. Analysis of its sequence suggests the existence of two post-transcriptional regulation steps: a translational mechanism due to the presence of an upstream ORF, and a maturation mechanism based on ON/OFF use of an intron encoding for a putative short leucine zipper. We also show that anoxia positively regulates splicing of *myb7* mRNA in rice roots and that uORF inhibits the expression of the downstream ORF in transiently transformed protoplasts.

2 Results and Discussion

2.1 Sequence Analysis of *myb7* cDNA Clone

Two million plaques of a cDNA library in lambda ZAP vector, were screened with a probe corresponding to a conserved domain of a *myb* cDNA from barley. One positive clone, *myb7*, was isolated. The isolated clone (1190 bp) was completely sequenced and found to correspond to the 5' untranslated region and part of the interrupted coding sequence of a novel *myb*-like gene. A schematic draft of *myb7* cDNA clone is shown in Fig. 1, where some peculiar features suggesting post-transcriptional regulation are also indicated. In Fig. 2 the nucleotidic and the putative amino acidic sequences are shown.

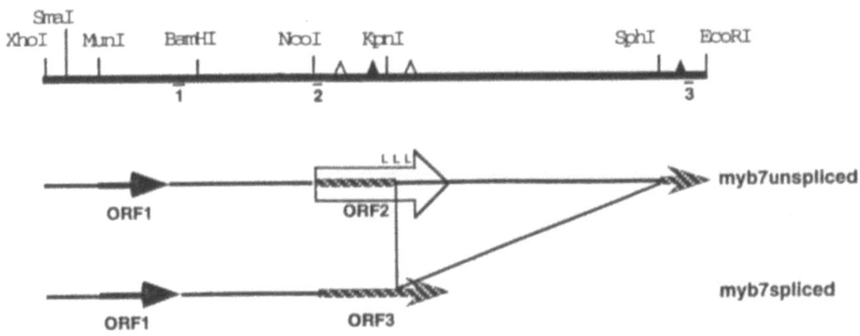

Fig. 1. Scheme of *myb7* cDNA clone. CK2 (△) and PKC (▲) putative phosphorylation sites are indicated; arrows indicate ORFs, while numbers indicate oligonucleotides used for PCR amplifications. Relevant restriction sites are also shown

An upstream ORF (uORF, position 118-228), indicated as ORF1 in Figs. 1 and 2, is present in the long leader sequence (at least 480 bp). Moreover, the *myb7* cDNA seems to be derived from an unspliced mRNA. The putative protein product of the longer ORF (ORF2, position 482-715) shows homology with MYB factors previously characterized in plants, but the homology drops in correspondence of a splicing site conserved in *myb* plant genes (arrows in Fig.3) (Paz-Ares et al., 1987; Grotewold et al., 1991; Urao et al., 1993). A canonical 3' acceptor site is present 508 bp downstream in *myb7* sequence; the ORF could be the result of splicing of this 508 bp (position 482-514 plus 1123-1190, Figs. 1, 2 and 3) that extends to the end of sequence, maintaining homology to *myb*-like sequences. Homology alignment of the nucleotide and amino acid sequences of several *myb* genes is shown in Fig. 3.

Translation of the unspliced mRNA would result in the synthesis of a polypeptide (ORF2, Fig 1, 2 and 3) containing an incomplete MYB repeat, followed by a short putative leucine zipper. Three leucine residues (capitalized in Fig. 2), separated by six amino acids and within a very hydrophobic context, are present in the aminoacidic sequence corresponding to the intron.

CK2 and PCK phosphorylation consensus sites are present both in the ORF2 and ORF3 (Figs. 1 and 2); the CK2 sites in ORF3 are in a conserved position in respect to other *myb* sequences. Regulation of MYB transcriptional factor activity by CK2 has been reported (Lusher et al., 1990).

2.2 The *myb7* uORF Inhibits Expression of Downstream ORF

The putative polypeptide encoded by the uORF (ORF1, Figs. 1 and 2) is 40 amino acids long and is rich in leucine and arginine residues (6 and 5 respectively). Searches in the GenBank, EMBL and SWISS-PROT data banks using the sequence of the uORF encoded polypeptide did not evidence significant homology to previously isolated sequences. The translational stop codon, present in this ORF, TGA, is the most widely used in plants (46%) and is followed by an adenine, which is frequently found in this position (41%) (Gallie, 1993).

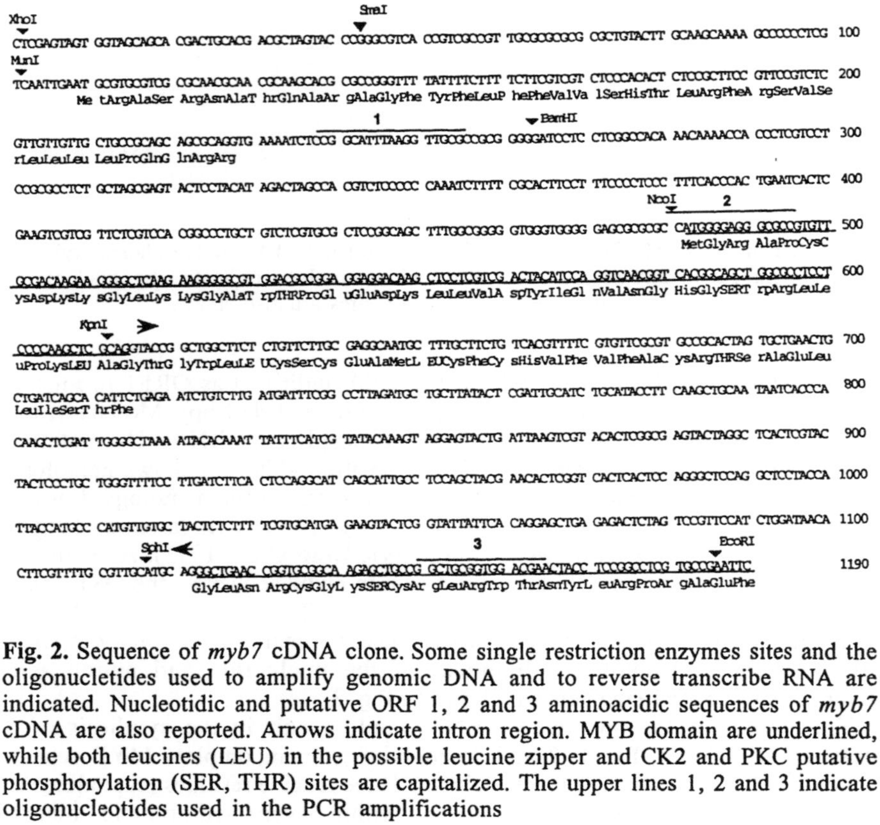

Fig. 2. Sequence of *myb7* cDNA clone. Some single restriction enzymes sites and the oligonucletides used to amplify genomic DNA and to reverse transcribe RNA are indicated. Nucleotidic and putative ORF 1, 2 and 3 aminoacidic sequences of *myb7* cDNA are also reported. Arrows indicate intron region. MYB domain are underlined, while both leucines (LEU) in the possible leucine zipper and CK2 and PKC putative phosphorylation (SER, THR) sites are capitalized. The upper lines 1, 2 and 3 indicate oligonucleotides used in the PCR amplifications

Most eukaryotic mRNAs encode a single polypeptide, starting at the first AUG codon; the introduction of AUG codons in the mRNA leader (uAUG), strongly reduces the expression of downstream coding sequences (Kozak, 1989). The large ribosomal subunit is supposed to bind the 5' end of transcript, to scan the leader, and to start translation at first AUG. Reduced initiation at downstream AUG depends on the inefficiency of internal re-initiation (Kozak, 1989). Nevertheless, many mRNAs, encoding critical regulatory proteins, appear to be designed for inefficient translation; translation is started at non-AUG codon or present uORFs in the leader, as in the cases of yeast GCN-4 (activator of amino acid biosynthesis) and of the maize O2 (regulator of zein biosynthesis) (Hinnebusch, 1990; Kozak, 1991; Lohmer et al., 1993).

```
MYB1_HORVU   .....S....  .E.AHTNK..  ..K...DR.T  A..KA..E.C  ..S...A...  L.........   56
MYB3_MAIZE   .....S....  .E.AHTN...  ..K...ER.V  A..RA..E.C  ..S...A...  L.........   56
MYBD_MAIZE   .....RA...  .A.E.V....  ..SK..DA.A  A.VKA..E.K  ..EV.QK...  R.........   56
MYBC_MAIZE   .....RA...  .A.E.V....  ..SK..DA.A  A.VKA..E.K  ..EV.QK...  R.........   56
MYBP_MAIZE   .....T....  .E.V.L...R  ..A...QL.A  N..AE..E.S  ..S...N...  L.........   56
MYB1_MAIZE   .GR..A....  .A.V.LN..S  ..PQ..MR.I  A..QK..HTN  ..A...Q...  L.........   58
MYB3_HORVU   .....PSSGA  VGQPKVRK.L  .SP...EK.Y  NH.IR..V.C  .SSV.RL.A.  N.........   58
MYB2_PHYPA   .....K....  .E.V.LR..P  ..S...QK.V  SH.TNN.LSC  ..AI..L...  L.........   56
GL1_ARATH    .RIR.RDE..  KENQEY.K.L  ..V...NI.M  D.VLN..T.Q  .NRIVRKT..  K.........   58
ORF2         .....A....  .D.K.L.K..  ..P...KL.V  D..QVN.H.S  ..L...L..T  GWLLC..---   53
ORF3         .....A....  .D.K.L.K..  ..P...KL.V  D..QVN.H.S  ..L...L...  N.........   56
Consensus    M--GR.PC--  C.K.G.KRGA  WT.EED..L.  .YI..HG.G.  WR.LPK.AGL  .RCGKSCRLR   60

MYB1_HORVU   .I........  .....HE.DE  ...K..S...  .K....GR.  ..........  ...THIRRKL  116
MYB3_MAIZE   .I........  ....TAD.D.  ..VK..S...  .K....AR.  ..........  ...THVRRKL  116
MYBD_MAIZE   .L.....NIR  ...I.YD...  ...R..R..  .......GR.  ....E.....  ...STLGRRA  116
MYBC_MAIZE   .L.....NIR  ...I.YD...  ...R..R..  .......GR.  ..........  ...STLGRRA  116
MYBP_MAIZE   .I....A.V.  ...I.KE...  I..K..AT..  .......SH.  ..........  ...S------  110
MYB1_MAIZE   .I........  ....TDE..E  A..R..G...  .K..K..AC.  ..........  V..THLKKKV  118
MYB3_HORVU   .I........  ..C..QQ...  H.VA..QI..  ....Q..SH.  ..........  F..SCIKKKL  118
MYB2_PHYPA   .T........  ..I..EA..N  ..LD..AT..  ....R..AQ.  ..........  ...TRLKKRL  116
GL1_ARATH    .M...S.NVN  K...TEQ...  ...R..K...  .......KRV  ......QV..  ...THLSKKL  118
ORF2         -----EAM.-  --C.C-----  ----------  ----------  ----------               60
ORF3         .T.....---  .AE.------  ----------  ----------  ----------  ----------   67
Consensus    W.NYLRPDLK  RGNFS..EED  LII.LH.LLG  NRWSLIA..L  PGRTDNEIKN  YWN.......  120
```

Fig. 3. Homology of the MYB7 putative protein products with other plant MYBs. Only differences in the consensus sequence are indicated. Dots indicate identity, dashes, gaps and arrows conserved positions of introns. Both proteins encoded by spliced (ORF3) and unspliced (ORF2) mRNA of *myb7* are reported. Other *myb* genes are of *Zea mays* (MAIZE), *Hordeum vulgare* (HORVU), *Physcomitrella patiens* (PHYPA) and *Arabidopsis thaliana* (ARATH). Dashes at the end of ORF3 correspond to the end of *myb7* clone, which is not full size

The expression of the MYB-like protein encoded by the *myb7* mRNA might be similarly regulated by the presence of the putative upstream ORF1. Its translation-start codon is in the very poor context TGA**ATG**C, where neither the nucleotide at -3 (T) nor the one at +4 (C) position corresponds to the one found in the eukaryotic sequence consensus (A>G)CC**AUG**G (Kozak, 1989; Gallie, 1993; Pantopoulos et al., 1994). However, in maize the presence of an upstream out of frame AUG embedded in a weak consensus context, as CAG**AUG**U, can block the expression of downstream luciferase coding sequence. This indicates that the first AUG is preferentially used although the second one is placed in a theoretically more favourable environment (Luehrsen and Walbot, 1994). Also the uORFs present in the leader sequence of GCN4 and O2 are in a sub-optimal consensus context (Hinnebusch, 1990; Lohmer et al., 1993). To test if uORF, contained in the *myb7* leader region, is able to inhibit translation of a downstream ORF, the fragment XhoI-BamHI was cloned in the puc8CaMVCATAT, upstream to the *cat* gene (pCaLead, Fig. 4). In an analogous construct the AUG of *myb7* uORF was mutagenized in AAT (X in pCALead*, Fig. 4). The CAT activity in *Phaseolus vulgaris* protoplasts, transiently transformed with both pCaLead, pCaLead* and pCaCat plasmids, indicates that *myb7* uORF inhibits the expression of the downstream *cat* gene, and that, at least in part, this inhibition depends on the AUG. *Myb7* is the first *myb* mRNA in which a putative uORF has been identified; the context of uORF initiation-codon is consistent with a translational regulatory function of this ORF. In condition of general reduction of mRNA translation (i.e. during stress) the inhibitory activity of the uORF might be by-passed.

2.3 Analysis of Expression of *myb7*

The DNA fragment used to screen the library, corresponding to the *myb* domain from barley, revealed, by hybridization on northern blot, the presence of two classes of messenger RNAs of about 1700 and 1200 bp, the difference between them being compatible with the size of the *myb7* intron (Fig. 5). In coleoptiles the relative abundance of the two classes of RNAs was independent of growth conditions, whereas in roots the higher band was predominant in aerobic and the lower one in anoxic tissues (Fig. 5). However, hybridization with the heterologous probe, followed by low stringency washes, evidenced the expression of a *myb* family in rice and the two bands could refer to different maturation forms deriving from the same genes or to different classes of genes.

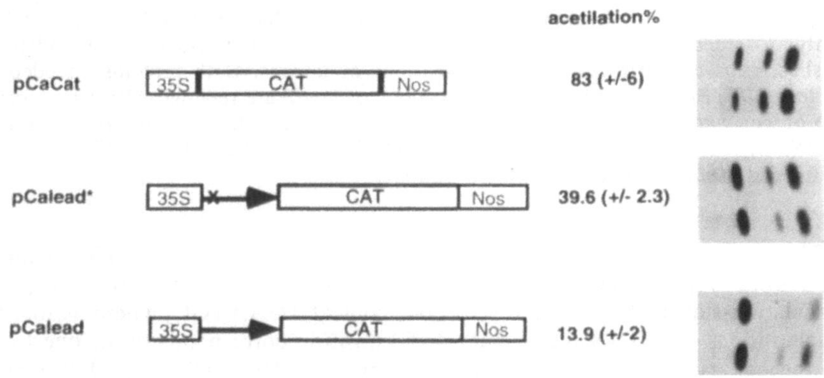

Fig. 4. *myb7* uORF inhibits expression of downstream CAT ORF. pCaCat= puc8CaMVCATAT plasmid; pCalead= puc8CaMVCATAT plasmid with *myb7* uORF inserted upstream to CAT ORF; pCalead*= analogous to pCalead, but with uORF-AUG mutagenized in AAT

Therefore, to study specifically the expression of *myb7* gene we analyzed the expression of this gene in rice roots and coleoptiles subjected or not to anoxia, using northern blot analysis with specific probes.

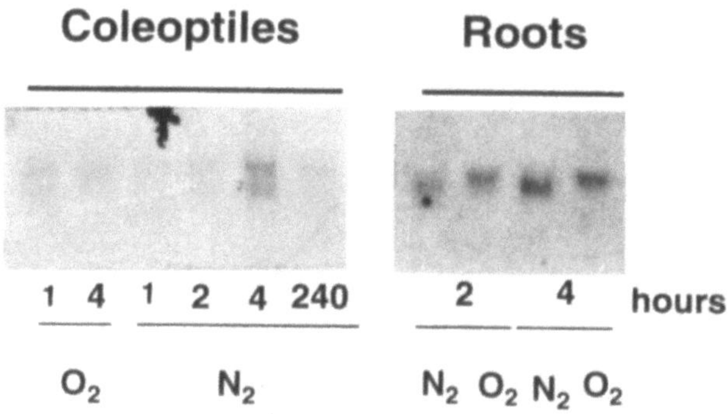

Fig. 5. *Myb* expression in rice roots is influenced by anoxia. Forty micrograms of total RNA, extracted from rice segments subjected to aerobic (O_2) or anaerobic (N_2) conditions were probed with a *myb* generic probe (*myb8*). Length of treatment is indicated. Two mRNA classes of about 1200 and 1700 bp are evident

However, even loading up 40 μg of total RNA or 5 μg of poly $(A)^+$, no signal was detected by using as probe either a KpnI-SphI fragment corresponding to the intron, or a XhoI-NcoI fragment corresponding to the untranslated 5' region, possibly due to a low amount of *myb7* mRNAs (data not shown). We therefore resorted to PCR analysis.

At first we analyzed the amplification products of genomic DNA amplified with oligo 2 and 3, both laying in the conserved *myb* region, on either side of the putative intron, or with oligo 1 and 3, the first of them lying in the untranslated 5' region. Amplified products were separated on agarose gel and blotted on nylon membrane. The results of hybridization with the 3' (fragment KpnI-SphI: intron region) or 5' (fragment NcoI-KpnI: *myb* conserved region) probes are shown in Fig. 6. When DNA was amplified with oligo 1 and 3 (lane oligo 1, Fig. 6) hybridization with both probes resulted in one band of 900 bp corresponding to the predicted amplified region of *myb7* cDNA. However, hybridization of DNA amplified with oligo 2 and 3 (lane oligo 2, Fig. 6) evidenced a band of 700 bp with the 3' probe , but several bands when probed with a general *myb* 5' domain. This indicates the presence of a family of *myb* genes, with a first intron in a conserved position, whose size ranges from almost 100 bp (corresponding to the lower of bands visible in Fig. 6 panel 5', lane oligo 2) to almost 500 bp, as in our *myb7* clone.

The presence of a single band hybridizing to the intron region in the genomic DNA amplified with oligo 2 and 3 (which lie in the conserved *myb* region) suggests that the *myb7* clone corresponds to a single gene. This is also supported by the presence of a single band hybridizing to conserved *myb* region within the amplification products of oligo 1 and 3. Additionally, these results indicate that oligo 1 and 3 can be used to amplify specifically the myb7 sequence. This allowed us to follow *myb7* expression in aerobic and anaerobic conditions.

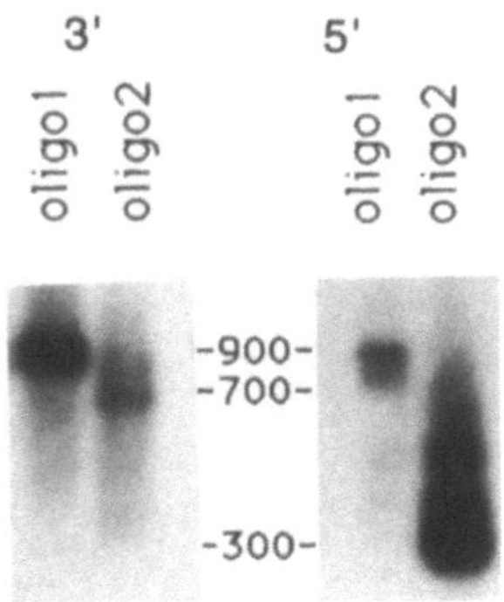

Fig. 6. Southern blot analysis of genomic DNA amplified by PCR. Genomic DNA was amplified using oligo 1 and 3 (oligo 1) or 2 and 3 (oligo 2). Blot analysis of amplified products hybridized with probe specific of *myb7* (3') or with a fragment containing a portion of *myb* domain (5') is shown. Size in bp are indicated

Total RNA, extracted from rice roots and coleoptiles, grown in aerobic conditions or subjected to 2 or 24 h of anoxia, was DNAse treated to eliminate genomic contamination and then reverse transcribed and amplified by using oligos 1 and 3. Amplification products were separated on 2% agarose gel, blotted on nylon membrane and hybridized to 5' *myb* generic probe, corresponding to first exon, to allow identification of both spliced and unspliced transcripts. At least three separate lots of tissues for each condition were analyzed. In the control reactions, without substrate DNA or without primers, amplified products were absent (data not shown).

Two bands of the expected size for the spliced and the unspliced forms were detected (Fig. 7); in coleoptiles the relative amount of fragments derived from unspliced and spliced messengers was slightly affected by anoxia, and the two forms seemed to be present in an almost equimolar amount. However, in stressed roots there was a dramatic increase in the ratio between the amount of amplified fragments corresponding to spliced and the unspliced mRNA. In fact, in aerobic roots the unspliced form was highly predominant, whereas in stressed roots the amplified fragment corresponding to the spliced transcript was the more abundant form.

In animal and fungal cells splicing is an accurate and efficient process and significant amounts of unspliced transcripts are detected only in cells subjected to severe stress (Yost, 1988). Most heat shock (*hsp*) genes in animals have no introns and this was supposed to be due to the evolutionary selection of failure of the splicing machinery during heat shock. In plants many *hsp* genes contain introns and the presence of unspliced RNAs has been also reported (Rochester et al., 1986). Splicing is inhibited by a variety of stresses such as treatment with heavy metals (Czarnecka et al., 1988; Frey et al., 1990) and hypoxia (Ortis and Strommer, 1990). The presence of unspliced RNAs has been also reported in plants not subjected to experimental stress (Paz-Ares et al., 1990; Arondel et al., 1991; Grotewold et al., 1991). However, field-grown plants in field are continually subjected to some degree of stress. Several mRNAs, encoding transcriptional factors, are regularly spliced and could result in mRNAs encoding different transcriptional factors with distinct functions (Green, 1991).

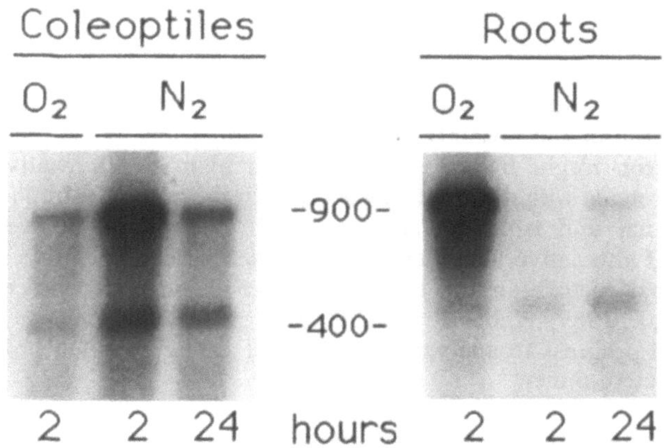

Fig. 7. Southern blot analysis of RT-PCR products. DNAse treated RNAs were reverse transcribed and amplified with oligo 1 and 2. Southern analysis with probe corresponding to 5' region of *myb7* indicate the presence of two mRNAs, the size of them corresponding to spliced (400 bp) and unspliced (900 bp) forms. In roots anoxia strongly induces splicing. O_2, aerobic; N_2, anaerobic conditions. Length of treatment (h) is indicated

The large amount of unspliced *myb7* RNA in respect to the spliced form, and the predominant presence of the spliced form of the *myb7* transcript in roots during anoxia, a stress situation that generally inhibits splicing, suggest that post-transcriptional regulation rather than a trivial escape is responsible for the observed switch between the prevalence of the unspliced and spliced forms. The observation that the 5' region of the intron could direct the synthesis of an in-frame short leucine zipper supports the idea of a regulatory role for *myb7* splicing.

Leucine zippers are involved in dimerization of several classes of transcriptional factors (Schmidt et al., 1990; Cole, 1991; Prendergast and Ziff, 1992; DeVetten and Ferl, 1995). Although most of them contain five or more regularly spaced leucine residues, only three or four have been found in several cases (Fornace et al., 1989; Katagiri et al., 1989; Weisshaar et al., 1991; Meier and Gruissem, 1994; DeVetten and Ferl, 1995).

The putative polypeptide, containing the leucine zipper, coded by ORF2, would be unable to bind DNA, because of the presence of an incomplete MYB domain. Moreover, no putative transcriptional activator region is present. However, a differentially-spliced FosB transcript in mouse 3T3 cells, in regenerating-liver cells and in hippocampal neurones encodes a truncated form of FosB able to dimerize with each of Jun proteins. The resulting heterodimers inhibits the transcriptional activity of both Jun and Fos (Nakabeppu and Nathans, 1991). Finally the functional antagonism between c-Jun and MyoD proteins in mammals seems to derive from the formation of heterodimers unable to activate the promoters, targets of the respective homodimers (Bengal et al., 1992).

The translation of the putative partial MYB-domain-leucine zipper polypeptide may provide the cell with a mechanism able to regulate the dimerization and hence activity of other transcriptional factors. We conclude that *myb7* gene is subjected to post-transcriptional regulation, at the level of mRNA maturation and possibly at the translational level. The ON/OFF use of the first intron might be an efficient mechanism to alternatively express regulatory proteins with different functions as in the case of murine c-*myb* gene. The functional and truncated forms of MYB proteins derived from the translation of alternatively-spliced c-*myb* transcript act as positive and negative transcriptional factors on the same set of genes in different cellular types (Ramsay et al., 1989; Weber et al., 1990).

Work is in progress to analyse the ability of ORF2 encoded protein to form homo- and hetero-dimers.

Acknowledgements. The authors thank Dr. W. Rhode for providing myb cDNA from barley. The research was supported by the National Research Council of Italy, special project RAISA, sub-project N. 2, Paper N. 2467.

References

Arondel V, Tchang F, Baillet B, Bignols F, Grellet F, Delseny M, Kader JC, Puigdomenech P (1991) Multiple mRNAs coding for phospholipid-transfer protein from *Zea* mays arise from alternative splicing. Gene **99**: 133-136

Avila J, Nieto C, Canas L, Benito MJ, Paz-Ares J (1993) *Petunia hybrida* genes related to the maize regulatory C1 gene and to animal myb proto-oncogenes. Plant J 3: 553-562

Baranowskij N , Frohberg C, Prat S, Willmitzer L (1994) A novel DNA binding protein with homology to Myb oncoproteins containing only one repeat can function as a transcriptional activator. EMBO J **13**: 5383-5392

Bengal Y, Ransone L , Scharfmann R, Dwarki Vj, Tapscott SJ, Weintraub H, Verma IM (1992) Functional antagonism between c-Jun and MyoD proteins: a direct physical association. Cell **68**: 507-519

Cole MD (1991) Myc meets Max. Cell **65**: 715-716

Czarnecka E, Nagao RT, Key JL, Gurley WB (1988) Characterization of Gmhsp26-A, a stress gene encoding a divergent heat shock protein of soybean: heavy-metal-induced inhibition of intron processing. Mol Cell Biol **8**: 1113-1122

DeVetten NC, Ferl RJ (1995) Characterization of a maize G-box binding factor that is induced by hypoxia. Plant J **7**: 589-601

England BP, Admon A, Tjian R (1992) Cloning of *Drosophila* transcription factor Adf-1 reveals homology to Myb oncoproteins. Proc Natl Acad Sci USA **89**: 683-687

Fornace AJ, Nebert DW, Hollander CM, Luethy JD, Papathanasiou M, Fargnoli J, Holbrook NJ (1989) Mammalian genes coordinately regulated by growth arrest signals and DNA-damaging agents. Mol Cell Biol **9**: 4196-4203

Frey M, Reinecke J, Grant S, Sadler H, Gierl A (1990) Excision of the En/Spm transposable element of *Zea* mays requires two element-encoded proteins. EMBO J **9**: 4037-4044

Gallie DR (1993) Post-trascriptional regulation of gene expression in plants. Annu Rev Plant Physiol Plant Mol Biol **44**: 77-105

Green MR (1991) Biochemical mechanisms of constitutive and regulated pre-mRNA splicing. *In* AR Inc, ed, Analytical Biochemistry, pp 559-599

Grotewold E, Athma P, Peterson T (1991) Alternatively spliced products of the maize P gene encode proteins with homology to the DNA-binding domain of myb-like transcription factors. Proc Natl Acad Sci USA **88**: 4587-4591

Guarante L, Berminghan-McDonogh O (1992) Conservation and evolution of transcriptional mechanism in eukaryotes. Trends in Genetics **8**: 27-32

Hinnebusch AG (1990) Involvment of an initiation factor and protein phosphorylation in translational control of GCN4 mRNA. Trends Biol Sci **15**: 148-152

Hsu T, Gogos JA, Kirsh SA, Kafatos FC (1992) Multiple zinc finger forms resulting from developmentally regulated alternative splicing of a transcription factor gene. Science **257**: 1946-1950

Jackson D, Culianez-Macia F, Prescott AG, Roberts K, Martin C (1991) Expression patterns of *myb* genes from *Antirrhinum* flowers. Plant Cell **3**: 115-125

Ju Q, Morrow BE Warner JR (1990) REB1, a yeast DNA-binding protein with many targets is essential for cell growth and bears some resemblance to the oncogene *myb*. Mol Cell Biol **10**: 5226-5234

Katagiri F, Lam E, Chua N-H (1989) Two tobacco DNA-binding proteins with homology to the nuclear factor CREB. Nature **340**: 727-729

Kozak M (1989) The scanning model for translation: an update. J Cell Biol **108**: 229-241

Kozak M (1991) Structural features in eukaryotic mRNAs that modulate the initiation of translation. J Biol Chem **266**: 19867-19870

Kuhn A, Voit R, Stefanovsky V, Evers R, Bianchi M, Grummt I (1994) Functional difference between the two splice variants of the nucleolar transcription factor UBF: the second HMG box determines specificity of DNA binding and transcriptional activity. EMBO J **13**: 416-424

Li L, Heller-Harrison R, Czech M, Olson EN (1992) FGF inactivates myogenic helix-loop-helix proteins through phosphorylation of a conserved protein kinase C site in their DNA-binding domains. Mol Cell Biol **12**: 4478-4485

Lohmer S, Maddaloni M, Motto M, Salamini F, Thompson RD (1993) Translation of the mRNA of the maize transcriptional activator Opaque-2 is inhibited by upstream open reading frames present in the leader sequence. Plant Cell **5**: 65-73

Luehrsen KR, Walbot V (1994) The impact of AUG start codon context on maize gene expression *in vivo*. Plant Cell Rep **13**: 454-458

Lusher B, Christenson E, Litchfield DW, Krebs EG, Eisenman RN (1990) Myb DNA binding inhibited by phosphorilation at a site deleted during oncogenic activation. Nature **344**: 517-521

Marocco A, Wissenbach M, Becker D, Paz-Ares J, Saedler H, Salamini F, Rohde W (1989) Multiple genes are transcribed in *Hordeum vulgare* and *Zea mays* that carry the DNA binding domain of the myb oncoproteins. Mol Gen Genet **216**: 183-187

Meier I, Gruissem W (1994) Novel conserved sequence motifs in plant G-box binding proteins and implications for interactive domains. Nucleic Acid Res **22**: 470-478

Nakabeppu Y, Nathans D (1991) A naturally occurring truncated form of FosB that inhibits Fos/Jun transcriptional activity. Cell **64**: 751-759

Ogata K, Hojo H, Aimoto S, Nakai T, Nakamura H, Sarai A, Ishii S, Nishimura Y (1992) Solution structure of a DNA-binding unit of Myb: a helix-turn-helix-related motif with conserved tryptophans forming a hydrophobic core. Proc Natl Acad Sci USA **89**: 6428-6432

Ortis DF, Strommer JN (1990) The *Mu1* maize transposable element induces tissue-specific aberrant splicing and polyadenylation in two Adh1 mutants. Mol Cell Biol **10**: 2090-2095

Pantopoulos K, Johansson HE, Hentze MW (1994) The role of the 5' untranslated region of eukaryotic messenger RNAs in translation and its investigation using antisense technologies. *In* WE Cohn, K Moldave, eds, Progress in Nucleic Acid Research and Molecular Biology, Academic Press, Inc, San Diego, pp 181-238

Paz-Ares J, Ghosal D, Wienand U, Peterson PA, Saedler H (1987) The regulatory c1 locus of *Zea mays* encodes a protein with homology to myb proto-oncogene products and with structural similarities to transcriptional activators. EMBO J **6**: 3553-3558

Paz-Ares J, Ghosal D, Sadler H (1990) Molecular analysis of the C1-I allele from *Zea mays*: a dominant mutant of the regulatory C1 locus. EMBO J **9**: 315-322

Prendergast GC, Ziff EB (1992) A new bind for myc. Trends in Genetics **8**: 91-96

Ramsay RG, Ishii S, Nishina Y, Soe G, Gonda TJ (1989) Characterization of alternate and truncated forms of murine c-myb proteins. Oncogene Research **4**: 259-269

Rochester DE, Winter JA, Shah DM (1986) The structure and expression of maize genes encoding the major heat shock, hsp70. EMBO J **9**: 315-322

Schmidt RJ, Burr FA, Auckerman MJ, Burr B (1990) Maize regulatory gene Opaque-2 encodes a protein with a 'leucine zipper' motif that binds to zein DNA. Proc Natl Acad Sci USA **87**: 46-50

Springhorn JP, Singh K, Kelly RA, Smith TW (1994) Post-transcriptional regulation of Id1 activity in cardiac muscle. J Biol Chem **269**: 5132-5136

Sun XH, Baltimore D (1991) An inhibitory domain of E12 transcription factor prevent DNA binding in E12 homodimers but not in E12 heterodimers. Cell **64**: 459-470

Urao T, Yamaguchi-Shinozaki K, Urao S, Shinozaki K (1993) An *Arabidopsis* myb homolog is induced by dehydration stress and its gene product binds to the conserved myb recognition sequence. Plant Cells: 1529-1539

Weber BL, Westin EH, Clarke MF (1990) Differentiation of mouse erythroleukemia cells enhanced by alternatively spliced c-myb mRNA. Science **249**: 1291-1293

Weisshaar B, Armstrong GA, Block A, DaCosta-Silva O, Hahlbrock K (1991) Light-inducible and constitutively expressed DNA-binding proteins recognizing a plant promoter element with functional relevance in light responsiveness. EMBO J **10**: 1777-1786

Yost HJL (1988) Translation of unspliced transcripts after heat shock. Science **242**: 1544-1548

Jasmonates - Secondary Messengers in Plant Defense and Stress Reactions

Christiane Reinbothe[1] and Steffen Reinbothe[2]

[1] Institute of Plant Biochemistry, D-06120 Halle/Saale, FRG
[2] Department of Genetics, Institute for Plant Sciences, ETH-Zentrum, CH-8092 Zurich, Switzerland

Abstract. Jasmonates influence plant gene expression in a pleiotropic manner. Whether released in plants upon pathogen attack and stress treatment or applied externally, jasmonates induce numerous plant defense and stress proteins (JIPs), and simultaneously lower or even shut down the expression of photosynthetic genes, including those for the small subunit of ribulose-1,5-bisphosphate carboxylase/oxygenase (the *rbcS* gene product) and a light-harvesting chlorophyll a/b binding protein (the *lhb* gene product). *jip* genes are rapidly transcriptionally activated, and the resulting *jip* transcripts are preferentially translated, as compared to *rbcS* and *lhb* mRNAs. However, rbcS and lhb transcripts are preserved in the non-polysomal fraction in the early stage of the jasmonate response, presumably to potentially allow recovery of the cell from stress treatment or pathogen attack. Within the plastid compartment, early changes in transcript functionality, as will be discussed for *rbcL*, the mRNA encoding the large subunit of RuBisCo, appear to be superimposed on delayed jasmonate effects on plastid transcription and RNA stabilities. Amino acids released from the degradation of chloroplast proteins are used for cytoplasmic JIP formation. In the final stage of the jasmonate response, protein biosynthesis is irreversibly inactivated. The underlying molecular mechanism requires the induction of JIP60, which is a novel ribosome-inactivating protein, and the specific marking of the plant ribosome. The coordinate interplay between JIP60 and its intracellular target, marked ribosomes, finally leads to localized cell death and thus may prevent the spread of bacteria, viruses and fungi beyond the site of infection.

Keywords. Barley, gene expression, *Hordeum vulgare*, jasmonates, photosynthetic genes, plant defense genes, post-transcriptional control, transcriptional control

1 Biosynthesis of Jasmonates

Jasmonates comprise a large family of related cyclopentanone compounds. Major physiologically active substances are jasmonic acid (-)-JA, and its methyl ester, methyl jasmonate (JaMe), as well as the amino acid conjugates of (-)-JA with isoleucine, valine or leucine. Besides these compounds, a large number of other (-)-JA derivatives have been identified (see Sembdner and Parthier, 1993 for review).

All of the different jasmonates are synthesized by a common biosynthetic pathway. Linolenic acid is converted through various steps into (+)-7-*iso*-JA (Vick and Zimmerman, 1984), which in most plant species is rapidly converted into (-)-JA by isomerization (Vick and Zimmerman, 1984), as shown in Fig. 1.

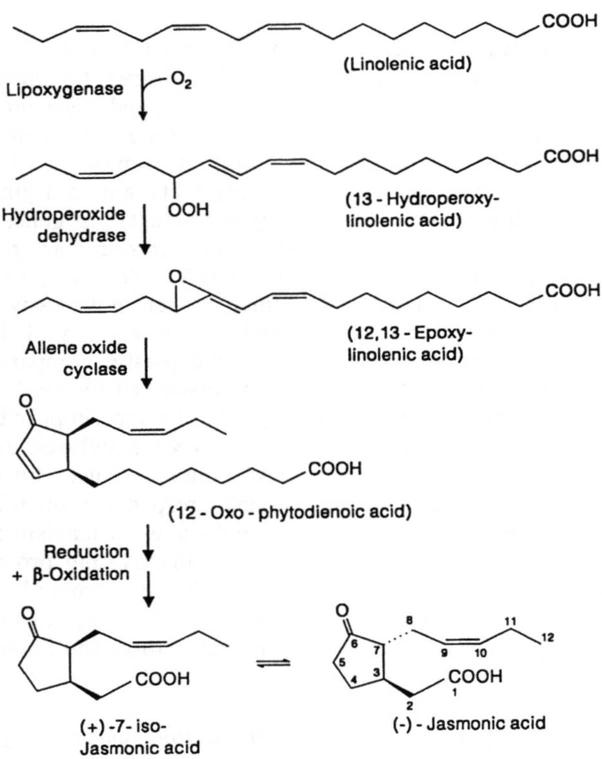

Fig. 1. Biosynthetic pathway of jasmonates

Subsequent modifications of either compound include hydroxylations at C-11 and/or C-12, the reduction of the keto group at C-6, and O-glucosylations of metabolites hydroxylated at C-11, C-12 and/or C-6. In addition, esterification or conjugation with aliphatic amino acids at C-1, giving rise to the above mentioned metabolites, have been reported (see Sembdner and Parthier, 1993, for review).

2 Physiological Effects of Jasmonates

Jasmonates affect plant growth and development in a pleiotropic manner. Processes influenced by jasmonates in diverse plant species include embryogenesis and seed germination, leaf senescence and leaf abscission, flower bud formation, and fruit ripening. Numerous other effects have previously been reviewed (Parthier, 1991; Staswick, 1992; Sembdner and Parthier, 1993).

Recently, jasmonates have been identified to be stress-related compounds. They are produced in plants upon wounding (Creelman et al., 1992), in plant cell cultures treated with fungal elicitors (Gundlach et al., 1992; Müller et al., 1993) and have also been proposed to be induced upon pathogen attack (Farmer and Ryan, 1992). Also under diverse stress conditions, such as during exposure of plants to high sugar concentrations or water deficit (desiccation), an increase in the concentration of endogenous jasmonates has been observed (Creelman and Mullet, 1995).

3 Barley Genes Affected by Jasmonates

In all plant species studied thus far, symptoms caused by exogenously applied or endogenously produced JaMe coincided with alterations in gene expression (summarized in Sembdner and Parthier, 1993). There are genes that responded positively or negatively to JaMe (see S. Reinbothe et al., 1994a, for a compilation of jasmonate-responsive genes). In barley, at least four groups of jasmonate-induced protein (*jip*) genes have been identified thus far based on cDNA cloning procedures (Andresen et al., 1992; Becker and Apel, 1992) and northern hybridization with gene- specific probes (Reinbothe et al., 1992a, b; C. Reinbothe et al., 1994). From the comparison of deduced amino acid sequences, putative functions could be inferred for all the encoded polypeptides with the exception of the 23 kD JIP (Andresen et al., 1992), which shares no similarity with other known protein sequences found in the databanks. One barley JIP, JIP6, which is synthesized as a 15 kD precursor (JIP15) (Andresen et al., 1992), was found to belong to the group of leaf thionins, fungitoxic molecules localized in the plant cell wall and in the vacuole (Bohlmann et al., 1988). Another JIP, JIP60 (Becker and Apel, 1992), was recently identified to be a ribosome-inactivating protein involved in the general down-regulation of protein biosynthesis in long-term JaMe-treated or long-term stressed leaf tissues (Chaudhry et al., 1994; Reinbothe et al., 1994b). Finally, some *jip* transcripts seem to encode "late embryogenesis abundant" (LEA) proteins (Reinbothe et al., 1992a, b; C. Reinbothe et al., 1994) that are thought to mitigate the physico-chemical changes encountered in cells of vegetative and generative tissues during cytosolic dehydration (Dure et al., 1989; Dure, 1993).

4 Temporal Pattern of Jasmonate-Induced Alterations in Plant Gene Expression Displayed by Barley Leaf Tissues

4.1 Transcriptional Activation of *jip* Genes

Detached leaf segments of barley respond to exogenously applied JaMe with a characteristic temporal pattern of change in gene expression. At least four overlapping stages of the jasmonate response can be distinguished. The first is a very rapid response of induction of cytoplasmic JIP synthesis, that is controlled transcriptionally. As shown in Fig. 2, *jip* genes are first transcribed 2 to 4 h after the onset of JaMe treatment, and their mRNAs start to accumulate shortly thereafter (see also Müller-Uri et al., 1988; Reinbothe et al., 1992b). In contrast to *jip* genes, genes encoding proteins involved in photosynthesis, such as the small subunit of ribulose-1,5-bisphosphate carboxylase/oxygenase (*rbcS* gene product of Rubisco) and a type III light harvesting chlorophyll *a/b* binding protein (*lhbC1* gene product), are rapidly turned off after the onset of JaMe treatment. These opposite changes suggest that *jip* and light-regulated genes in barley may contain shared as well as distinct *cis*-elements within their promoters that cause the observed differences in transcriptional control. Interestingly, both *jip* and photosynthetic genes have been shown to contain similar G-box motifs. It is hence tempting to speculate that these motifs might bind transcription factors that could activate *jip* genes but would depress transcription of photosynthetic genes. Alternatively, JaMe could cause *de novo* formation of factors required for transcription of *jip* genes or could inhibit synthesis of proteins operating in transcription of photosynthetic genes. Both jasmonate effects have been documented for numerous other polypeptides in barley leaf tissues (Reinbothe et al., 1993a, b, c).

4.2 Preferential Translation of *jip versus* Photosynthetic mRNAs

During the second stage of the jasmonate response, beginning after approximately 8 h, most preexisting (control) mRNAs are displaced from polysomes in favor of *jip* mRNAs. Previous results have identified translation initiation as the site at which protein synthesis is selectively depressed by JaMe (Reinbothe et al., 1993a, 1993b). It has been suggested that *jip* transcripts may contain similar elements within their 5' untranslated leaders that enhance their translation, as compared to that of most preexisting mRNAs.

Taking a previous paper of Goldsbrough et al. (1993) into account, a conserved motif has been found in stress-induced transcripts from diverse plant species during database searchings; it shares the consensus sequence:

5' U C A U C U U C U U 3'

In some cases, this sequence element occurred not only in the 5' untranslated regions of the mRNAs but also in the corresponding promoter regions of the genes studied (Siebertz et al., 1989; Castresana et al., 1990; Binet et al., 1991; Douglas et al., 1991). While this sequence has previously been shown to act as a *cis*-regulatory element within the promoter of a β-glucanase of barley (Goldsbrough et al., 1993), its function as a translational enhancer remains to be determined. It is at least conceivable, however, that nuclear proteins which recognize the element in the promoter may also be able to bind to the corresponding mRNA leader sequence. Protein binding to both sites could cause heteroduplex formation and thus might be part of an autoregulatory mechanism of coupled transcriptional and post-transcriptional control of stress-responsive genes.

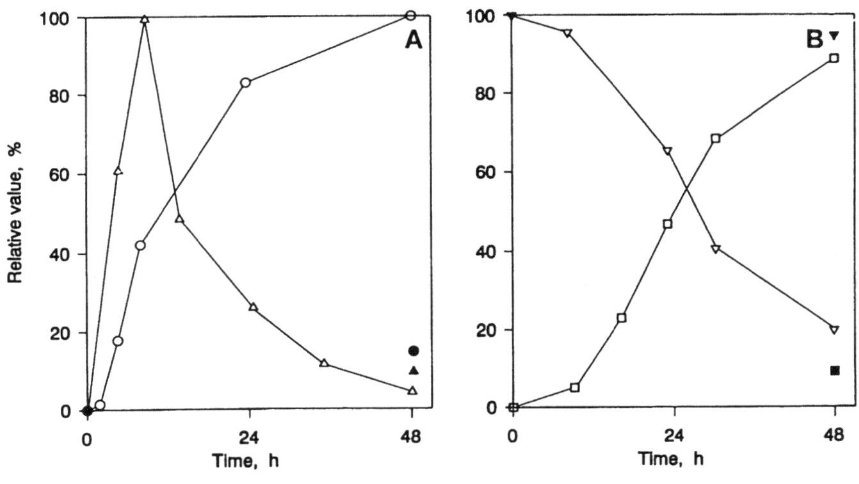

Fig. 2. JIP60 expression in JaMe- and water treated barley leaf tissues. Changes in *jip60* gene transcription (△,▲) and *jip60* steady state transcript level (○,●)(A), are compared with JIP60 protein concentrations (□,■) as well as the overall rates of protein synthesis (▽,▼) (B) in JaMe- (△,○,□,▽) and water treated (▲,●,■,▼) leaf tissues

4.3 Preservation of Photosynthetic Transcripts

After approximately 24 h of JaMe treatment, control but not JIP protein synthesis has ceased, although mRNAs such as *rbcS* and *lhbC1* are still present at pretreatment levels *in vivo* (Müller-Uri et al., 1988; Reinbothe et al., 1993a, b). Despite the drastic reduction in transcription of the *rbcS* and *lhbC1* genes, the corresponding transcripts are maintained during this stage of the jasmonate response, suggesting that such mRNAs may be preserved in specialized subcellular structures, such as messenger ribonucleoprotein particles (Larson and Sells, 1987; Scherrer, 1990), in JaMe-treated leaf tissues. For example, control mRNAs in JaMe-treated leaf tissues may become sequestered onto the intermediate filament network from which they could later be mobilized for protein synthesis after metabolization of the externally applied or internally produced JaMe under stress conditions (S. Reinbothe et al., 1994b). This interpretation could also explain why expression of elements of the cytoskeleton, such as tubulin and actin, is maintained at a much higher level than that of photosynthetic genes during the early stage of the jasmonate response. Actin and tubulin genes were neither turned off immediately after exposing leaf tissues to JaMe nor were their transcripts displaced from polysomes in favor of *jip* mRNAs during this stage of the jasmonate response (Reinbothe et al., 1993a and unpublished results).

4.4 Destabilization of Photosynthetic Transcripts

During the third stage of the jasmonate response, beginning after approximately 36 h, a massive decrease of control mRNAs, such as *rbcS* and *lhbC1*, can be observed. We assume that this rapid degradation of *rbcS* and *lhbC1* transcripts might be due to their release from the above mentioned storage structures, such as mRNP. However, the half-life times of these transcripts appear to be much shorter than those in water treated leaf tissues, in which the levels of these transcripts also decline, suggesting a selective mRNA destabilization to occur in JaMe-treated leaf tissues. It is as yet undetermined whether specific modifications in their 5' and 3' untranslated regions, which both have been shown to contain determinants for mRNA stability (summarized in Sachs, 1993; Takagi et al., 1993), may destine *rbcS* and *lhbC1* transcripts to be degraded by either constitutive or JaMe-induced ribonucleases. In contrast to these transcripts, *jip* transcripts remain stable for up to 48 h (Fig. 2).

4.5 Decay of Cytoplasmic Leaf Polysomes and Cell Death

Superimposed on the effects at the post-transcriptional level, a decay of cytoplasmic polysomes, indicative of a general depression of translation initiation can be observed in long-term JaMe-treated leaf tissues (Fig. 2, see also S. Reinbothe et al., 1994a). This effect is likely caused by the interaction of the recently identified ribosome-inactivating protein JIP60 with ribosomes marked specifically for cleavage into their ribosomal subunits by this protein (S. Reinbothe et al., 1994a, b).

JIP60 has a bipartite structure that consists of an N-terminal domain reminiscent of type I ribosome-inactivating proteins (RIPs) and a C-terminal domain similar to eukaryotic translation initiation factors of type eIFγ. Based on these sequence relationships, two modes of action have been proposed (Chaudhry et al., 1994; S. Reinbothe et al., 1994b). JIP60 might operate as a classical RIP with *N*-glycosidase activity and thus would cleave a particular adenine residue in the α-sarcin/ ricin loop, a highly conserved region of the 28S (26S) rRNA of eukaryotic (prokaryotic) ribosomes. However, our previous studies do not support such a mode of JIP60 action. Although JIP60 is able to bind to the α-sarcin/ ricin loop, it does not cleave the *N*-glycosidic bond between adenine and ribose. JIP60 as a functional entity of its two domains does not display RIP activity. Rather, JIP60 dissociates ribosomes into their ribosomal subunits (S. Reinbothe et al., 1994b). This mode of action might be due to the presence of the eIF4γ domain. However, initiation factors of this type normally recruit mRNAs onto polysomes and do not cause dissociation of ribosomes (see Chaudhry et al., 1994, for discussion).

During the final stage of the jasmonate response, beginning after approximately 48 h, the cytoplasmic ribosomes and their ribosomal subunits, including their 25S and 18S rRNAs, decline drastically (Reinbothe et al., 1993b). In concert with those changes in gene expression taking place in the plastid compartment (Reinbothe et al., 1993c), these alterations cause characteristic senescence symptoms, such as loss of chlorophyll and degradation of Rubisco, and ultimately lead to cell death.

5 Jasmonates as Putative Signalling Molecules in Local and Systemic Plant Defense Reactions

Farmer and Ryan (1992) previously put forth a model of pathogen- and wound-induced signal transduction involved in the systemic activation of proteinase inhibitor gene expression in tomato and potato. Accordingly, wounding causes the expression of systemin (Pearce et al., 1991; McGurl et al., 1992). This peptide is transported locally and systemically within wounded plants (McGurl et al., 1994; Narvaez-Vasquez et al., 1995), and induces or activates linolenic acid synthesis after its binding to putative receptors in the plasma membrane of non-wounded cells (Schaller and Ryan, 1994). Linolenic acid is subsequently converted into (-)-JA and JaMe (cf. Fig. 1), which in turn activate proteinase inhibitor genes (Farmer and Ryan, 1992).

Although the model proposed by Farmer and Ryan (1992) might explain how plants acquire systemic resistance against pathogens and feeding insects, it does not address the question of why plant tissues respond to certain pathogens with rapid cell death in the first place. The results summarized in this review might offer an explanation, however. We propose that JaMe released *in situ* upon pathogen attack or wounding might cause the induction of the entire defense gene expression program at the end of which cell death occurs. Plant defense genes would become rapidly activated, their mRNAs would accumulate and could direct synthesis of an array of diverse defense proteins. Simultaneously, photosynthetic genes would be turned off; their transcripts would be maintained, however, to potentially allow recovery from pathogen attack. A final response could be the down-regulation of protein synthesis, subsequent degradation of ribosomes, and depression of other key metabolic functions. In particular, the interaction between proteins with activities similar to that of JIP60 and ribosomes marked for cleavage by theses proteins might be one of the triggers through which irreversible changes would become manifested in pathogen-infected cells.

In distant plant organs and tissues, the suicidal activity of JIP60 would remain silent, however. The target of JIP60, the ribosome, would probably not become modified at a certain, subthreshold jasmonate concentration. The imported jasmonate would likely be destined for rapid metabolization (Sembdner et al., 1990; Gundlach et al., 1992; Müller et al., 1993). However, the transient increase in the jasmonate concentration would probably be sufficient to induce the expression of plant defense genes. Their encoded products, in concert with other defense-related compounds, such as secondary metabolites, could then help to confer systemic resistance to plant tissues against subsequent pathogen attack.

Programmed cell death could also be a defense strategy to deter pathogenic plant viruses. Such viruses have to reprogram the metabolism of their hosts in order to insure their propagation. They do so by translating their transcripts or parts therefrom at the plant ribosome (see Zaitlin and Hull, 1987, for review). Because of the above average rate of translation initiation of viral mRNAs (Herson et al., 1979; Gallie et al., 1987), plants appear to have no possibility to protect their ribosomes against virally-encoded mRNAs. However, one may speculate that the pathogen-induced, likely JaMe-mediated, down-regulation of protein biosynthesis might be a defense strategy to specifically depress translation of viral RNAs and thus to prevent the spread of such RNA viruses beyond the infection site. JIP60 as the functional entity of its N-terminal and C-terminal domains, or JIP60's RIP domain might interact with the plant ribosome. If tested *in vitro*, the RIP domain of JIP60 cleaves the universally conserved *N*-glycosidic bond in the α-sarcin/ricin loop (Chaudhry et al., 1994). Because a processing site reminiscent of that of virally-encoded proteases is present between the N-terminal RIP domain and the C-terminal eIF4γ domain, it has been suggested that viral infection may cause the RIP domain to be released (Chaudhry et al., 1994). In turn, inactivation of the plant protein biosynthetic machinery could occur. This would presumably lead to cell death and could thus prevent the spread of viruses within the plant.

In addition to jasmonates, other signals have been shown to be produced during pathogenesis, such as plant cell wall-derived oligouronides (Ryan and Farmer, 1991), or compounds such as ethylene, abscisic acid and salicylic acid (see Lamb et al., 1989; Enyedi et al., 1992; Klessig and Malamy, 1994, for reviews). This diversity of secondary messengers may be part of a complex set of stress and defense response systems through which plants acquire the capability to deter a variety of pathogens in their environment.

References

Andresen I, Becker W, Schlüter K, Burges J, Parthier B, Apel K (1992) The identification of leaf thionin as one of the main jasmonate-induced proteins of barley (*Hordeum vulgare*). Plant Mol Biol **19**: 193-204

Becker W, Apel K (1992) Isolation and characterization of a cDNA encoding a novel jasmonate-induced protein of barley (*Hordeum vulgare* L.). Plant Mol Biol **19**: 1065-1067

Binet M-N, Weil J-H, Tessier, L-H (1991) Structure and expression of sunflower ubiquitin genes. Plant Mol Biol **17**: 395-407

Bohlmann H, Clausen S, Behnke S, Giese H, Hiller C, Reimann-Philipp U, Schrader G, Barkholt V, Apel K (1988) Leaf-specific thionins of barley - a novel class of cell wall proteins toxic to plant-pathogenic fungi and possible involvement in the defense mechanism of plants. EMBO J **7**: 1559-1565

Castresana C, De Carvalho F, Gheysen G, Habets M, Inze D, van Montagu M (1990) Tissue-specific and pathogen-induced regulation of *Nicotiana plumbaginifolia* β-1,3-glucanase gene. Plant Cell **2**: 1131-1143

Chaudhry B, Müller-Uri F, Cameron-Mills V, Gough S, Simpson D, Skriver K, Mundy J (1994) The barley 60 kDa jasmonate-induced protein (JIP60) is a novel ribosome-inactivating protein. Plant J **6**: 815-824

Creelman RA, Mullet JE (1995) Jasmonic acid distribution and action in plants: regulation during development and response to biotic and abiotic stress. Proc Natl Acad Sci USA **92**: 4114-4119

Creelman RA, Tierney ML, Mullet JE (1992) Jasmonic acid/methyl jasmonate accumulate in wounded soybean hypocotyls and modulate wound gene expression. Proc Natl Acad Sci USA **89**: 4938-4941

Douglas CJ, Hauffe KD, Ites-Morales M-E, Ellard M, Paszkowski U, Hahlbrock K, Dangl JL (1991) Exonic sequences are required for elicitor and light activation of a plant defense gene, but promoter sequences are sufficient for tissue specific expression. EMBO J **10**: 1767-1775

Dure III L, Crouch M, Harada J, Ho T-HD, Mundy J, Quatrano R, Thomas T, Sung ZR (1989) Common amino acid sequence domains among the LEA proteins of higher plants. Plant Mol Biol **12**: 475-486

Dure III L (1993) A repeating 11-mer amino acid motif and plant desiccation. Plant J **3**: 363-369

Enyedi AJ, Yalpani N, Silverman P, Raskin I (1992) Signal molecules in systemic plant resistance to pathogens and pests. Cell **70**: 879-886

Farmer EE, Ryan CA (1992) Octadecanoid precursors of jasmonic acid activate the synthesis of wound-inducible proteinase inhibitors. Plant Cell **4**: 129-134

Gallie DR, Sleat DE, Watts JW, Turner PC, Wilson TMA (1987) The 5'-leader sequence of tobacco mosaic virus RNA enhances the expression of foreign gene transcripts *in vitro* and *in vivo*. Nucl Acids Res **15**: 3257-3272 .

Goldsbrough A, Albrecht H, Stratford R (1993) Salicylic acid-inducible binding of a tobacco nuclear protein to a 10 bp sequence which is highly conserved among stress-inducible genes. Plant J **3**: 563-571

Gundlach H, Müller M, Kutchan TM, Zenk MH (1992) Jasmonic acid is a signal transducer in elicitor-induced plant cell cultures. Proc Natl Acad Sci USA **89**: 2389-2393

Herson D, Schmidt A, Seal SN, Marcus A, Vloten-Doting L (1979) Competitive mRNA translation in an *in vitro* system from wheat germ. J Biol Chem **254**: 8245-8249

Klessig DF, Malamy J (1994) The salicylic acid signal in plants. Plant Mol Biol **26**: 1439-1458

Lamb CJ, Lawton MA, Dron M, Dixon RA (1989) Signals and transduction mechanisms for activation of plant defenses against microbial attack. Cell **56**: 215-224

Larson DE, Sells BH (1987) The function of proteins that interact with mRNA. Mol Cell Biochem **74**: 5-15

McGurl B, Pearce G, Orozco-Cardenas M, Ryan CA (1992) Structure, expression and antisense inhibition of the systemin precursor gene. Science **255**: 1570-1573

McGurl B, Orozco-Cardenas ML, Pearce G, Ryan CA (1994) Over-expression of the prosystemin gene in transgenic tomato plants generates a systemic signal that constitutively induces proteinase inhibitor synthesis. Proc Natl Acad Sci USA **91**: 9799-9802

Müller MJ, Brodschelm W, Spannagl E, Zenk MH (1993) Signalling in the elicitation process is mediated through the octadecanoid pathway leading to jasmonic acid. Proc Natl Acad Sci USA **90**: 7490-7494

Müller-Uri F, Parthier B, Nover L (1988) Jasmonate-induced alteration of gene expression in barley leaf segments analyzed by *in vivo* and *in vitro* protein synthesis. Planta **176**: 241-248

Narvaez-Vasquez J, Pearce G, Orozco-Cardenas ML, Franceschi VR, Ryan CA (1995) Autoradiographic and biochemical evidence for the systemic translocation of systemin in tomato plants. Planta **195**: 593-600

Parthier B (1991) Jasmonates, new regulators of plant growth and development: Many facts and few hypotheses on their actions. Bot Acta **104**: 446-454

Pearce G, Strydom D, Johnson S, Ryan CA (1991) A polypeptide from tomato leaves induces wound-inducible proteinase inhibitor proteins. Science **253**: 895-898

Reinbothe S, Machmudova A, Wasternack C, Reinbothe C, Parthier B (1992a) Jasmonate-induced proteins in cotton: immunological relationship to the respective barley proteins and homology of transcripts to late embryogenesis abundant (Lea) mRNAs. J Plant Growth Regul **11**: 7-14

Reinbothe S, Reinbothe C, Lehmann J, Parthier B (1992b) Differential accumulation of methyl jasmonate-induced mRNAs in response to abscisic acid and desiccation in barley (*Hordeum vulgare*). Physiol Plant **86**: 49-56

Reinbothe S, Reinbothe C, Parthier B (1993a) Methyl jasmonate represses translation initiation of a specific set of mRNAs in barley. Plant J **4**: 459-467

Reinbothe S, Reinbothe C, Parthier B (1993b) Methyl jasmonate-regulated translation of nuclear-encoded chloroplast proteins in barley. J Biol Chem **268**: 10606- 10611

Reinbothe S, Reinbothe C, Heintzen C, Seidenbecher C, Parthier B (1993c) A methyl jasmonate-induced shift in the length of the 5' untranslated region impairs translation of the plastid *rbcL* transcript in barley. EMBO J **12**: 1505-1512

Reinbothe C, Tewes A, Lehmann J, Parthier B, Reinbothe S (1994) Induction by methyl jasmonate of embryogenesis-related proteins and mRNAs in *Nicotiana plumbaginifolia*. Plant Sci **104**: 59-70

Reinbothe S, Mollenhauer B, Reinbothe C (1994a) JIPs and RIPs: The regulation of plant gene expression by jasmonates in response to environmental cues and pathogens. Plant Cell **6**: 1197-1209

Reinbothe S, Reinbothe C, Lehmann J, Becker W, Apel K, Parthier B (1994b) JIP60, a methyl jasmonate-induced ribosome-inactivating protein involved in plant stress reactions. Proc Natl Acad Sci USA **91**: 7012-7016

Ryan CA, Farmer EE (1991) Oligosaccharide signals in plants: A current assessment. Annu Rev Plant Physiol Plant Mol Biol **42**: 651-674

Sachs AB (1993) Messenger RNA degradation in eukaryotes. Cell **74**: 413-421

Schaller A, Ryan CA (1994) Identification of a 50-kDa systemin-binding protein in tomato plasma membranes having Kex2p-like properties. Proc Natl Acad Sci USA **91**: 11802-11806

Scherrer K (1990) Prosomes, subcomplexes of untranslated mRNP. Mol Biol Reports **14**: 1-9

Sembdner G, Meyer A, Miersch O, Brückner C (1990) Metabolism of jasmonic acid. *In* RP Pharis, SB Rood, eds, Plant Growth Substances 1988. Springer, Berlin, pp 374-379

Sembdner G, Parthier B (1993) The biochemistry and the physiological and molecular actions of jasmonates. Annu Rev Plant Physiol Plant Mol Biol **44**: 569-589

Siebertz B, Logemann J, Willmitzer L, Schell J (1989) *Cis*-analysis of the wound-inducible promoter *wun1* in transgenic tobacco plants and histochemical localisation of its expression. Plant Cell **1**: 961-968

Staswick PE (1992) Jasmonates, genes, and fragrant signals. Plant Physiol **99**: 804-807

Takagi M, Taylor CB, Newman TC, Green P (1993) The effect of sequences with high AU content on mRNA stability in tobacco. Proc Natl Acad Sci USA **90**: 11811-11815

Vick BA, Zimmerman DC (1984) Biosynthesis of jasmonic acid by several plant species. Plant Physiol **75**: 458-461

Zaitlin M, Hull R (1987) Plant virus-host interactions. Annu Rev Plant Physiol **38**: 291-315

Signal Perception and Intracellular Transduction in the *Phytophthora Sojae*/Parsley Interaction

Dirk Nennstiel[1], Klaus Hahlbrock[2], Thorsten Jabs[3], Dierk Scheel[1], Wolfgang Wirtz[1], Sabine Zimmermann[4] and Thorsten Nürnberger[1]

[1] Department of Stress and Developmental Biology, Institut für Pflanzenbiochemie, D-06120 Halle, Germany
[2] Department of Biochemistry, Max-Planck-Institut für Züchtungsforschung, D-50829 Köln, Germany
[3] Department of Phytopathology, Agricultural University, NL-6700 EE Wageningen, The Netherlands
[4] Institute of Plant Sciences, CNRS, F-91198 Gif-sur-Yvette Cedex, France

Abstract. Cultured parsley (*Petroselinum crispum*) cells or protoplasts respond to treatment with a crude cell wall preparation from *Phytophthora sojae* with the transcriptional activation of the same set of defense-related genes as are activated in parsley leaves upon infection with fungal spores. A 42 kD glycoprotein was purified from the fungal culture filtrate and its elicitor activity was found to reside exclusively in the protein moiety. A core sequence consisting of 13 amino acids (Pep-13) in the glycoprotein was identified, which stimulates the same responses as the crude cell wall elicitor, namely Ca_2^+ and H^+-influxes, effluxes of K^+-and Cl^- ions, an oxidative burst, ethylene production, defense-related gene activation, and formation of phytoalexins. Over-expression in *E. coli* of the full-length cDNA encoding the protein portion of the 42 kD glycoprotein, as well as of clones in which the 13 amino acids representing Pep-13 were replaced by either 2 or 6 unrelated amino acids, showed that this region is necessary and sufficient for the elicitor activity of the intact glycoprotein. Using (^{125}I)Tyr-Pep-13 as ligand in binding assays a single-class high-affinity binding site in parsley microsomal membranes and protoplasts could be detected. Binding was specific, saturable and reversible. By chemical crosslinking a 91 kD parsley plasma membrane protein was identified to be the receptor of the peptide elicitor. In patch-clamp analyses a calcium-permeable ion channel (LEAC) of large unitary conductance was identified, the activity of which could be modified by elicitor. A series of peptides was tested for their ability to compete for binding of Pep-13 to parsley microsomes, to activate LEAC as well as macroscopic ion fluxes, oxidative burst, and phytoalexin formation. Findings demonstrated a functional link between the various plant responses to elicitor treatment. The use of a series of ion channel inhibitors as well as inhibitors of the oxidative burst revealed that the observed ion fluxes and active oxygen species are an absolute requirement for defense-related gene activation and subsequent phytoalexin formation. Furthermore, the same ion channel inhibitors were found to efficiently inhibit the elicitor-induced oxidative burst. These findings allow us to establish a sequence of events that may constitute part of a signalling cascade triggering pathogen defense in plants.

Keywords. Defense gene activation, elicitor, parsley, pathogen defense, *Petroselinum crispum*, phytoalexin

1 Introduction

Plants are resistant to the vast majority of potentially pathogenic microorganisms they encounter in their environment (species or non-host resistance). Only in a minority of cases have true host/pathogen interactions evolved in which the pathogens have developed effective mechanisms for circumventing the defense machinery of a particular plant species, resulting in colonization of the plant. Individual cultivars of these species, however, are capable of mounting a successful resistance response against the invading pathogen (cultivar or host resistance) (Bailey, 1983; Ellingboe, 1984).

Defense reactions elicited in plants undergoing either type of resistance are very similar and result from both transcriptional activation of specific genes, such as phytoalexin biosynthesis (Hahlbrock and Scheel, 1989; Dixon and Lamb, 1990), or allosteric enzyme activation, such as reinforcement of the cell wall (Kauss et al., 1989) or the oxidative burst (Levine et al., 1994). These reactions are frequently associated with hypersensitive cell death, although a causal link has yet to be demonstrated (Atkinson, 1993; Jakobek and Lindgren, 1993).

Initiation of defense responses requires recognition of either plant- or pathogen-derived signals, termed elicitors, by the plant. This is believed to be mediated by receptors which specifically bind these signal molecules and thereby initiate intracellular transmission of the signal. Subsequent conversion of this signal through a signal transduction cascade results in induction of a particular cell response(s), such as defense gene activation.

The non-host resistance response of parsley leaves infected with zoospores of the soybean pathogen, *Phytophthora sojae*, comprises hypersensitive lesion formation, cell wall apposition of callose, accumulation of furanocoumarin phytoalexins, and transcriptional activation of defense-related genes (Jahnen and Hahlbrock, 1988; Schmelzer et al., 1989). A part of this complex response, namely fungus-induced activation of defense-related genes and phytoalexin production, can be also observed in suspension-cultured parsley cells or protoplasts upon treatment with elicitors derived from the fungal cell wall. We have used this system to study signal perception mechanisms and very early elicitor-induced signalling events.

2 Elicitor Characterization

A prerequisite for signal perception studies at the plant plasma membrane is the availability of a pure, molecularly defined ligand preparation. A 42 kD glycoprotein elicitor was purified from *P. sojae* culture filtrate whose ability to stimulate phytoalexin production in parsley protoplasts resided in the protein moiety (Parker et al., 1991). Heat denaturation of this protein did not affect its elicitor activity, indicating that this activity is determined by primary sequence information rather than secondary structure motifs. A peptide consisting of 13 amino acids (Pep-13) within the intact glycoprotein was determined to be sufficient to stimulate the same responses as the fungal cell wall elicitor and the glycoprotein elicitor (Nürnberger et al., 1994a). These responses comprise transcriptional activation of genes encoding phenylalanine ammonia lyase and 4-coumarate:CoA ligase, the reduction of transcript levels of genes encoding parsley histones, cdc2 and cyclin, stimulation of ethylene and phytoalexin biosynthesis, and an increase in extracellular chitinase activity (Nürnberger et al., 1994b; Logemann et al., 1996). Additionally, simultaneous treatment of parsley cells with UV light and either elicitor resulted in the abolition of light-induced vacuolar accumulation of flavonoids (Lozoya et al., 1991; Nürnberger et al., 1994b). We have not yet detected any response of parsley cells that could be stimulated by the fungal cell wall elicitor, and not by Pep-13.

The isolation of a cDNA encoding the protein portion of the glycoprotein elicitor from *P. sojae* (Sacks et al., 1995) allowed localization of this peptide to the C-terminal half of the protein. It was not associated with the single glycosylation site predicted within the protein and its amino acid sequence did not show any significant homology to sequences contained in several data bases. To ascertain whether Pep-13 was the sole determinant of elicitor activity within the mature glycoprotein, the full-length cDNA, subclones spanning different regions, as well as subclones in which the 13 amino acids representing Pep-13 were replaced by either two or six amino acids unrelated to the peptide, were expressed in *E. coli* and expression products assayed for elicitor activity. Only those expression products which contained the Pep-13 sequence were elicitor-active, indicating that Pep-13 is not only sufficient but also necessary for the elicitor activity of the intact glycoprotein (Fig. 1). Deletion of one N- and one C-terminal amino acid yielded the minimum peptide with full elicitor activity (Pep-11) (Nürnberger et al., 1994a). Substitution analysis, in which individual amino acids of Pep-11 were progressively replaced by alanine, identified two residues (arrows in Fig. 1) critical for activity. All other exchanges exerted little or no effect on the elicitor activity of this peptide.

Recently, an 85 kD glycoprotein was purified from the culture filtrate of *Phytophthora infestans* that induced the synthesis of the same pattern of phytoalexins in parsley cells as is elicited by the *P. sojae*-derived Pep-13. Similarly, the elicitor activity of this protein resided in (glyco)peptides smaller than 10 kD. Comparison of peptide sequences derived from tryptic fragments of this protein or from the corresponding cDNA with the respective sequences of the *P. sojae* glycoprotein revealed no homologies between the two elicitor proteins.

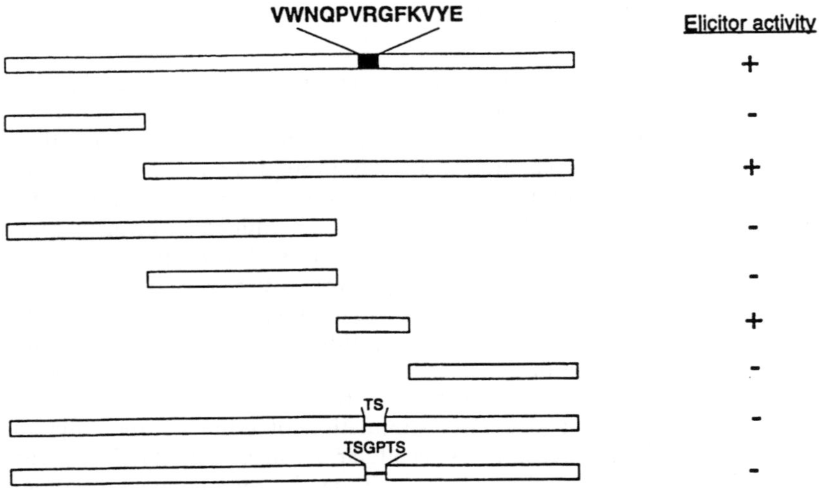

Fig. 1. Elicitor activities of portions of the *P. sojae* elicitor expressed as fusion proteins with glutathione S-transferase in *E. coli.* The full-length cDNA encoding the protein portion of the 42 kD elicitor, subclones spanning different regions of the protein, and subclones, in which the Pep-13-encoding sequence was replaced with nucleotide sequences coding for TS or TSGPTS, respectively, were transformed into *E. coli*. After IPTG-treatment, proteins accumulating in inclusion bodies were solubilized and assayed for their ability to stimulate phytoalexin formation in parsley protoplasts (+: full activity relative to that of the native glycoprotein elicitor; –: no activity detectable). The bars represent the expressed portions of the elicitor polypeptide, with the N-terminus on the left. The closed bar denotes the 13 amino acid region representing Pep-13

3 Elicitor Perception

A large number of chemically diverse elicitors derived from phytopathogenic and non-pathogenic fungi have been identified that stimulate defense responses in intact plants or cultured plant cells (Ebel and Cosio, 1994; Boller, 1995, for review). The cultivar or species specificity of elicitors, their ability to induce plant defense responses at low nanomolar concentrations, and the precisely defined signal structures required for elicitor-mediated activation of these reactions suggest the involvement of receptors in elicitor recognition and subsequent intracellular signal generation.

Incubation of parsley protoplasts with a fluoresceinated Pep-13 derivative allowed the visualization of receptor/ligand complexes on the protoplast surface by means of SEIG-EPOM (silver enhanced immunogold labeling with anti-fluorescein antibodies as viewed by epipolarization microscopy) (Diekmann et al., 1994). To obtain kinetic parameters of elicitor binding we performed ligand binding assays with parsley microsomal membrane preparations or protoplasts and radioiodinated Pep-13 (^{125}I-Pep-13). The existence of a saturable, single-class binding site with high affinity for this ligand (K_d=2.4nM) was demonstrated in both receptor preparations (Nürnberger et al., 1994a). Binding was reversible and specific with respect to both structural features of the ligand and the plant species that recognized this ligand.

In chemical cross-linking experiments using parsley microsomal or plasma membranes, radioiodinated Pep-13 could be covalently bound to its binding site with each of three homo-bifunctional reagents (Nürnberger et al., 1995). Analysis by SDS-PAGE and autoradiography of solubilized membrane proteins demonstrated labeling of a 91 kD protein, which could not be detected after either addition of excess unlabeled Pep-13 or omission of the cross-linker from the reaction mixture. The competitor concentration found to half-maximally reduce the intensity of the cross-linked band was 6 nM, in good agreement with the IC_{50} value of 4.7 nM obtained in ligand binding assays. Coupling of ^{125}I-Pep-13 to the 91 kD protein required the same structural elements within the ligand as found for binding of ^{125}I-Pep-13 to parsley microsomal membranes. These findings strongly suggest that the 91 kD protein represents the Pep-13 binding site or an integral part of it.

In an attempt to isolate the receptor/ligand complex by means of affinity chromatography, we employed a biologically active, biotinylated derivative of ^{125}I-Pep-13 in chemical cross-linking experiments. Detergent-solubilized microsomal membrane proteins were incubated with avidin agarose, and, after thoroughly washing the matrix, bound material was eluted with SDS-PAGE sample buffer containing 500 mM biotin. SDS-PAGE/autoradiography analysis of bound proteins demonstrated the potential of this experimental strategy for isolation of the Pep-13 binding site (Fig. 2).

4 Signal Transduction

The effects of elicitors on ion fluxes across the plasma membrane have been considered to be part of elicitor-specific signal transduction leading to the induction of plant defense responses in several plant/pathogen interactions (Ebel and Cosio, 1994). Changes in ion permeability of the plasma membrane (influxes of H^+ and Ca^{2+}, effluxes of K^+ and Cl^-) are among the earliest reactions detectable after treatment of parsley cells with Pep-13. Four lines of evidence strongly suggest that oligopeptide-stimulated ion fluxes and phytoalexin formation in parsley cells are functionally inseparable. (i) Omission of Ca^{2+} from the culture medium blocked defense-related gene activation and phytoalexin formation, as was also shown for the crude cell wall elicitor at the level of transcript accumulation (Dietrich et al., 1990) and transcript synthesis (Ebel and Scheel, 1992). Furthermore, elicitor treatment of cultured parsley cells in the presence of Ca^{2+} channel blockers, such as flunarizine, significantly inhibited both the targeted ion flux and phytoalexin production (Nürnberger et al., 1994b). (ii) Initial changes in ion fluxes were detectable within 2-5 min after addition of the elicitor and peaked after 20-30 min, thereby clearly preceding transcriptional activation of defense-related genes and phytoalexin formation. (iii) Similar elicitor concentrations were needed for half-maximal stimulation of the various plant responses. (iv) Identical structural properties of the oligopeptide were found to be required for efficient stimulation of all four ion fluxes and furanocoumarin synthesis, as well as for binding of the elicitor to its binding site. This strongly suggests that the Pep-13 binding site may, indeed, function as a receptor mediating transformation of a pathogen-derived signal into a plant defense response. Furthermore, these data provide evidence that rapidly induced ion fluxes may constitute elements of the signal transduction cascade triggering pathogen defense in parsley.

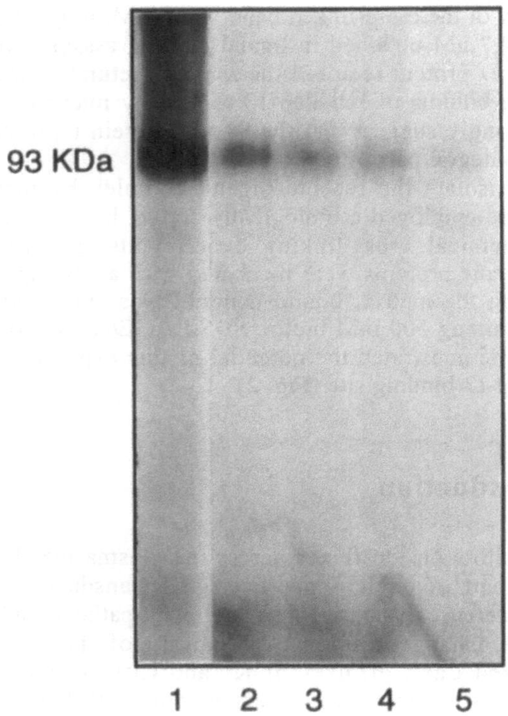

Fig. 2. Affinity chromatography isolation of the Pep-13/receptor complex.
Chemical crosslinking of biotinylated ^{125}I-Pep-13 to parsley microsomal membranes was performed as described (Nürnberger et al., 1995). After treatment of the membranes in 1% SDS (10 min at 37°C) solubilized material was subjected to avidin agarose, the matrix thoroughly washed, bound proteins eluted (with standard SDS-PAGE sample buffer containing 500mM biotin) and subsequently analyzed by SDS-PAGE/autoradiography. Size of a molecular mass standard is indicated on the left. lane 1, solubilized membrane proteins applied to affinity chromatography; lanes 2-5, fractions containing the Pep-13/receptor complex as eluted from avidin agarose

To electrophysiologically identify and characterize ion channels that contribute to the observed elicitor-induced ion fluxes, patch-clamp studies using parsley protoplasts were performed. In both "whole-cell" configuration and excised membrane patches in "out side-out" configuration, an inward-rectifying Ca^{2+} channel was detected. This channel exhibited very little, voltage-independent activity in the resting state. The unitary conductance of this channel at extracellular Ca^{2+} concentrations of 1 mM was 140 pS, unusually large for plant Ca^{2+} channels. More interestingly, addition of either cell wall or peptide elicitor drastically increased the activity of this channel (Fig. 3). This activation was due solely to an increase in the number of channels that opened simultaneously, since no changes in the single-channel conductance were observed upon elicitor application.

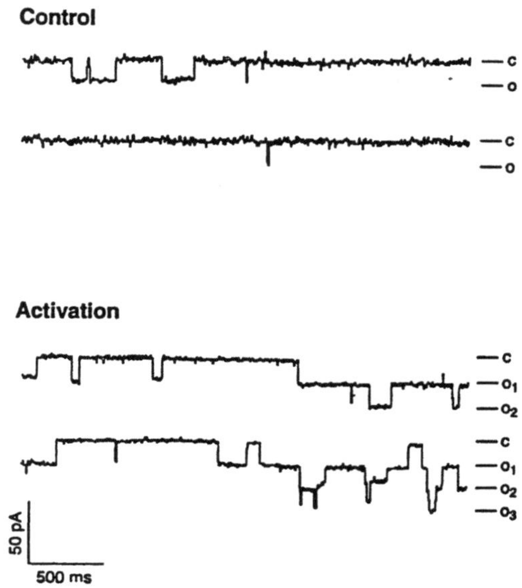

Fig. 3. Elicitor stimulation of a plasma membrane Ca^{2+} inward rectifying ion channel in parsley. Patch-clamp analysis of parsley protoplasts was performed as described (Zimmermann et al., 1994). Whole-cell recording of a parsley protoplast (bath solution 240mM $CaCl_2$, pH 5.2; pipette solution 150mM KCl, 1mM $MgCl_2$, 1mMATP, pH 6.8) either in the absence (Control) or presence of 100nM Pep-13 (Activation) is shown

Another response of parsley cells to elicitor treatment that is stimulated within 2-5 min, is the synthesis and extracellular accumulation of hydrogen peroxide (Nürnberger et al., 1994b). This reaction, known as the oxidative burst, was dependent on extracellular Ca^{2+} and could be prevented by the same ion channel inhibitors that were found to inhibit both elicitor-induced ion fluxes and phytoalexin formation. Additionally, the non-selective ionophore, amphotericin

B, stimulated not only the transcriptional activation of the same pattern of defense genes as did the elicitor, but also the oxidative burst. These findings indicate that elicitor-induced ion fluxes are required for the induction of the elicitor-induced oxidative burst in cultured parsley cells as well.

To address the question of whether reactive oxygen species are involved in defense gene activation and phytoalexin production in elicitor-treated parsley cells, we employed an inhibitor of NADPH oxidase, diphenylene iodonium (DPI). NADPH oxidase catalyzes the formation of superoxide anions which are subsequently converted into hydrogen peroxide by the action of superoxide dismutases. In our system, DPI was found to inhibit not only the elicitor-induced oxidative burst, but also transcriptional activation of defense genes and phytoalexin production, whereas ion channel activities, constitutive gene expression and cell viability remained unaffected. Catalase, a scavenger of hydrogen peroxide, did not interfere with the stimulation of plant defense responses when added concomitantly with the elicitor. It is therefore very likely that superoxide anions rather than hydrogen peroxide represent the reactive oxygen species that are involved in signalling leading to defense gene activation.

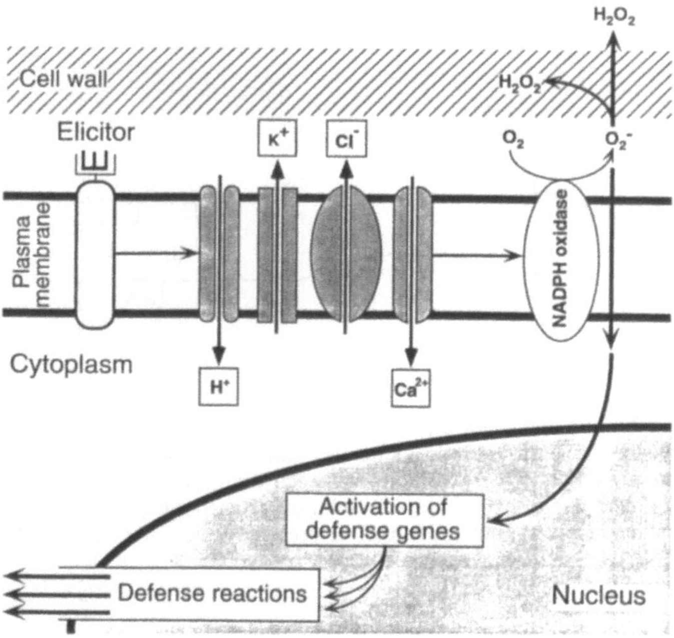

Fig. 4. Hypothetical model of elicitor signal transduction in parsley

5 Conclusion

The model shown in Fig. 4 summarizes our current knowledge of elicitor signal perception and transduction in parsley. Binding of a peptide originating from a fungal cell wall constituent (Hahlbrock et al., 1995) to a 91 kD parsley plasma membrane protein initiates changes in membrane ion channel activities. An elicitor-responsive, inward-rectifying Ca^{2+}channel was identified that may be responsible for the increase in the cytosolic Ca^{2+} concentration observed in elicitor-treated parsley cells (Sacks et al., 1993). Subsequently, through the action of a plasma membrane NADPH oxidase, superoxide anions are generated that mediate further transduction of the intracellular signal. Elicitor-specific, Ca^{2+}-dependent *in vivo* phosphorylation of several proteins has also been demonstrated in parsley (Dietrich et al., 1990). These proteins may then be involved directly or indirectly in regulation and expression of defense-related genes.

Acknowledgements. We thank Magdalena Jung and Lydia Bollenbach for excellent technical assistance. The support of Rainer Hedrich, Jean-Marie Frachisse and Jean Guern in patch-clamp analyses is gratefully acknowledged. We are grateful to Wendy Sacks for critical reading of the manuscript. This work was supported by the Deutsche Forschungsgemeinschaft (Sche 235/3-1 and 3-2), by a doctoral fellowship from Fond der Chemischen Industrie to D.N., and by a postdoctoral fellowship from Bayer Chemical Company, Leverkusen, to T.N.

References

Atkinson MM (1993) Molecular mechanisms of pathogen recognition by plants. Adv Plant Pathol **10**: 35-64

Bailey JA (1983) *In* JA Bailey, BJ Deverall, eds, The Dynamics of Host Defense. Academic Press, New York, pp 1-32

Boller T (1995) Chemoperception of microbial signals in plant cells. Annu Rev Plant Physiol Plant Mol Biol **46**: 189-214

Diekmann W, Herkt B, Low PS, Nürnberger T, Scheel D, Terschüren C, Robinson DG (1994) Visualization of elicitor-binding loci at the plant cell surface. Planta **195**: 126-137

Dietrich A, Mayer JE, Hahlbrock K (1990) Fungal elicitor triggers rapid, transient, and specific protein phosphorylation in parsley cell suspension cultures. J Biol Chem **265**: 6360-6368

Dixon RJ, Lamb CJ (1990) Molecular communication in interactions between plants and microbial pathogens. Annu Rev Plant Physiol Plant Mol Biol **41**: 339-367

Ebel J, Cosio EG (1994) Elicitors of plant defense responses. Int Rev Cytol **148**: 1-36

Ebel J, Scheel D (1992) Elicitor recognition and signal transduction. *In* T Boller, F Meins, eds, Genes Involved in Plant Defense. Springer-Verlag, Wien, pp 184-205

Ellingboe HA (1984) Genetics of host-parasite relations: an essay. Adv Plant Pathol **2**: 131-151

Hahlbrock K, Scheel D (1989) Physiology and molecular biology of phenylpropanoid metabolism. Annu Rev Plant Physiol Plant Mol Biol **40**: 347-369

Hahlbrock K, Scheel D, Logemann E, Nürnberger T, Parniske M, Reinold S, Sacks WR, Schmelzer E (1995) Oligopeptide elicitor-mediated defense gene activation in cultured parsley cells. Proc Natl Acad Sci USA **92**: 4150-4157

Jahnen W, Hahlbrock K (1988) Cellular localization of nonhost resistance reactions of parsley (*Petroselinum crispum*) to fungal infection. Planta **173**: 197-204

Jakobek JL, Lindgren PB (1993) Generalized induction of defense responses in bean is not correlated with the induction of the hypersensitive reaction. Plant Cell **5**: 49-56

Kauss H, Waldmann T, Jeblick W, Euler G, Ranjeva R, Domard A (1989) Ca^{2+} is important but not the only signal in callose synthesis induced by chitosan, saponins and polyene antibiotics. *In* BJJ Lugtenberg, ed, Signal Molecules in Plants and Plant-Microbe Interactions. Springer-Verlag, Berlin, pp 107-116

Levine A, Tenhaken R, Dixon RA, Lamb CJ (1994) H_2O_2 from the oxidative burst orchestrates the plant hypersensitive disease resistance response. Cell **79**: 583-593

Logemann E, Wu S-C, Schröder J, Schmelzer E, Somssich IE, Hahlbrock K (1996) Gene activation by UV-light, fungal elicitor or fungal infection in Petroselinum crispum is correlated with repression of cell cycle-related genes. Plant J **8**: 865-876

Lozoya E, Block A, Lois R, Hahlbrock K, Scheel D (1991) Transcriptional repression of light-induced flavonoid synthesis by elicitor treatment of cultured parsley cells. Plant J **1**: 227-234

Nürnberger T, Nennstiel D, Jabs T, Sacks WR, Hahlbrock K, Scheel D (1994a) High affinity binding of a fungal oligopeptide elicitor to parsley plasma membranes triggers multiple defense responses. Cell **78**: 449-460

Nürnberger T, Colling C, Hahlbrock K, Jabs T, Renelt A, Sacks WR, Scheel D (1994b) Perception and transduction of an elicitor signal in cultured parsley cells. Biochem Soc Symp **60**: 173-182

Nürnberger T, Nennstiel D, Hahlbrock K, Scheel D (1995) Covalent cross-linking of the *Phytophthora megasperma* oligopeptide elicitor to ist receptor in parsley membranes. Proc Natl Acad Sci USA **92**: 2338-2342

Parker JE, Schulte W, Hahlbrock K, Scheel D (1991) An extracellular glycoprotein from Phytophthora megasperma f.sp. glycinea elicits phytoalexin biosynthesis in cultured parsley cells and protoplasts. Mol Plant Microbe Interact **4**: 19-27

Sacks WR, Ferreira P, Hahlbrock K, Jabs T, Nürnberger T, Renelt A, Scheel D (1993) Elicitor recognition and intracellular signal transduction in plant defense. *In* EW Nester, DPS Verma, eds, Advances in Molecular Genetics of Plant-Microbe Interactions. Kluwer Academic Publishers, Dordrecht, pp 485-495

Sacks W, Nürnberger T, Hahlbrock K, Scheel D (1995) Molecular characterization of nucleotide sequences encoding the extracellular glycoprotein elicitor from *Phytophthora megasperma*. Mol Gen Genet **246**: 45-55

Schmelzer E, Krüger-Lebus S, Hahlbrock K (1989) Temporal and spatial patterns of gene expression around sites of attempted infection on parsley leaves. Plant Cell **1**: 993-1001

Zimmermann S, Thomine S, Guern J, Barbier-Brygoo H (1994) An anion current at the plasma membrane of tobacco protoplasts shows ATP-dependent voltage regulation and is modulated by auxin. Plant J **6**: 707-716

Author Index

Subject Index